HSIUNG'S DIAGNOSTIC VIROLOGY

AUTHORS AND CONTRIBUTORS

JOSEPH R. BOVE, M.D., professor emeritus, Laboratory Medicine, Yale University School of Medicine

VERONICA F. CHAN, PH.D., professor, College of Public Health, University of the Philippines, Manila; director, WHO National Influenza Center, Philippines; visiting research scientist, Laboratory Medicine, Yale University School of Medicine

CAROLINE K. Y. FONG, PH.D., research scientist, Laboratory Medicine, Yale University School of Medicine; chief, General Diagnostic Section, Virology Reference Laboratory, Veterans Affairs Medical Center, West Haven, Connecticut

G. D. HSIUNG, PH.D., professor emeritus, Laboratory Medicine, Yale University School of Medicine; director emeritus, Virology Reference Laboratory, Veterans Affairs Medical Center, West Haven, Connecticut

NICK KARABATSOS, PH.D., chief, Diagnostic Section, Arbovirus Diseases Branch, Division of Vector-borne Infectious Diseases, National Center for Infectious Diseases, Centers for Disease Control, Fort Collins, Colorado

MARIE L. LANDRY, M.D., associate professor, Laboratory Medicine and Internal Medicine (Infectious Diseases), Yale University School of Medicine; director, Virology Reference Laboratory, Veterans Affairs Medical Center, West Haven, Connecticut

FRANK J. MICHALSKI, PH.D., director, Diagnostic Virology Laboratory, Met Path, Inc., Corning Laboratory Services, Teterboro, New Jersey

GARY STACK, M.D., PH.D., assistant professor, Laboratory Medicine, Yale University School of Medicine; chief, Pathology and Laboratory Medicine Service, Veterans Affairs Medical Center, West Haven, Connecticut

FOURTH EDITION

HSIUNG'S DIAGNOSTIC VIROLOGY

AS ILLUSTRATED BY LIGHT AND ELECTRON MICROSCOPY

G. D. HSIUNG,
CAROLINE K. Y. FONG,
and MARIE L. LANDRY

YALE UNIVERSITY PRESS
New Haven and London

Designed by Jill Breitbarth and set in Gill Sans and New Baskerville types by The Composing Room of Michigan, Inc., Grand Rapids, Michigan. Printed in the United States of America by Edwards Brothers, Inc., Ann Arbor, Michigan.

Library of Congress Cataloging-in-Publication Data

Hsiung, G. D. (Gueh Djen), 1918–
Hsiung's diagnostic virology : as illustrated by light and electron microscopy / G. D. Hsiung, Caroline K. Y. Fong, and Marie L. Landry — 4th ed.
p. cm.
Rev. ed. of: Diagnostic virology. 3d ed. c1982.
Includes bibliographical references and index.
ISBN 0-300-05845-4 (alk. paper)
1. Diagnostic virology—Handbooks, manuals, etc. 2. Viruses—Identification—Handbooks, manuals, etc. 3. Medical microscopy—Handbooks, manuals, etc. I. Fong, Caroline K. Y. II. Landry, Marie L., 1948– . III. Hsiung, G. D. (Gueh Djen), 1918– Diagnostic virology. IV. Title. V. Title: Diagnostic virology.
[DNLM: 1. Virus Diseases—diagnosis—laboratory manuals. 2. Virology—laboratory manuals. QW 25 H873d 1994] QR387.H74 1994
616′.0194—dc20
DNLM/DLC
for Library of Congress 93-6338
CIP

A catalogue record for this book is available from the British Library.

The paper in this book meets the guidelines for permanence and durability of the Committee on Production Guidelines for Book Longevity of the Council on Library Resources.

10 9 8 7 6 5 4 3 2 1

CONTENTS

PREFACE TO THE FOURTH EDITION

The first edition of this book was published as a laboratory manual in 1964 and was designed for use by students in a two-week postgraduate workshop in diagnostic virology. In the ensuing years, the book has had two editions. It has been used in various virology laboratories as a standard handbook for the recognition and characterization of virus-induced cellular changes seen by light microscopy, as well as virus morphology as revealed by electron microscopy. Despite rapid changes in diagnostic virology, the photographs and illustrations are still indispensable as a visual reference at the laboratory bench.

The enormous expansion of virology over the past ten years has served as the impetus for the fourth edition of the book, which contains several new chapters and considerable revisions and updates for the other chapters. The revisions of certain chapters were made possible by the expertise and rich experience in clinical virology laboratories of my two coauthors and the contributors. The emergence of the human immunodeficiency virus (HIV) and the resultant epidemic of acquired immunodeficiency syndrome (AIDS) have prompted the inclusion of a chapter on human retroviruses with illustrative electron micrographs of human retroviruses.

The book, as in its previous editions, is designed primarily as a visual aid to the changes in cultured cells induced by viruses commonly encountered in human diseases. In the chapters on representative viruses, we describe the various techniques available, emphasizing the use of selected cell culture systems for isolation and identification. The book is intended to assist individuals involved in a clinical virology laboratory; the descriptive information serves as a guide to the illustrations. In-depth discussions on the theories and mechanisms of virus replication are limited in order to maintain the ease of use that a handbook demands.

It is our experience that the photographs and descriptions of each virus group can be applied to the viruses of other animal species. The procedures may be useful to those engaged in diagnostic and experimental virology in both human and veterinary medicine.

ACKNOWLEDGMENTS

The authors are grateful to Dr. Veronica F. Chan for her devotion in editing the entire book. Without her efforts to research, collect, and assemble the materials presented in the various chapters, it would have been impossible to complete this book in time for publication.

Our special thanks are extended to Lillemore Wallmark for the skillful photographic reproduction of the various illustrations and to Sigrid Klein for preparing the photographs of both light and electron microscopy. We are extremely grateful to Georgi Danley and Jennie Annatone for their assistance in word processing and their patience in putting the ever-changing manuscripts in order.

We are indebted to Dr. Daniel Medina of Yale University Medical School and Doctors Mingxu Guan and F. Diaz-Mitoma of the Children's Hospital of Eastern Ontario, Canada, who provided their unpublished light and electron micrographs of human retroviruses and human herpesvirus-6, respectively. Permissions granted by publishers for the reproduction of photographs are acknowledged.

The support for the reproduction of color plates by Bion Enterprises, Ltd., is greatly appreciated. The continued interest of Yale University Press in the development of this book over the past thirty years and the support from the National Institutes of Health and the West Haven Veterans Affairs Medical Center are also deeply appreciated.

G.D.H.
New Haven
March 1993

GUIDE TO READERS

This handbook is divided into six parts and is designed to provide a comprehensive visual aid to virus-induced cellular changes and virus morphology.

Part I comprises four chapters that summarize the overall discipline of diagnostic virology, guidelines for specimen collection, transport and processing, and safety precautions.

Part II consists of nine chapters that describe the practical methods and protocols for virus isolation and identification, viral antigen and genome detection, and the determination of virus-induced antibodies.

Parts III and IV consist of chapters on the recognition and characterization of representative RNA and DNA viruses. Each chapter describes a virus family, focusing on representative viruses that may be encountered in a clinical virology laboratory. The distinct growth characteristics of each virus are illustrated by photomicrographs of virus-induced cellular changes and electron micrographs of virus morphology in infected cells.

Part V comprises chapters on two groups of viruses, hepatitis and arboviruses, that are not commonly assayed in a routine virology laboratory. Detailed information is provided, however, for those laboratories in which tests for these viruses are performed.

The last three chapters that form part VI describe briefly cell culture preparations, endogenous viral and mycoplasmal contaminants, and culture media.

To keep the book a handy size, bibliographic references are not cited in the text but are presented in abridged lists of review papers and selected articles of original reports at the end of each section or chapter.

ABBREVIATIONS

AAV adeno-associated virus
ACIF anticomplement immunofluorescence
ACV acyclovir
AGMK African green monkey kidney
AIDS acquired immunodeficiency syndrome
ALF allantoic fluid
ALT alanine aminotransferase
AMF amniotic fluid
AO acridine orange staining
ARD acute respiratory disease
A549 cell line derived from human lung carcinoma
BAL bronchoalveolar lavage
BGMK Buffalo green monkey kidney cell line
BHK-21 cell line derived from baby Syrian hamster kidney tissue
BKV a human polyomavirus
BL Burkitt's lymphoma
BME basal medium, Eagle's
BPL beta-propiolactone
BSA bovine serum albumin
BSS balanced salt solution
CAP College of American Pathologists
CAM chorioallantoic membrane
CE chicken embryo
CER hamster cells
CF complement-fixation
CIE counter immunoelectrophoresis
CMV cytomegalovirus
CNS central nervous system
CPE cytopathic effect
CPM counts per minute
CS calf serum
CSF cerebrospinal fluid
CV-1 cell line derived from African green monkey kidney
DMSO dimethylsulfoxide
DNA deoxyribonucleic acid
EB-3 cell line derived from Burkitt lymphoma harboring a herpeslike virus particle
EBNA Epstein-Barr virus nuclear antigen
EBV Epstein-Barr virus
ECHO enteric cytopathogenic human orphan virus
ED_{90} effective dose, 90% reduction
EDTA ethylene diaminetetraacetic acid
EEE Eastern equine encephalitis virus
EES Eagle's basal medium prepared with Earle's BSS
EHS Eagle's basal medium prepared with Hanks' BSS
EIA enzyme immunoassay
ELISA enzyme-linked immunosorbent assay
EM electron microscope, electron microscopy

FA fluorescent antibody
FAMA fluorescent antimembrane antibody
FBS fetal bovine serum
FITC fluorescein isothiocyanate
GMK green monkey kidney
GPE guinea pig embryo
H&E hematoxylin-eosin staining
HA hemagglutinin, hemagglutination
HAd hemadsorption
HAd-I hemadsorption-inhibition
HAV hepatitis A virus
HB_cAg hepatitis B core antigen
HB_eAg hepatitis B e antigen
HB_sAg hepatitis B surface antigen
HBV hepatitis B virus
HCV hepatitis C virus
HDAg hepatitis D antigen
HDF human diploid fibroblast
HDV hepatitis D virus
HEF human embryonic fibroblast
HEK human embryonic kidney
HeLa cell line derived from epithelioid carcinoma of human cervix
HEP-2 cell line derived from epidermoid carcinoma of human larynx
HEV hepatitis E virus
HI hemagglutination-inhibition
HIV human immunodeficiency virus
HL a heteroploid cell line
HPV human papilloma virus
HSV-1 herpes simplex virus type 1
HSV-2 herpes simplex virus type 2
HTLV-I human T-cell lymphotropic virus type I
HTLV-II human T-cell lymphotropic virus type II
IAHA immune adherence hemagglutination
IC intracerebral
IEM immunoelectron microscopy
IF immunofluorescent
IFA indirect fluorescent antibody
IgG immunoglobulin G
IgM immunoglobulin M
IL-2 interleukin 2
IM infectious mononucleosis
IP immunoperoxidase
JCAHO Joint Commission for the Accreditation of Healthcare Organizations
JCV a human polyomavirus
KB cell line derived from carcinoma of human nasopharynx
LA latex agglutination
LBCf Laboratory Branch complement-fixation test (Centers for Disease Control)
$LLC\text{-}MK_2$ cell line derived from rhesus monkey kidney
MDCK Madin-Darby canine kidney cell line
MEM minimum essential medium
MK monkey kidney
ML mink lung cell line
MOI multiplicity of infection
MRC-5 cell line derived from normal fetal lung tissue
MVM minute virus of mice
NDV Newcastle disease virus
NPC nasopharyngeal carcinoma
NPS nasopharyngeal secretion
NS-1 cell line derived from mouse plasmacytoma
NT neutralization test
PAP papanicolaou
PBMC peripheral blood mononuclear cells
PBS phosphate-buffered saline
PCR polymerase chain reaction
PEG polyethylene glycol
PFU plaque-forming unit
PHA phytohemagglutinin
PHFG primary human fetal glial cells
PML progressive multifocal leukoencephalopathy
PS cloned porcine stable cell line
PTA phosphotungstic acid
Raji cell line derived from Burkitt lymphoma, does not contain herpesvirus particle
RBC red blood cells
RD cell line derived from a rhabdomyosarcoma tissue
RDE receptor-destroying enzyme
RhMK rhesus monkey kidney
RIA radioimmunoassay
RK rabbit kidney
RNA ribonucleic acid

RS	rectal swab
RSV	respiratory syncytial virus
SA	simian agent
SSPE	subacute sclerosing panencephalitis
SV	simian virus
SV_5	simian virus 5 (a parainfluenza virus)
SV_{40}	simian virus 40 (vacuolating virus of monkey), a polyomavirus
TC	tissue culture
TCGF	T-cell growth factor
$TCID_{50}$	tissue culture infectious dose 50% endpoint
TS	throat swab
URI	upper respiratory illness
UV	ultraviolet
VB	Veronal buffer solution
VCA	viral capsid antigen of EBV
Vero	cell line derived from African green monkey kidney tissue
VSV	vesicular stomatitis virus
VTM	viral transport medium
VZV	varicella-zoster virus
WB	Western immunoblot
WEE	Western equine encephalitis virus
WI-38	diploid cell line derived from human embryonic lung

PART ONE

GENERAL ASPECTS

1

GENERAL PRINCIPLES AND CURRENT TRENDS IN DIAGNOSTIC VIROLOGY

G. D. HSIUNG

Undeniably, much progress has been made in diagnostic virology over the past decade. With the advent of more advanced technology, the laboratory diagnosis of viral infections has become more rapid and accurate, yet for many viruses, virus isolation remains the most sensitive procedure and the gold standard in diagnostic virology. With the availability of specific antiviral chemotherapeutic agents in recent years, clinicians are demanding prompt laboratory diagnosis so that appropriate management can be instituted for their patients. Today, diagnostic virology has become an integral part of clinical medicine.

RAPID VIRAL DIAGNOSIS: DETECTION OF VIRAL ANTIGENS AND VIRAL GENOMES

Although the isolation of viruses remains an essential approach in diagnostic virology, recently developed techniques for detecting viral antigens and viral nucleic acid are in many instances helpful supplementary tools for rapid laboratory diagnosis. For example, the detection of noncultivable viruses has been made possible by the availability of immunologic tests. Enzyme immunoassays (EIA) are of particular value in recognizing viruses directly in clinical specimens without cultivation, as in the case of rotavirus. The use of commercial kits with high-quality monoclonal antibodies has enhanced tremendously the sensitivity and specificity of the detection of viral antigens. Likewise, viruses with slow growth rates are detectable earlier by the monoclonal antibody staining of infected cell cultures. A good example is that of cyto-

ISBN 0-300-05845-4.

megalovirus (CMV); the observation of a cytopathic effect (CPE) in a conventional culture requires two to three weeks, but the detection of CMV early antigens in a shell vial culture takes only 16 hours. In addition, the development of genomic probes has made a rapid laboratory diagnosis possible for certain noncultivable viruses; for example, a human papillomavirus infection is detected in clinical specimens by means of nucleic acid hybridization. The amplification of viral sequences by polymerase chain reaction (PCR) has introduced an even more sensitive technology for virological diagnosis. Nevertheless, molecular amplification techniques are as yet not routinely applicable for clinical laboratories. These sophisticated techniques for the detection of viral antigens or genomes are, however, directed at specific pathogens. Thus, only the expected agents will be detected, but not those that are unexpected.

MONOCLONAL ANTIBODIES FOR RAPID IDENTIFICATION OF VIRAL ISOLATES

In the past, only polyclonal antisera produced in animals were available for the identification of virus isolates; variations in quality were problematic, and sensitivity as well as specificity varied. For example, typing herpes simplex virus type 1 and type 2 (HSV-1, -2) by antisera was limited by extensive cross-reactions between the two viruses. At present, a variety of high-quality monoclonal antibodies is available commercially. These antibodies include those against influenza viruses A and B, parainfluenza virus types 1–4, mumps, measles, HSV-1 and -2, adenovirus, varicella-zoster virus (VZV), CMV, and respiratory syncytial virus (RSV). As a result, the use of a monoclonal antibody in an immunofluorescence (IF) assay to identify these viruses has become routine. The identification of virus isolates can usually be made on the same day that CPE or hemadsorption (HAd) occurs in cultures, which greatly benefits patient management.

RECOGNITION OF VIRAL AGENTS UNDER ELECTRON MICROSCOPY

Like cell culture isolation, direct visualization of virions under an electron microscope (EM) also allows the recognition of unexpected viral agents. The initial recognition of Epstein-Barr herpesvirus in cultured lymphoblastic cells obtained from Burkitt's lymphoma patients was made possible by EM, as were the discovery of two kinds of particles—surface antigen and Dane particles—in the sera of patients with hepatitis B; the observation of small virus particles in stool samples of hepatitis A patients; and the discovery of a rotavirus in the stool samples of children with infantile diarrhea. More recently, the confirmation of a retrovirus in the peripheral blood mononuclear cells of patients with acquired immunodeficiency syndrome (AIDS) and the discovery of human herpesvirus type 6 are examples of the value of EM examination. At times when the viral isolate eludes identification by other methods, direct visualization by EM remains a most valuable tool for viral identification.

CHANGING INTEREST IN VIRUS GROUPS

As new viruses continue to surface, changes in interest in the different virus groups become apparent. The waves of interest, the surges of research activity, and the demand for virological diagnosis occur year by year, as reflected in the number of publications in the *Index Medicus* and as depicted in figures 1.1 and 1.2. This variation is dictated by need as well as the availability of new techniques at the time of the need. In the early 1950s, the paralytic polio-

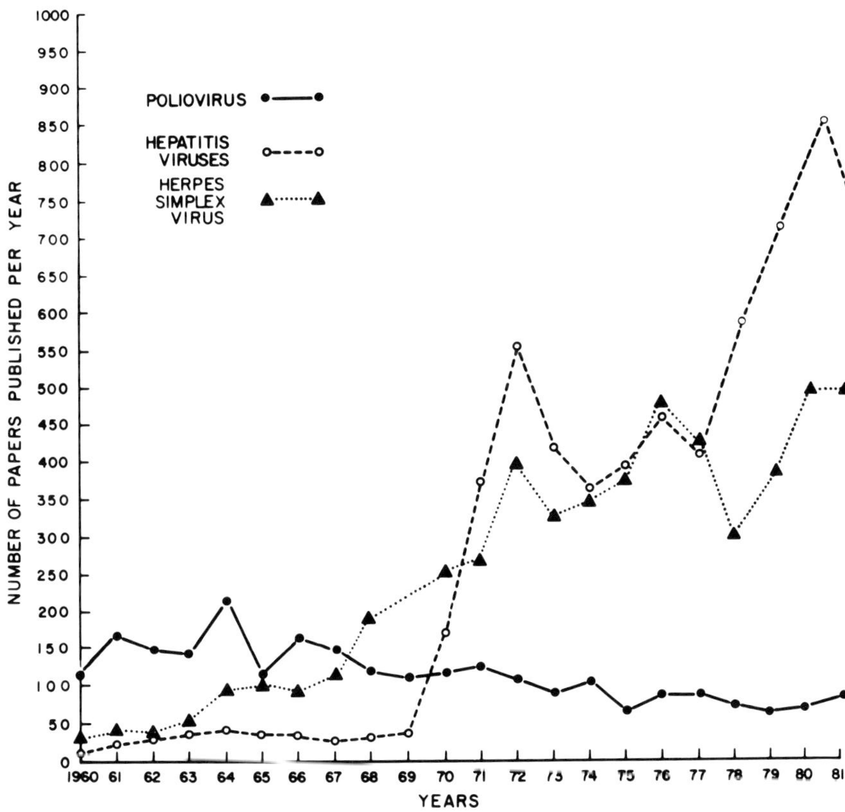

Figure 1.1 Number of publications on three representative groups of viruses reported annually in the *Index Medicus*, 1960–81. (Modified from G. D. Hsiung, *Yale J. Biol. Med.* 53:2 [1980])

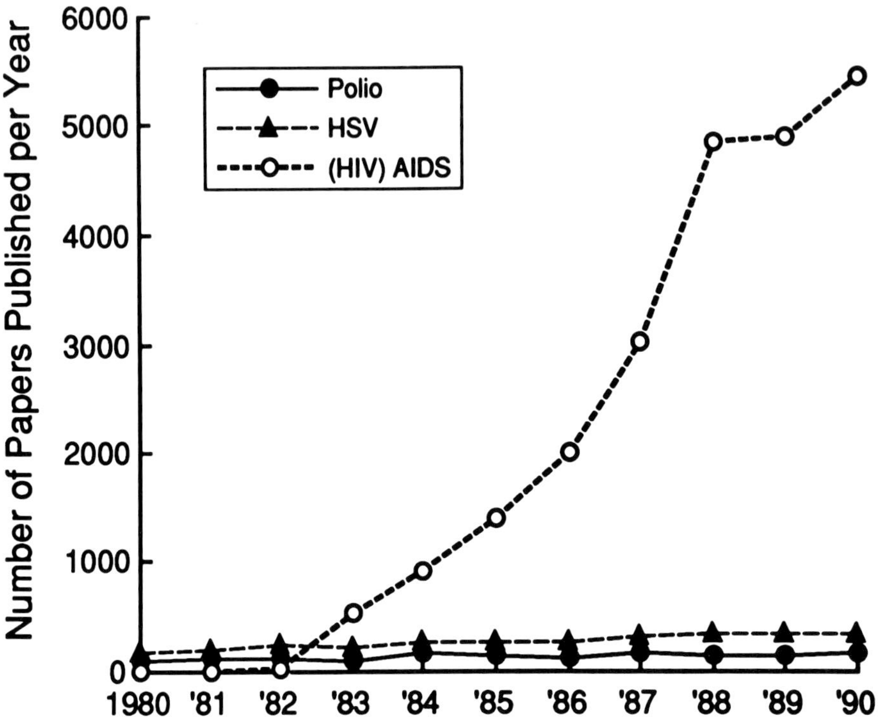

Figure 1.2 Number of publications on three representative groups of viruses recorded annually in *Medline*, 1980–90. (Modified from G. D. Hsiung, *Yale J. Biol. Med.* 62:86 [1986])

myelitis attack rate was at its peak in the United States, hence the interest in studying the poliovirus intensified. The development of cell culture, a new method for the isolation and cultivation of viruses, allowed studies of pathogenesis and the development of a safe and effective vaccine. With the use of the inactivated poliovirus vaccine in 1955, there was a noticeable decrease in the number of reported cases of poliovirus infection. A further decrease in the number of cases of poliomyelitis was noted beginning in 1961 when a more effective means of prevention, the live oral poliovirus vaccine, became available. Thus, the number of reports on poliovirus studies gradually declined during the period 1960–80 (figure 1.1).

Few studies on herpesviruses, including herpes simplex virus, were published in the early 1960s—perhaps because these agents were not considered to cause major human disease other than an occasional case of herpes encephalitis. Greater interest in the herpesviruses in the early 1970s was stimulated by the epidemiological association between HSV-2 and cervical carcinoma, the successful treatment of herpes simplex encephalitis first by adenine arabinoside and then acyclovir, and the recognition of genital herpesvirus as an important cause of sexually transmitted disease. More recently, owing to the ability to treat CMV infection among organ transplant and AIDS patients with ganciclovir, the interest in the study of CMV has increased. In addition, the success of the rapid detection of CMV early antigen in a shell vial culture with a monoclonal antibody (as noted above) has greatly enhanced these studies.

Although viral hepatitis (HBV) has long been recognized as an important clinical entity, until recently the number of papers published on hepatitis was small. The discovery of hepatitis B surface antigens in patients' sera in the early 1970s; the recognition of HBV virions by EM; and the application of radioimmunoassay and other immunologic methods for detecting the hepatitis B virus antigen and antibody were undoubtedly responsible for the sudden increased interest and published papers on hepatitis B in the 1970s. With the widespread use of the recombinant hepatitis B vaccine, further interest in studying the hepatitis B virus may eventually decline, although studies on other types of hepatitis viruses may increase.

The number of cases of human immunodeficiency virus (HIV) infection has attained epidemic proportions since AIDS was first recognized in the early 1980s. Investigators therefore continue to devote their time to the study of HIV, and the number of papers published on HIV increased rapidly between 1980 and 1990 (figure 1.2).

Much progress has been made in diagnostic virology. Though the detection of viral antigen, antibody, and nucleic acid determination are essential implements for rapid viral diagnosis, only suspected viral agents are being sought, therefore, which limits the recognition of unsuspected viral agents. Virus isolation in cell cultures, however, remains open minded and the gold standard in viral diagnosis. Undoubtedly owing to the influx of chemotherapeutic agents for the treatment of viral diseases in recent years, clinicians are demanding a more rapid and accurate viral diagnosis for the more effective management of their patients.

Viruses have their own agenda; one infection may be controlled while another, "new" viral agent may emerge. The quest for a more effective means of laboratory diagnosis must continue. Thus, for optimal results, the combination of conventional methods and newly developed techniques should be employed for a rapid and accurate viral diagnosis.

2

PROPERTIES AND CLASSIFICATION OF VIRUSES

G. D. HSIUNG

Viruses are infectious agents generally characterized by their submicroscopic size and their requirement to replicate within living cells. The nature of these agents can best be understood from their physicochemical properties and their means of replication in a susceptible host. Although many systems have been proposed for the classification of animal viruses, there is general agreement that the structure and composition of viruses such as nucleic acid type and configuration, capsid symmetry, size, and shape of a virion and presence or absence of an envelope are important characteristics for developing a taxonomic scheme. Physical and chemical methods are not routinely used in a clinical diagnostic laboratory. There are, however, instances when a newly isolated virus cannot be identified by standard serologic techniques. In such instances, it may be necessary to study the agent's basic properties—including nucleic acid composition and size and shape—by electron microscopy, in order to characterize the virus more specifically.

Viruses may contain either ribonucleic acid (RNA viruses) or deoxyribonucleic acid (DNA viruses), but never both as their genome (table 2.1). A complete virion is composed of a nucleic acid core enclosed in a protein coat—the *capsid*—together forming a nucleocapsid. In some viruses, their nucleocapsids are surrounded by envelopes.

The size and shape of a virion can be determined directly by electron microscopic examination (see chapter 13). The morphology of major virus families, arranged according to size, are illustrated in figure 2.1. These include the DNA viruses: Poxviridae, Herpesviridae, Adenoviridae, Papovaviridae, and Parvoviridae; and the RNA viruses: Paramyxoviridae, Rhabdoviridae, Retroviridae, Reoviridae, and Enteroviridae.

ISBN 0-300-05845-4.

Table 2.1 Physicochemical Properties of Major Virus Families Infecting Humans[a]

Nucleic Acid Composition	Virus Family	Genome Structure[b]	Virion: Diameter (nm)	Virion: Nucleocapsid Symmetry	Virion: Envelope (E or N)[c]	Viral Assembly Site[d]	Representative Viruses Encountered in a Clinical Laboratory
RNA	Picornaviridae	SS	25–30	Icosahedral	N	C	Enterovirus (polio-, coxsackie-, echo-) Rhinovirus
	Orthomyxoviridae	SS	80–120	Helical	E	C	Influenza virus
	Paramyxoviridae	SS	150–300	Helical	E	C	Parainfluenza virus Mumps virus Measles virus RS virus
	Reoviridae	DS	70–80	Icosahedral	N	C	Reovirus Rotavirus
	Retroviridae	SS	100–120	Icosahedral	E	C	HTLV-I[e] HIV-1
	Rhabdoviridae	SS	60–180	Helical	E	C	Rabies virus
DNA	Herpesviridae	DS	150–200	Icosahedral	E	Nu	HSV—1&2 CMV, VZV, EBV HHV-6
	Adenoviridae	DS	70–80	Icosahedral	N	Nu	Adenovirus
	Papovaviridae	DS	45–55	Icosahedral	N	Nu	HPV, JC, BK
	Parvoviridae	SS	18–20	Icosahedral	N	Nu	B19
	Poxviridae	DS	230–300	Complex	E	C	Vaccinia virus

a. Other viruses not listed in this table are Coronaviridae and Arenaviridae, both RNA viruses, ranging in size from 100–160 nm with helical nucleocapsids, each surrounded by an envelope.

b. SS = single-stranded; DS = double-stranded

c. E = enveloped, sensitive to ether treatment; N = naked, resistant to ether treatment

d. C = cytoplasm; Nu = nucleus

e. See list of abbreviations.

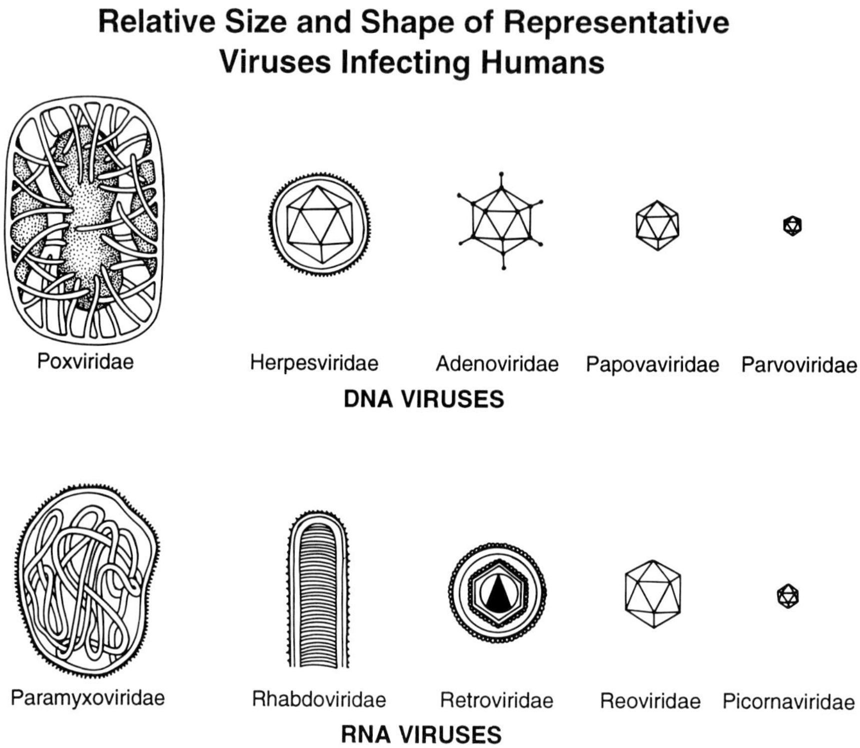

Figure 2.1 Relative size and shape of representative virus families infecting humans: *Top row,* DNA viruses; *bottom row,* RNA viruses

Table 2.1 also lists some of the common viruses encountered in a clinical laboratory. Detailed growth characteristics of each RNA and DNA virus family are discussed separately in the following chapters. For hepatitis viruses and those viruses transmitted by arthropods (arboviruses), readers are referred to part V, chapters 26 and 27, since these two groups include both RNA and DNA viruses.

3

SPECIMEN COLLECTION, TRANSPORT, AND PROCESSING FOR VIROLOGICAL STUDIES

C. K. Y. FONG AND M. L. LANDRY

COLLECTION OF SPECIMENS

Successful isolation of viruses from clinical materials depends largely on the proper collection and handling of specimens. Ideally, specimens for virus studies should be collected as early as possible in the course of the disease (figure 3.1)—that is, within the first three days after the onset of illness. All samples should be properly labeled and accompanied by such information as the patient's name, age, sex, date of onset of illness, date of collection, and presumptive clinical diagnosis. This latter information is essential for the selection of the most sensitive test system in the laboratory. Specimens for virus isolation should be transported immediately to the laboratory. If delays are unavoidable, you may keep specimens refrigerated at 4°C for no more than three days. For longer periods, store frozen at −70°C, but never at −20°C.

Selection of Specimen

The selection of appropriate specimens for viral culture depends on the clinical disease and the virus suspected. Since one virus may cause several different clinical diseases and a clinical disease entity may be associated with several different viruses, it is important to select the most appropriate specimen for each case.

For some viral diseases, several different specimens may be taken. For example, for a patient with a central nervous system (CNS) disease that is suspected to be due to an enterovirus, cerebrospinal fluid (CSF) and throat

ISBN 0-300-05845-4.

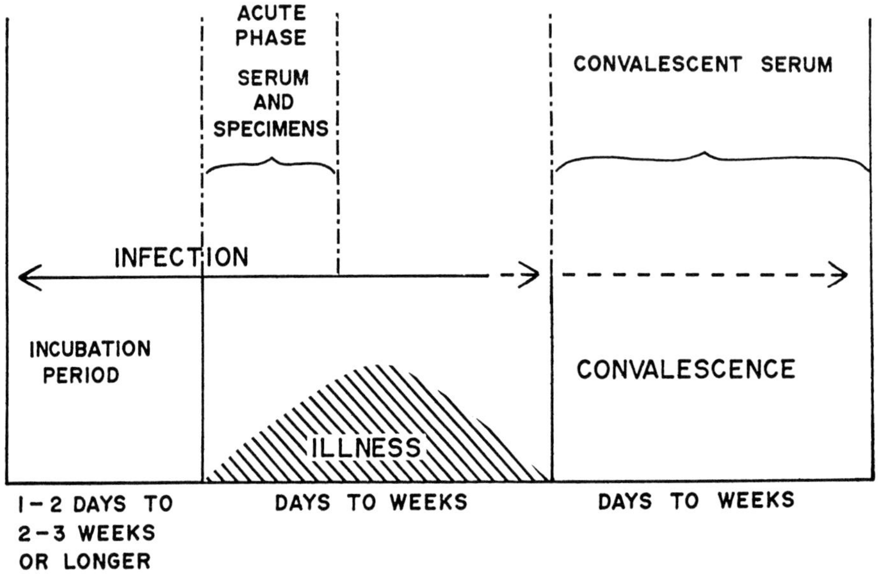

Figure 3.1 Schedule for collecting specimens for virus studies (from G. D. Hsiung and R. H. Green, *Virology and Rickettsiology*, 1978, vol. 1, pt. 1, 4)

and rectal swabs should be considered. For CNS disease following parotitis, urine and CSF specimens should be taken for mumps virus isolation. A summary of specimen selection for suspected viral disease is given in table 3.1.

Collection Procedures

It is important to choose not only the most appropriate site for specimen collection, but also to collect an adequate amount for virus studies. Guidelines for specimen collection are provided in table 3.2.

Place all swabs and tissues in a viral transport medium (VTM) (see below) to prevent drying. Although any type of swab may be used satisfactorily for most specimens, calcium alginate swabs may inactivate herpes simplex virus and chlamydiae and should be avoided. Some laboratories prefer metal or plastic swab shafts over wooden shafts since the latter may be toxic to certain viruses.

Place body fluids in leakproof, screw-cap, sterile containers. Keep all specimens at 4°C until inoculation into a culture, except whole blood for HIV or a CMV culture, which should be kept at room temperature.

SPECIMEN TRANSPORT

Viral Transport Medium

Viral transport medium usually consists of a broth medium or balanced salt solution, supplemented with 0.5% gelatin, serum, or bovine serum albumin

Table 3.1 Selection of Specimens for Viral Studies*

Clinical Syndrome	Viruses Suspected	Examples of Specimens to Be Collected	Laboratory Test
Respiratory			
Pneumonia			
Normal adult	Influenza A, B, adenovirus	BAL, lung tissue	Culture
Immunosuppressed	CMV, adenovirus, VZV, HSV, RSV	BAL, lung tissue	Culture
Upper respiratory illness or pharyngitis	Rhinovirus, RSV, adenovirus, parainfluenza, influenza A, B, enterovirus	Nasal wash, NP swab, TS	Culture
	EBV	Serum	Serology (IF)
Pleurodynia	Enterovirus	TS	Culture
Ocular			
Conjunctivitis	Enterovirus, adenovirus	Conjunctival swab, TS	Culture
Keratitis	HSV, VZV, and adenovirus	Lesion swab, corneal swab	Culture
Infectious mononucleosis	EBV	Serum	Serology (IF)
	CMV, adenovirus	Saliva, urine, blood	Culture
	HIV	Serum, blood	Serology (ELISA, IF, WB)
Cardiovascular			Culture
Myocarditis, pericarditis	Enterovirus	TS, stool, endocardial biopsy	Culture
	Influenza	TS, endocardial biopsy	Culture
	CMV	TS, urine, endocardial biopsy	Culture

(*continued*)

Table 3.1 (Continued)

Clinical Syndrome	Viruses Suspected	Examples of Specimens to Be Collected	Laboratory Test
Digestive tract			
Gastroenteritis	Rotavirus	Stool	ELISA, EM
	Norwalk agent	Stool, acute and convalescent serum	EM, IEM
	Adenovirus	Stool	EM, ELISA, culture
	Enterovirus	Stool	Culture
Colitis	CMV	Colon biopsy, stool	Culture
Proctitis	HSV	Lesion swab, RS	Culture
Hepatitis	Hepatitis A, B, C, D	Serum	Serology (ELISA, IAHA, IF, RIA)
	CMV, HSV, and adenovirus	Liver tissue	Culture
	EBV	Serum	Serology (IF)
Cutaneous or mucous membrane			
Vesicular	HSV, VZV, vaccinia	Lesion swab	Culture
	Enterovirus	TS, stool	Culture
Exanthematous	Measles	TS, serum	Culture, Serology (ELISA, IF)
	Rubella	Serum	Serology (ELISA, LA)
	Enterovirus, adenovirus	TS, stool	Culture
	CMV	TS, urine	Culture

Central Nervous System			
Encephalitis	HSV	Brain biopsy	Culture
	VZV	CSF	Serology (ELISA, FAMA, IF)
	Alphavirus (EEE, WEE, etc.)	Serum	Serology (ELISA, HI)
	Adenovirus, mumps	CSF, TS, urine	Culture
	Influenza	TS, CSF	Culture
	Enterovirus	CSF, TS, stool	Culture
	HIV	Serum, CSF	ELISA (for antigen and antibody)
		CSF, blood	Culture
Meningitis	Enterovirus	CSF, TS, stool	Culture
	HSV	CSF, lesion swab	Culture
	Mumps	CSF, urine, serum	Culture
PML	Polyomavirus	Brain tissue	EM

*Acute and convalescent sera can be collected for antibody studies. Serologic testing, however, is not practical for enteroviruses, rhinoviruses, or polyomaviruses.

BAL	bronchoalveolar lavage
CSF	cerebrospinal fluid
ELISA	enzyme-linked immunosorbent assay
EM	electron microscopy
FAMA	fluorescent antimembrane antibody
IAHA	immune adherence hemagglutination
IEM	immune electron microscopy
IF	immunofluorescence
NP	nasopharyngeal
NS	nasal swab
RIA	radioimmunoassay
TS	throat swab
WB	Western immunoblot

Table 3.2 Collection of Specimens for Viral Culture

Specimen	Methods for Collection
Throat swab	Swab posterior pharynx and tonsils vigorously. Place swab in VTM.
Nasal swab	Rotate swab gently in nostril to obtain secretions. Place swab in VTM.
Nasopharyngeal aspirate	Insert catheter into posterior nasopharynx. Suction mucus into a mucus trap, then wash mucus through suction catheter and into trap with viral transport medium, a balanced salt solution, or saline.
Nasopharyngeal swab	Insert flexible fine shaft into nasopharynx, just past the point of resistance. Rotate gently. Remove and place in VTM.
Bronchoalveolar lavage	Wedge bronchoscope into subsegmental bronchus, insert four 50-ml boluses of sterile saline into the suction port with immediate return suction after the insertion of each sample. Submit 5–10 ml for viral studies.
Pleural fluid	Collect 2 ml into a sterile vial.
Eye swab	Rub premoistened swab gently several times over involved surface of the eye or obtain conjunctival scrapings.
Lesion swab	Unroof vesicles or crusts to obtain fluid; if ulcer, swab base of ulcer. Do not just swab surface. Clean lesions with sterile saline before and with antiseptic after obtaining specimen.
Vesicle fluid	Aspirate fluid with a 26- or 27-gauge needle-syringe. Rinse the syringe with VTM.
Urine	Obtain clean catch midstream urine (5–10 ml) in sterile container; first voided A.M. urine is best. Collect three specimens for CMV to increase yield.
CSF	Collect 2 ml in sterile container.
Stool	Collect 2–4 g in sterile container. Fill no more than halfway.
Rectal swab	Rotate swab in rectum to obtain stool. For proctitis, swab rectal mucosa.
Whole blood	Collect blood (5–10 ml) in tube with anticoagulant, keep at room temperature, and deliver promptly to the virology laboratory. Specimens should be processed within 2–5 hours of collection (24 hours maximum).
Bone marrow	Collect at least 2 ml into a tube with anticoagulant (EDTA or heparin).
Tissues (brain, liver, lung, etc.)	Collect aseptically, using separate sterile instruments for each tissue sample. Place in balanced salt solution or saline.

(BSA) to protect the viability of viruses in transit and antibiotics to prevent bacterial and fungal growth (for formulae, see the end of this chapter). Many types of VTM are commercially available. Virocult[R] is a commercial specimen collection kit (Medical Wire & Equipment Company, Cleveland, Ohio). It consists of a swab and a plastic transtube containing a 1-ml viral transport medium absorbed in a sponge. After specimen collection, the swab is inserted into the transtube for transport.

Culturette(TM) (Marion Scientific, Kansas City, Missouri), a common specimen collection device, consists of a plastic tube containing a sterile, rayon-tipped applicator and an ampule of modified Stuart's transport medium without a protein stabilizer or antibiotics. The advantage of Culturette is that it can be used for specimen collection for many different microorganisms in addition to viruses. Since Culturette contains neither protein nor antibiotics, its use is limited to short transport times.

Another type of viral transport system is called Transporter(TM) (Bartel Immunodiagnostics, Baxter Scientific Products), which contains a monolayer of human fibroblast cells grown on the side of the lower conical portion of a plastic centrifuge tube that contains culture medium. Owing to the presence of viable cells, the shelf-life of Transporter is usually short (up to one month) and should be examined for intact cell monolayer before use. Specimens collected by Transporter should be held at room temperature until they are transported to a laboratory. A higher virus recovery rate has been reported for Transporter.

Method of Transport

Label specimen containers, place in a plastic bag accompanied by a requisition with pertinent clinical information, and transport as soon as possible to a nearby virology laboratory. If a delay is anticipated, keep all specimens, except whole blood for HIV and CMV culture, at 4°C.

Shipping Specimen

When it is necessary to ship a specimen, pack it in leakproof containers enclosed in a second watertight container with absorbent material (figure 3.2). Place the specimen container in a styrofoam box with commercial refrigerant packs. Avoid using wet ice, because it will melt and leak from the package. Use dry ice only if the delay in transport is longer than seven days. Never freeze whole blood. Some enveloped viruses, such as CMV and VZV, are inactivated by freezing and thawing.

Pack all specimens in a manner to avoid leakage or breakage, and to withstand the trauma of mailing. Ship by rapid overnight delivery services. Sera for antibody studies can be sent by regular mail. Display label on each package stating: "DIAGNOSTIC SPECIMEN."

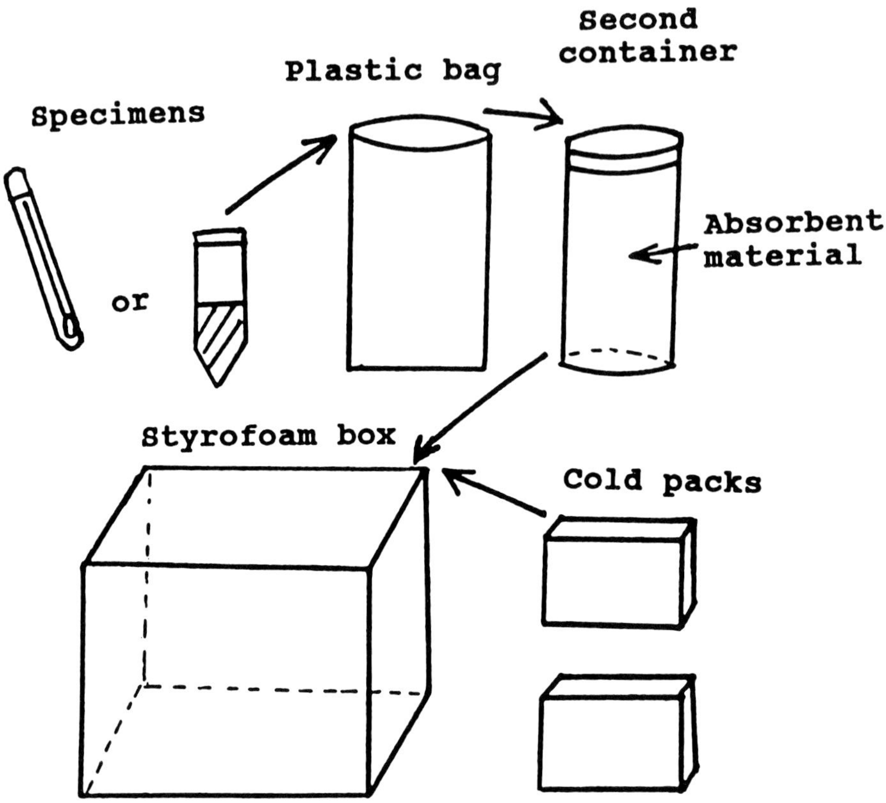

Figure 3.2 Proper packaging of specimens for shipping

PROCESSING SPECIMENS FOR VIRUS ISOLATION

Swabs (throat, eye, rectal, lesion, and genital, etc.)

1. Place swab in a vial containing 2.0 ml of VTM, rub swab against side of vial, and break off the swab tip into the vial.
2. Vortex with the swab in the vial for 10–30 seconds and inoculate 0.2 ml into each culture tube.

Urine

1. Mix 2 ml urine with 2 ml VTM containing antibiotics in a sterile vial to reduce urine toxicity.
2. Adjust the pH to 7.0–7.4 with 7.5% sodium bicarbonate containing phenol red and inoculate 0.4 ml into each cell culture tube.

Bronchoalveolar Lavage (BAL) or Bronchial Washing

Centrifuge BAL at 1,400 × *g* for 20 minutes at 4°C if the total volume is greater than 5 ml and resuspend the pellet with 5 ml supernatant. Add antibiotics concentrate and glass beads to the centrifuge tube and vortex the content. Inoculate 0.2–0.3 ml into each culture tube. If the total volume of BAL is less than 5 ml, treat with antibiotics concentrate, then inoculate into culture tubes directly.

Sputum

Sputums are generally too viscous to be pipetted. There are two effective ways to handle sputums:

1. Pick up sufficient sputum using a sterile swab and place the swab in a vial of VTM, add beads to the vial, and vortex to break up the mucus.
2. Alternatively, add about 3–4 ml VTM into the sputum container; then add beads, vortex the container, and inoculate the suspension into culture tubes.

Autopsy or Biopsy Tissues

The standard method for processing tissues is by homogenization or grinding to break cells and release intracellular viruses. If the virus suspected is incomplete or latent, however, or is an enveloped virus (for example, herpesvirus [HSV] or varicella-zoster virus [VZV]) that has not been released from the cells, better results may be obtained by mincing the tissue, thus leaving some cells intact for cocultivation. Theoretically, this would allow the virus to continue its replication cycle and release infectious virus.

MINCED TISSUE

1. Place tissue in small sterile beaker or petri dish. Add prewarmed 0.25% trypsin in Hanks' balanced salt solution (BSS) to the tissue and mince tissue finely with sterile scissors.
2. Transfer tissue or cell suspension to a centrifuge tube containing growth medium with 10% fetal bovine serum (FBS).
3. Centrifuge the cell suspension and resuspend cells in VTM to make a 20% v/v suspension for inoculation into cell cultures.

HOMOGENIZATION

1. Soft Tissue (for example, brain tissue): Add VTM to container holding the tissue; and place sterile glass beads in the container, and vortex for 30–60 seconds to break down the tissue. Inoculate the cell suspension into cell cultures.

2. Fibrous Tissue (such as liver, lung, colon, esophagus)
   a. Use glass homogenizer: Pipette 2–3 ml VTM into a sterile glass homogenizer, and transfer the tissue to the homogenizer. Apply 20 strokes with the plunger of the homogenizer until no tissue is left. Transfer the cell suspension to a specimen vial for inoculation. If there is too much cellular debris, clarify by centrifugation. Inoculate the supernatant into cell culture.
   b. Use "Stomacher": Place 2–3 ml of VTM in a Stomacher specimen bag, add tissue (1 g is sufficient) into the bag, and roll the top of the bag down to close. Place the first specimen bag inside a second bag, and heat to seal the outer bag. Place the entire bag in a Stomacher and turn it on for 30–60 seconds to homogenize specimen. Open the bag under a hood, and transfer the tissue suspension into a specimen vial for culture inoculation.

Stools

Make a 20% stool suspension in phosphate-buffered saline (PBS) containing antibiotics, then process as follows: with cap secured, vortex the tube to break the stool, and centrifuge at 1,400 × *g* for 20 minutes. Collect supernatant into a sterile glass vial and inoculate supernatant into cell cultures.

Cerebrospinal Fluid

Use CSF undiluted and without antibiotics. Adequate volume is often a problem with CSF specimen. If adequate specimen is not available, inoculate cell cultures, prioritized according to the virus suspected.

Whole Blood

In general, when whole blood with anticoagulant is submitted, viruses harbored in the leukocyte fraction are sought, such as cytomegalovirus. Less toxicity and a higher isolation rate are obtained when the leukocyte fraction is separated prior to inoculation. For good leukocyte separation, fresh blood should be kept at room temperature and, ideally, processed within two to five hours. Methods for separating whole blood are as follows:

1. Dextran Separation: This method yields a combined leukocyte fraction (lymphocytes, mononuclear, and polymorphonuclear cells) from the whole blood. It is the simplest and fastest method, and for CMV, a combined leukocyte fraction is most desirable. Fresh blood samples, however, are essential.
   a. Mix 2 parts of whole blood to 1 part of dextran 6% (mol wt 70,000), dissolved in sterile normal saline, in a centrifuge tube. Final ratio

of blood to dextran is 2:1. The mixture is incubated at 37°C for 30 minutes to allow the RBC to settle.

b. Remove the leukocyte-enriched supernatant from the sedimented erythrocytes and transfer to a centrifuge tube containing 5–7 ml culture medium (Eagle's minimum essential medium, MEM).

c. Centrifuge the tube for 10 minutes at 900 × *g*; discard the supernatant.

d. Repeat step C using 5–7 ml MEM: If a significant amount of red blood cells (RBC) are present in the leukocyte pellet, remove the supernate after the second wash. Add 5 ml sterile distilled water, quickly resuspend the cells with a pipette, and vortex 5 seconds to lyse RBC. Add 10 ml medium containing 2% calf serum immediately to the leukocyte suspension to stop the lysing effect. Centrifuge the tube for 10 minutes at 900 × *g*, remove the supernatant and resuspend the pellet with 2 ml VTM, and inoculate into cell cultures, 0.3 ml per tube or shell vial.

2. Ficoll-Hypaque Separation: Ficoll-Hypaque separation yields a mononuclear fraction (lymphocytes and monocytes) from whole blood.
   a. Mix equal volumes of whole blood with sterile normal saline in a sterile test tube.
   b. Pipette sufficient Ficoll-Hypaque reagent to a centrifuge tube so that the final ratio of diluted blood to Ficoll-Hypaque is 3:1.
   c. Overlay carefully three parts of diluted blood over the Ficoll-Hypaque and centrifuge for 30 minutes at 400 × *g* at room temperature.
   d. Remove and discard the top layer (the plasma layer).
   e. Remove carefully the mononuclear cell band (about 2 ml) and transfer to a centrifuge tube containing 5 ml MEM.
   f. Centrifuge for 10 minutes at 900 × *g*, remove and discard the supernatant.
   g. Add 7 ml of MEM to the tube and repeat the centrifugation-wash procedure.
   h. Add 2 ml of VTM to the cell pellet and inoculate into cell cultures.

Bone Marrow

1. Bone Marrow Aspirate: If the specimen is an aspirate in an adequate amount, it should be processed like whole blood sample and separated by either the dextran or the Ficoll-Paque-Macrodex separation method.
2. Bone Marrow Biopsy: If the specimen is a biopsy, it should be processed like a fibrous biopsy (that is, homogenized using a glass homogenizer). (Note: Bone marrow specimens tend to be extremely

bloody, hence, they are very toxic to cell monolayers. It is necessary to rinse inoculated cell monolayers (either roller tubes or shell vials) after the initial 1-hour adsorption time to remove the excess inoculum.)

Serum (or Plasma)

Serum (clotted blood) or plasma (anticoagulated blood) can be obtained by centrifuging the blood for 10 minutes at $900 \times g$.

The serum or plasma is then inoculated into appropriate cell cultures.

SPECIMEN INOCULATION FOR VIRUS ISOLATION

For virus isolation in conventional cell culture and centrifugation culture, each specimen is inoculated into a variety of different cell cultures that are selected based on optimum sensitivity to the suspected virus or to a wide range of viruses, if a comprehensive viral study is required. (For detailed procedures, see chapter 5).

SEROLOGIC TESTS

Serologic tests are necessary to diagnose acute infection when virus isolation is impractical or to validate the significance of the isolation of certain viruses. When serologic tests are used for diagnosis, paired sera are required, one obtained in the acute stage (3–5 days after onset) of the illness and another 10–21 or more days after onset. Occasionally, single serum taken shortly after illness can be used for antibody detection in the IgM fraction. The procedures for various serologic tests are described in part II.

For more detailed procedures on specimen collection and handling, virus isolation, and serologic tests, the supplementary reading for part I listed at the end of the section is recommended.

VIRUS TRANSPORT MEDIUM COMMONLY USED IN A CLINICAL VIROLOGY LABORATORY

1. Hanks' Balanced Salt Solution with Gelatin

Hanks' balanced salt solution	100 ml
Gelatin	0.5 g

Autoclave at 15 lb pressure for 15 minutes. Before use, add antibiotics to desired concentration and dispense aliquots of 2 or 3.5 ml in screw-cap vials. Store at 4°C.

2. Buffered Tryptose Phosphate Broth with Gelatin

Tryptose phosphate broth	2.95 g
$Na_2HPO_4 \cdot 7\ H_20$	2.06 g
$NaH_2PO_4 \cdot H_20$	0.08 g
Gelatin	0.50 g
Phenol red (0.5%)	0.40 ml
Distilled water	100.00 ml

Autoclave at 15 lb pressure for 15 minutes. Before use, add antibiotics to desired concentration, and dispense aliquots of 2 or 3.5 ml to a screw-cap vial. Store at refrigerated temperature (4°C).

3. Veal Infusion Broth with Bovine Serum Albumin

Veal infusion broth	2.5 g
Phenol red (0.5%)	0.4 ml
Distilled water	100.0 ml

Autoclave at 15 lb pressure for 15 minutes. After cooling, add 0.5 g bovine serum albumin, dissolve, and filter to sterilize through 0.45 μm millipore filter. Before use, add antibiotics (vancomycin, gentamicin, and Fungizone) to desired concentration and dispense aliquots of 2 or 3.5 ml to each screw-cap vial. Store at refrigerated temperature.

4

QUALITY ASSURANCE AND LABORATORY SAFETY

M. L. LANDRY

QUALITY ASSURANCE IN THE CLINICAL LABORATORY

Over the years, laboratories have become more complex, with an ever-increasing variety of assays, complex data handling and reporting systems, large test volumes, and greater pressure for rapid turnaround. The central role the laboratory has assumed in the diagnosis and treatment of patients has led to concern over the serious impact of laboratory errors.

To reduce and prevent errors in laboratory testing, federal regulations have been developed and have recently been revised. The College of American Pathologists (CAP) and the Joint Commission for the Accreditation of Healthcare Organizations (JCAHO) have incorporated these guidelines into their inspection and accreditation procedures. Quality control records for reagents, tests performed, equipment maintenance and calibration, written procedures for specimen collection and laboratory analysis, and participation in proficiency testing have all become standard elements of laboratory quality control (table 4.1). A qualified and well-trained workforce, however, is the foundation of an effective quality control program.

In recent years, the emphasis has changed to a broader concept of quality, now referred to as *quality assurance* or *quality assessment*. Quality assurance extends beyond internal laboratory processes to include the selection and ordering of tests by physicians, specimen transport, the reporting and interpretation of results, and the effect of test results on patient outcomes (figure 4.1). Thus, quality assurance programs should serve to enhance communication and collaboration between the laboratory and other health care workers,

ISBN 0-300-05845-4.

Table 4.1 Standard Elements of Laboratory Quality Assurance

Standard Elements	Items Implemented or Provided by Laboratory
Well-trained work-force	High standards and continuing education
Written procedures	Specimen collection and processing Laboratory test procedure manual
Monitoring and recordkeeping	Source of cell cultures and media Reagents and kits Test controls Virus stocks Equipment maintenance and calibration
Proficiency testing	Performed at regular time intervals
Communication	Ongoing dialogue with other health care workers to identify and correct problems promptly

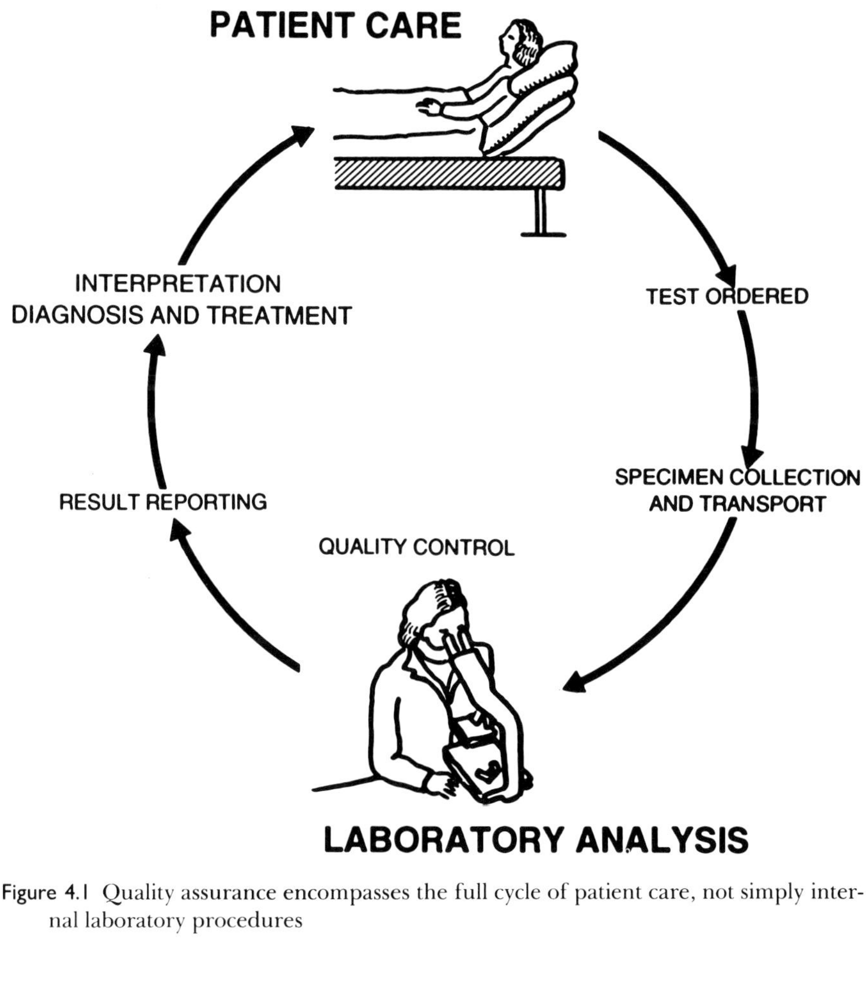

Figure 4.1 Quality assurance encompasses the full cycle of patient care, not simply internal laboratory procedures

including physicians, nurses, phlebotomists, transporters, clerical workers, and computer programmers.

Laboratory directors should obtain copies of federal regulations, CAP, and JCAHO standards. Implementing a quality assurance program is time-consuming, but the goal is worthwhile: achieving error-free testing that provides relevant information within a time frame that benefits patient care.

LABORATORY SAFETY

Laboratory-acquired infections have been recognized for many years. Consequently, CAP as well as JCAHO include an evaluation of safety issues in their standards for inspection and accreditation. In the past decade, however, the risk of HIV infection has heightened the concerns of laboratory workers. This concern has led to the introduction of "universal precautions" based on recommendations from the Centers for Disease Control (CDC) and the guidelines on occupational exposure to blood-borne pathogens by the Occupational Safety and Health Administration (OSHA). Laboratory directors should obtain copies of all relevant federal regulations, as well as inspection and accreditation standards, in order to design and implement an appropriate safety program for their laboratories.

Goals of a Laboratory Safety Program

The goal of a laboratory safety program is to provide a safe place in which to work. Procedures and proper equipment must be in place to minimize accidents and prevent exposure to hazardous materials. Personnel must understand the hazards and characteristics of all materials they will handle or be exposed to in the laboratory setting and be trained in proper techniques.

Understanding Risks

There are four categories of laboratory hazards: physical, microbiological, chemical, and radiological. In a virology laboratory, infection can occur from aerosols, contamination of the skin, eyes, or mucous membranes, inoculation by puncture, and by ingestion. Representative viruses and methods to prevent laboratory transmission are listed in table 4.2. Laboratory infections can occur when handling clinical specimens, cultures, or virus stocks. Even uninoculated cell cultures may contain endogenous or adventitious agents that may cause human infection. It should be remembered that although enveloped viruses are usually inactivated by ethanol and acetone, nonenveloped viruses are not, but require chlorox or formalin. Thus, it is extremely important for personnel to adhere strictly to specific laboratory safety rules (see the end of this section).

Table 4.2 Safety Precautions to Avoid Virus Transmission in a Virology Laboratory

Mode of Transmission	Representative Viruses	Safety Precautions
Aerosol inhalation	Respiratory viruses, VZV	Avoid aerosol-generating activities;* use containment devices; work in biological safety cabinet; use centrifuge safety cups
Hand contamination with self-inoculation	Enterovirus, rhinovirus, adenovirus, HSV, CMV, RSV	Wear gloves Frequent handwashing Keep hands away from face, mouth, and eyes Use proper technique
Contamination of abrasions or cuts	HBV, HSV, HIV	Cover all cuts and abrasions Receive HBV vaccine
Injection or splash	HBV, HIV, HSV, CMV	Handle and dispose of sharps safely Use proper technique Wear protective clothing and face shield Work in biological safety cabinet or behind shield Receive HBV vaccine

*Examples: Centrifuging, sonicating, blending, forcefully expelling contents of pipettes or syringes, and popping tops off Vacutainer tubes.

Elements of Laboratory Safety

BIOSAFETY LEVELS

Guidelines for handling infectious agents have been developed and published by the CDC and the National Institutes of Health (NIH). Four biosafety levels have been delineated, with recommended safety facilities, practices, and equipment appropriate to the agent handled (table 4.3). The four levels correspond to the severity of illness associated with the infectious agent and the potential for transmission from an aerosol. Clinical virology laboratories follow biosafety level (BSL) 2 guidelines. The in vitro cultivation of human retroviruses from clinical specimens requires BSL 3 practices, although a BSL 3 facility is not necessary unless large-scale virus production is conducted.

PROCEDURES

Laboratory procedures should be evaluated for potential risks and specific safety precautions should be instituted to prevent the opportunity for trans-

Table 4.3 Summary of Recommended Biosafety Levels for Infectious Agents

Biosafety Level	Practices and Techniques	Safety Equipment	Facilities
1	Standard microbiological practices	None: primary containment provided by adherence to standard laboratory practices during open bench operations.	Basic
2	Level 1 practices, plus: laboratory coats, decontamination of all infectious wastes, limited access, protective gloves, and biohazard warning signs as indicated.	Partial containment equipment (i.e., class I or II biological safety cabinets) used to conduct mechanical manipulative procedures that have high aerosol potential that may increase the risk of exposure to personnel.	Basic
3	Level 2 practices, plus: special laboratory clothing, controlled access.	Partial containment equipment used for all manipulations of infectious material.	Containment
4	Level 3 practices, plus: entrance through changeroom where street clothing is removed and laboratory clothing put on, shower on exit; all wastes are decontaminated on exit from the facility.	Maximum containment equipment (i.e., class III biological safety cabinet or partial containment equipment in combination with full-body, air-supplied, positive-pressure personnel suit) used for all procedures and activities.	Maximum containment

Source: *Biosafety in Microbiological and Biomedical Laboratories*, 1988, U.S. Public Health Service, CDC, and NIH.

mission. Safe laboratory practices and techniques are the most important defense against laboratory acquired infections.

FACILITY DESIGN AND PROTECTIVE EQUIPMENT

Facility design with proper ventilation, equipment, chemical and radioisotope storage areas, is essential. Properly functioning biological safety cabinets (BSC) and centrifuges with sealed buckets provide protection from aerosols. Pipette-aids protect from ingestion which can occur with mouth-pipetting. Protective clothing and gloves provide a barrier between the worker and the work. Eye and face protection should be readily accessible, if splash or spatter is possible.

BIOLOGICAL WASTE, CHEMICAL, AND RADIOISOTOPE HANDLING AND DISPOSAL

Careful attention must be given to waste handling and decontamination: a detailed chemical hygiene program as well as procedures for safe handling and disposal of radioisotopes should be in place. Guidelines for handling and disposal of virological waste are listed below.

General Laboratory Safety Procedures for Handling and Disposal of Virological Waste

1. All contaminated materials, including contaminated pipettes, needles and syringes, should be placed in appropriate containers, properly labeled and autoclaved before being washed or disposed of. No infectious substances should be allowed to enter the building drainage system.
2. Stock solutions of suitable disinfectants, including 10% chlorox and/or 70% ethanol, should be available at the work site.
3. Special attention should be given to personnel in contact with infected animals. Water bottles or cages in which infected animals are kept should be autoclaved after use. Care must be taken to discard infected animal carcasses and tissues that are sent to the incinerator for disposal. Infected embryonic eggs should be treated in the same manner.
4. Autoclave or steam heat, when it can be applied, is the most effective means of sterilization. Where autoclave cannot be applied, chlorox is effective in decontaminating liquid waste and surfaces.

Specific Laboratory Safety Rules for Personnel Working in a Virology Laboratory

1. Wear laboratory coats at all times when working in a virology laboratory, and use gloves when contact with potentially infectious material is possible.
2. Keep all vials, culture tubes, flasks, etc., containing virus *tightly capped* when not being used.
3. Avoid aerosol generating activities or work only with proper containment equipment, such as a biological safety cabinet and centrifuge safety cups.

4. Do not mouth pipette: all dilutions and inoculations of virus are to be done with pipetting devices. Discard contaminated pipettes and waste into containers with disinfectant.
5. *Never* put a contaminated pipette on the surface of your workbench. *Never* walk around with a contaminated pipette in your hand. *Never* discard any virus-containing fluids in the sink.
6. No eating, drinking, or smoking in the laboratory and no storage of food or beverages in laboratory cabinets or refrigerators are allowed.
7. Scrub down the bench tops with disinfectant at the end of each day.
8. Remove gloves and wash hands thoroughly before leaving the laboratory.

TRAINING OF PERSONNEL

A key element in any laboratory safety program is personnel training, which is an ongoing task since safety consciousness requires continual reinforcement. The training of personnel in safety procedures has now been mandated by OSHA and must be done on a regular basis and documented.

Immunization with the hepatitis B virus vaccine is strongly encouraged if contact with human blood is anticipated. Immunization with other virus vaccines is recommended when available and indicated.

ACCIDENT CONTINGENCY PLAN

An accident contingency plan should be developed and in place for biological, chemical, and radioisotope spills. A tray or kit containing appropriate materials to contain and decontaminate spills should be readily available.

INSPECTIONS

Regular safety inspections within the laboratory by an internal safety officer are desirable. A safe workplace is worth the effort.

SUPPLEMENTARY READING, PART ONE

Chapter 1: General Principles and Current Trends in Diagnostic Virology

Chernesky, M. A. An evolutionary change in diagnostic virology. *Yale J. Biol. Med.* 62:89–92 (1989).

Evans, A. S. *Viral Infections of Humans: Epidemiology and Control.* 3d ed. New York: Plenum, 1989.

Fields, B. N., and Knipe, D. M., eds. *Fields Virology.* 2d ed. New York: Raven, 1990.

Gardner, P. S., and McQuillin, J. *Rapid Viral Diagnosis, Application of Immunofluorescence.* 2d ed. London: Butterworth, 1980.

Hsiung, G. D. The impact of cell culture sensitivity on rapid viral diagnosis: A historical perspective. *Yale J. Biol. Med.* 62:79–88 (1989).

Hsiung, G. D.; Fong, C. K. Y.; and August, M. The use of electron microscopy in diagnosis of viral infection. *Prog. Med. Virol.* 25:133–59 (1979).

James, J., and Bell, G. T. Human monoclonal antibody production. *J. Immun. Method.* 100:5–40 (1987).

Kohler, G., and Milstein, C. Continuous culture of fused cells secreting antibody of predefined specificity. *Nature* 256:495–97 (1975).

Lennette, D. A.; George Ray, C.; and Menegus, M. A.; sec. eds. Sec. 2, Viruses. In Balows, A.; Hausler, W. J.; Herrmann, K. L.; Isenberg, H. D.; and Shadomy, H. J., eds., *Manual of Clinical Microbiology,* 5th ed., pp. 811–1032. Washington, D.C.: American Society for Microbiology, 1991.

Richman, D. D. Developments in rapid viral diagnosis. *Infect. Dis. Clin. N. Am.* 1:311–22 (1987).

Schmidt, N. J., and Emmons, R. W. *Diagnostic Procedures for Viral, Rickettsial and Chlamydial Infections,* 6th ed. Washington, D.C.: APHA, 1989.

Specter, S., and Lancz, G. J., eds. *Clinical Virology Manual,* 2d ed. Norwalk, Conn.: Appleton and Lange, 1992.

Spector, S. A., and Dankuer, W. M. Rapid viral diagnostic techniques. *Adv. Ped. Infect. Dis.* 1:37–59 (1986).

Subbarao, E. K.; Griffis, J.; and Warner, J. L. Detection of multiple viral agents in nasopharyngeal specimens yielding respiratory syncytial virus (RSV): An assessment of diagnostic strategy and clinical significance. *Diag. Microbiol. Infect. Dis.* 12:327–32 (1989).

Yolken, R. H.; Coutlee, F.; and Viscidi, R. P. New prospects for the diagnosis of viral infections. *Yale J. Biol. Med.* 62:131–39 (1989).

Chapter 2: Properties and Classification of Viruses

Andrewes, C. H. Nomenclature of viruses. *Nature* (London) 173:260–61 (1954).

Buck, C., and Paulino, G., eds. *American Type Culture Collection, Catalogue of Animal Viruses and Antisera, Chlamydiae and Rickettsiae,* 6th ed. Rockville, Md.: American Type Culture Collection, 1990.

Francki, R. I. B.; Faquet, C. M.; Knudson, D. L.; and Brown, F.; eds., *Classification and Nomenclature of Viruses.* Fifth Report of the International Committee on Taxonomy of Viruses, Archives of Virology, supp. 2. New York: Springer-Verlag, 1990.

Karabatsos, N. *International Catalog of Arboviruses including Certain Other Viruses of Vertebrates,* 3d ed. San Antonio: American Society for Tropical Medicine and Hygiene, 1985.

Kingsbury, D. W. Biological concepts in virus classification. *Intervirology* 29:242–53 (1988).

Melnick, J. L. Taxonomy and nomenclature of viruses. *Intervirology* 17:1–200 (1982).

Murphy, F. A., and Kingsbury, D. W. Virus taxonomy. In Fields, B. N., and Knipe, D. M., eds., *Fields Virology*, 2d ed., pp. 9–35. New York: Raven, 1990.

White, D. O. Virus classification. In Dulbecco, R., ed., *Encyclopedia of Human Biology* (1991), s.v. "virology, medical."

Chapter 3: Specimen Collection, Transport, and Processing for Virological Studies

Bettoli, E. J.; Brewer, P. M.; Oxtoby, M. J.; Zaidi, A. A.; and Guinan, M. E. The role of temperature and swab materials in the recovery of herpes simplex virus from lesions. *J. Infect. Dis.* 145:399 (1982).

Crane, L. R.; Gutterman, P. A.; Chapel, T.; and Lerner, A. M. Incubation of swab material with herpes simplex virus. *J. Infect. Dis.* 141:531 (1980).

Howell, C. L.; Miller, M. J.; and Martin, W. J. Comparison of rates of virus isolation from leukocyte populations separated from blood by conventional and Ficoll-Paque/Macrodex methods. *J. Clin. Microbiol.* 10:533–37 (1979).

Huntoon, C. J.; House, R. F., Jr.; and Smith, T. F. Recovery of viruses from three transport media incorporated into culturettes. *Arch. Path. Lab. Med.* 105:436–37 (1981).

Johnson, F. B.; Leavitt, R. W.; and Richards, D. F. Evaluation of the Virocult transport tube for isolation of herpes simplex virus from clinical specimens. *J. Clin. Microbiol.* 20:120–22 (1984).

Perez, T. R.; Mosman, P. L.; and Juchau, S. V. Experience with Virocult as a viral collection and transportation system. *Diag. Microbiol. Infect. Dis.* 2:7–9 (1984).

Rodin, P.; Hare, M. J.; Barwell, C. F.; and Withers, M. J. Transport of herpes simplex virus in Stuart's medium. *Br. J. Vener. Dis.* 47:198–99 (1971).

Schmidt, N. J., and Emmons, R. W. General principles of laboratory diagnostic methods for viral, rickettsial and chlamydial infections. In Schmidt, N. J., and Emmons, R. W., eds., *Diagnostic Procedures for Viral, Rickettsial and Chlamydial Infections*, 6th edition, pp. 1–35. Washington, D.C.: American Public Health Association, 1989.

Yeager, A. S.; Morris, J. E.; and Prober, C. G. Storage and transport of cultures for herpes simplex virus type 2. *Am. J. Clin. Path.* 72:977–79 (1979).

Chapter 4: Quality Assurance and Laboratory Safety

Centers for Disease Control. Guidelines for prevention of transmission of human immunodeficiency virus and hepatitis B virus to health-care and public-safety workers. *Morbid. Mortal. Weekly Rep.* 38 (supp. 6): 1–37 (1989).

———. Recommendations for prevention of HIV transmission in health-care settings. *Morbid. Mortal. Weekly Rep.* 36:3S–18S (1987).

———. Update: Universal precautions for prevention of transmission of human immunodeficiency virus, hepatitis B virus, and other blood-borne pathogens in health-care settings. *Morbid. Mortal. Weekly Rep.* 37:377–82, 387–88 (1988).

Commission on Laboratory Accreditation. Inspection checklist. Sec. 1, Laboratory General. Skokie, Ill.: College of American Pathologists, 1988.

Committee on Hazardous Substances in the Laboratory, National Research Council. Prudent practices for handling hazardous chemicals in laboratories. Washington, D.C.: National Academy Press, 1980.

Department of Labor, Occupational Safety and Health Administration. 29 CFR 1910. 1030. Occupational exposures to blood-borne pathogens: Final rule. *Fed. Regist.* 56:64175–65182 (1991).

———. Occupational exposures to hazardous chemicals in laboratories: Final rule. *Fed. Regist.* 56:3300–35 (1990).

Joint Commission for the Accreditation of Healthcare Organizations. *Accreditation Manual for Hospitals*. Chicago: Joint Commission for the Accreditation of Healthcare Organizations, 1989.

National Committee for Clinical Laboratory Standards. Tentative Guideline M29-T: Protection of laboratory workers from infectious disease transmitted by blood and tissue. Villanova, Pa., 1989.

U.S. Postal Service. 39 CFR, pt. 3. Mailability of etiologic agents. *Fed. Regist.* 54:11970–972 (1989).

U.S. Public Health Service, Centers for Disease Control, and National Institutes of Health. Biosafety in microbiological and biomedical laboratories, 2d ed. Publication (NIH) no. 88–8395, Washington, D.C.: Department of Health and Human Services, 1988.

PART TWO

METHODS COMMONLY USED FOR VIRUS ISOLATION AND IDENTIFICATION, VIRAL ANTIGEN AND GENOME DETECTION, AND ANTIBODY DETERMINATION

5

VIRUS ISOLATION AND IDENTIFICATION METHODS

G. D. HSIUNG

The methods commonly used for virus isolation vary from laboratory to laboratory. These variations depend not only on the availability of specialized resources and equipment of each setting, but also on the availability of trained personnel. In many instances, the isolation of a viral agent is the most sensitive and accurate method for the diagnosis of a viral infection and is considered the gold standard with a few exceptions. Since viruses replicate only in living cells, a constant supply of cell cultures and experimental animals is necessary (figure 5.1). Newborn mice and embryonated chicken eggs are useful alternate host systems, but they are used only in selected laboratories. Figure 5.2 outlines the methods of isolating and identifying viruses encountered in a clinical laboratory. Cell cultures remain the most commonly used system in diagnostic virology laboratories today.

CELL CULTURE SYSTEMS

Since the discovery that the poliovirus multiplies in cultured nonneural tissues, cell cultures have become the most convenient system for the isolation of many viruses. A variety of primary cell cultures and passaged cell lines are used. Although most cell types are available commercially, many can be readily prepared by individual laboratories. Ideally, only freshly prepared cell cultures should be used. Aged cells are frequently less sensitive to virus infection. In table 5.1, the four cell culture systems commonly used in the isolation of viruses of the major virus groups are given. Primary rabbit kidney or guinea pig embryo cells are especially useful in the rapid identification of

ISBN 0-300-05845-4.

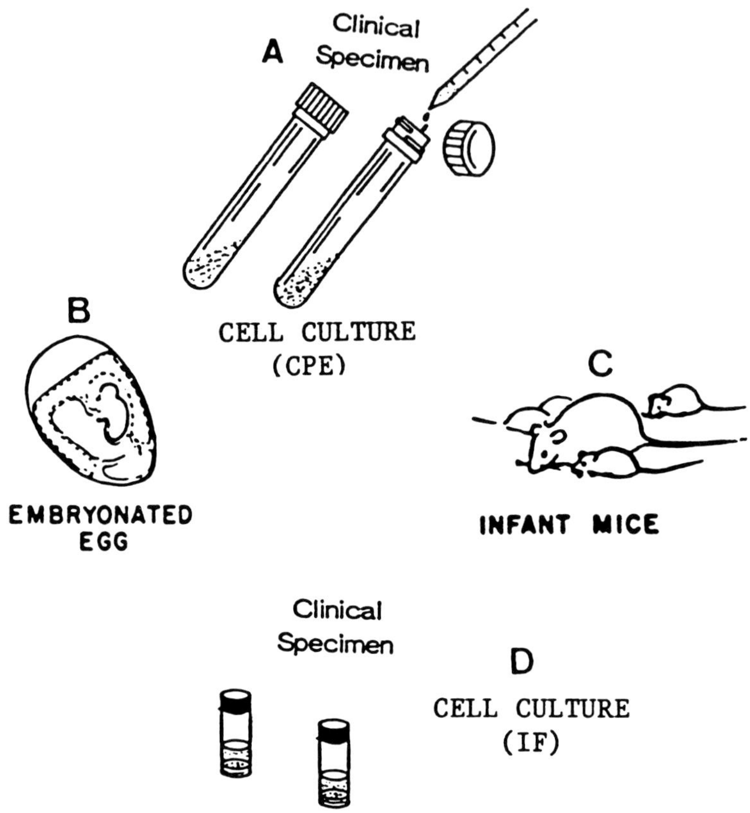

Figure 5.1 Host systems for virus isolation: *(A)* Cell culture in roller tubes for observation of CPE; *(B)* Embryonated eggs; *(C)* Newborn mice; *(D)* Cell cultures on coverslips in shell vial for immunocytochemical staining

HSV since other viruses do not induce visible changes in these cell systems. (The details of cytopathology induced by individual viruses are given in separate chapters dealing with specific virus groups.) In general, the following procedures apply:

Conventional Tube Culture (figure 5.3, *A* and *B*)

1. Remove medium, inoculate specimens—0.1–0.3 ml into each culture tube—and incubate inoculated cultures at 37°C for one hour in stationary racks. Add maintenance medium. Incubate cultures in a roller drum or stationary racks.
2. Check inoculated culture tubes daily for cytopathic effects. (Certain cell lines [for example, A549 or Hep-2 cultures, derived from human carcinoma tissues] require frequent medium changes.)
3. When CPE occurs, subculture into a fresh culture of the same cell type to ensure recovery of sufficient virus for identification of the isolate.

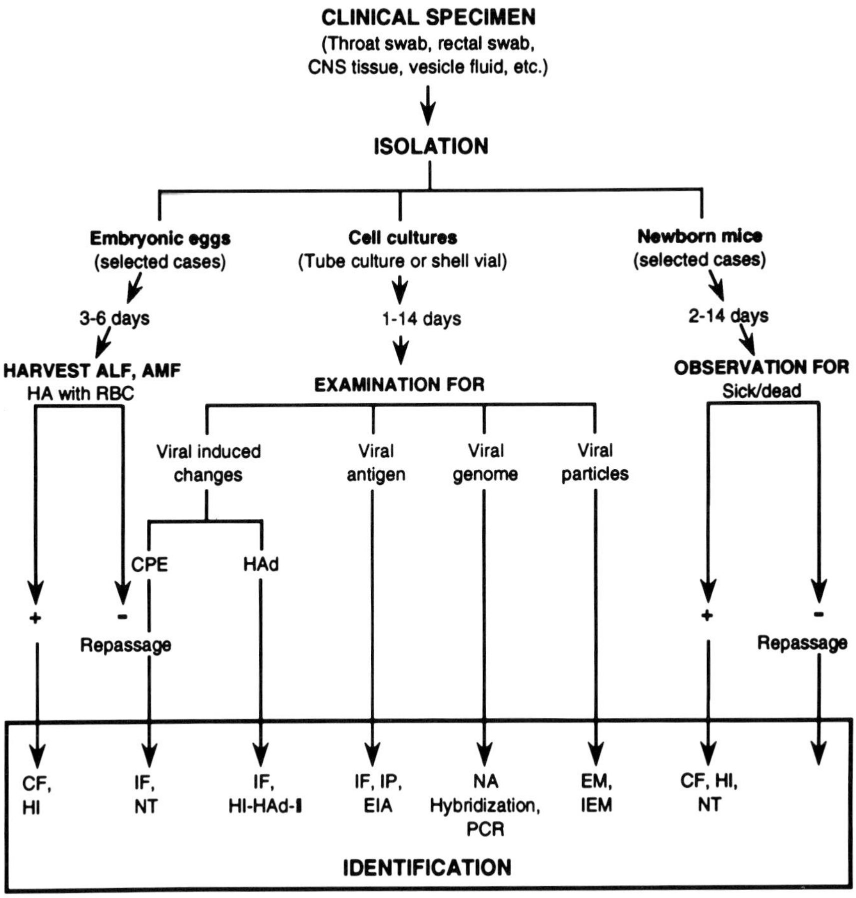

Figure 5.2 Flowchart for virus isolation and identification

ALF allantoic fluid
AMF amniotic fluid
CF complement fixation test
CNS central nervous system
CPE cytopathic effect
EIA enzyme immunoassay
EM electron microscopy
HA hemagglutinin
HAd hemadsorption test
HI hemagglutination-inhibition test
IEM immunoelectron microscopy
IF immunofluorescent staining
IP immunoperoxidase test
NA nucleic acid hybridization
NT neutralization test
PCR polymerase chain reaction
RBC red blood cells

Table 5.1 Different Levels of Sensitivity of Cell Culture Systems to Viruses Commonly Encountered in a Clinical Virology Laboratory

Virus	CPE in Cell Cultures[a] Primary Monkey Kidney	Human Embryo Fibroblast	A549/ HEP-2 Cell Lines	PRK or GPE[b]
RNA virus				
Enterovirus (except coxsackie A)	+++	++	++/−	−
Rhinovirus	+	+++	+	−
Influenza and para-influenza viruses	(HAd+++)	(HAd+)	(HAd+)	−
Respiratory syncytial virus	(SYN.+)	+	(SYN.+++)	−
DNA virus				
Adenoviruses	+/−	++	+++	+/−
Herpes simplex viruses	+/−	++	++	+++
Varicella-zoster virus	+/−	+++	+[c]/−	+/−
Cytomegalovirus	−	+++	−	−

a. Level of sensitivity by CPE: − nonsensitive; + low sensitive; ++ moderately sensitive; +++ highly sensitive

b. When HSV is expected, primary rabbit kidney or guinea pig embryo cells are used.

c. A549 cells are occasionally susceptible to VZV.

4. For some viruses, such as the influenza and parainfluenza viruses, where CPE may not be evident, the hemadsorption test is commonly used (see chapter 15).

Centrifugation Culture in Shell Vial (figure 5.3, *C* and *D*)

Detection of CPE in tube cultures may require only one day for viruses such as poliovirus or herpes simplex virus but 3–14 days for other viruses such as cytomegalovirus. It has been recognized that centrifugation of clinical specimens containing viruses or chlamydia onto cell monolayers enhances infectivity. For the inoculation of shell vial cultures, the following procedures apply:

1. Remove medium from each shell vial culture containing coverslip with cell monolayer.
2. Inoculate specimens—0.1–0.3 ml into each shell vial.
3. Centrifuge the inoculated shell vial cultures at 700 × *g* for 30–60 minutes.
4. Wash the infected shell vial culture once with PBS when inoculated with blood, urine, or stool sample to reduce toxicity.

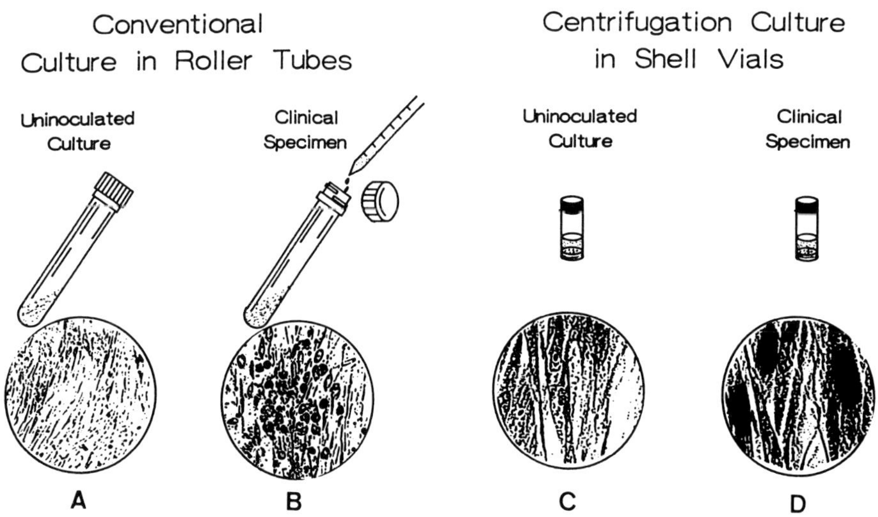

Figure 5.3 Conventional culture in roller tubes vs. centrifugation culture in shell vials: *(A)* Uninfected human embryo fibroblast (HEF) cell monolayer in a roller tube; *(B)* HEF cells showing CMV-induced cell rounding (*A* and *B*, 100×); *(C)* Uninfected HEF cells on a coverslip, fixed, and stained; *(D)* HEF cells fixed and stained with immunoperoxidase showing CMV-induced antigen in the nuclei of infected cells (dark-stained areas) (*C* and *D*, 400×). (Reproduced with permission, M. L. Landry and G. D. Hsiung, *Clinical Virology Manual,* 2d ed., S. Specter and G. J. Lancz, eds., Norwalk, Conn.: Appleton and Lange, 1992, 47)

5. Add 1-ml maintenance medium to each shell vial and then incubate at 37°C for 24–48 hours.
6. Examine the shell vial monolayer for CPE or other evidence of cellular changes.
7. Remove the supernate. Add 1-ml PBS to wash the infected cells on the coverslip.
8. Add 1 ml of cold acetone to fix the infected cells on the coverslip for 10 minutes.
9. For identification, remove the coverslip containing fixed cells from the shell vial, and stain with specific antibody, using either direct or indirect procedures with a fluorescein or peroxidase label, as desired (see chapter 7).

IDENTIFICATION OF VIRUSES ISOLATED IN CELL CULTURE SYSTEMS

Examples of cellular changes induced by major groups of viruses in different cell cultures are represented in figure 5.4. Although a virus may be able to

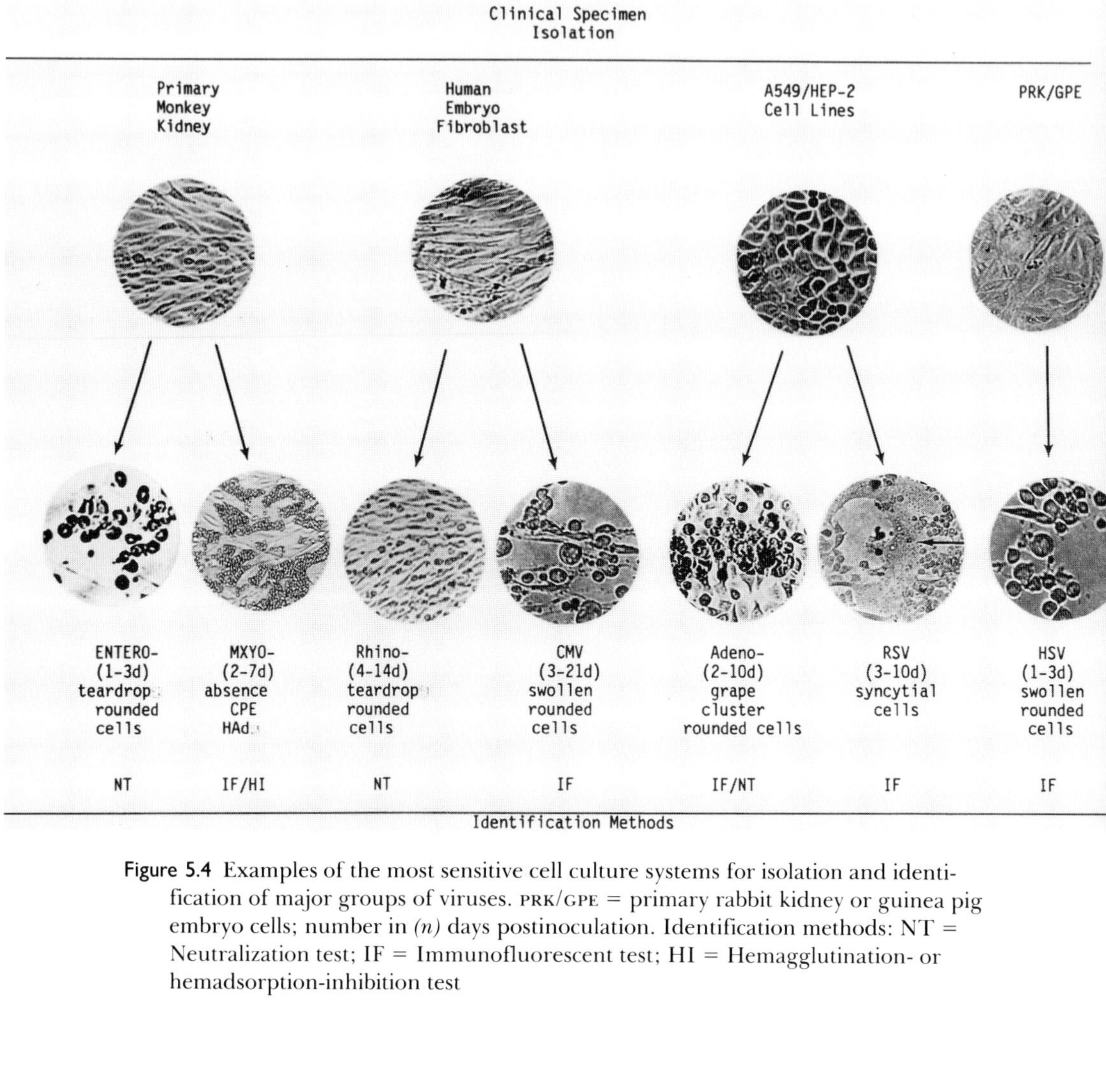

Figure 5.4 Examples of the most sensitive cell culture systems for isolation and identification of major groups of viruses. PRK/GPE = primary rabbit kidney or guinea pig embryo cells; number in *(n)* days postinoculation. Identification methods: NT = Neutralization test; IF = Immunofluorescent test; HI = Hemagglutination- or hemadsorption-inhibition test

infect and replicate in several cell systems, some cell cultures are more sensitive than others (see table 5.1).

When a virus isolate is obtained after the inoculation of several types of cell cultures, a presumptive identification can often be made based upon the type of cell culture that exhibits cytopathic effects, the characteristic CPE observed, and the specimen source. A final identification is made by the use of a virus-specific antiserum to detect viral antigens or to inhibit virus-induced changes in infected cultures. With the availability of monoclonal antibodies for most of the commonly isolated viruses in the clinical laboratory, immunofluorescent (IF) staining (see chapter 7) has been used routinely as soon as CPE-HAds occurs, with a few exceptions (see figure 5.4). For certain viruses, such as the enterovirus group, a neutralization test (NT) in the cell culture is necessary to obtain final identification. For other viruses, immunologic tests such as hemadsorption- or hemagglutination-inhibition and complement-fixation can also be used.

APPENDIX: EMBRYONATED EGGS AND INFANT MICE

In selected cases, the following two techniques may be used when available.

Newborn Mouse Inoculation. Most group A coxsackieviruses and some arboviruses are isolated best in newborn mice. Specimens suspected of harboring these viruses should be inoculated into this animal system if available. Caution should be taken to avoid errors related to the method of inoculation (figure 5.5).

1. Inoculate specimens into newborn mice within 24–48 hours of birth with 0.01–0.02 ml/mouse intracerebrally and/or 0.03–0.05 ml/mouse intraperitoneally, using a 25-gauge 3/8″-long needle and a 1-cc syringe.
2. Check the inoculated newborn mice twice daily for signs of illness, paralysis, or death.
3. Harvest mouse brain or skeletal muscle when the animals are paralyzed or when other symptoms appear. Make a 10% tissue suspension and inoculate it into the appropriate monolayer cell culture or mice for further study.

Embryonated Egg Inoculation. The chick embryo is a highly sensitive host for the primary isolation of several virus types. For certain strains of influenza virus A and for the mumps virus, the amniotic route of inoculation is preferable. Chorioallantoic membranes, however, are highly susceptible to pock formation by herpes simplex virus and poxviruses.

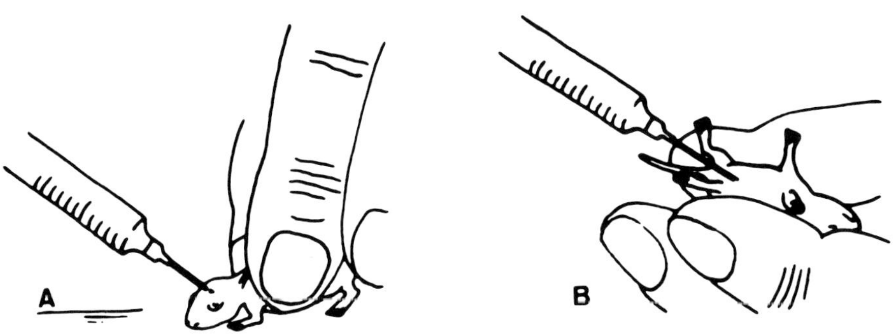

Figure 5.5 Mouse inoculation: *(A)* Intracerebral route; *(B)* Intraperitoneal route

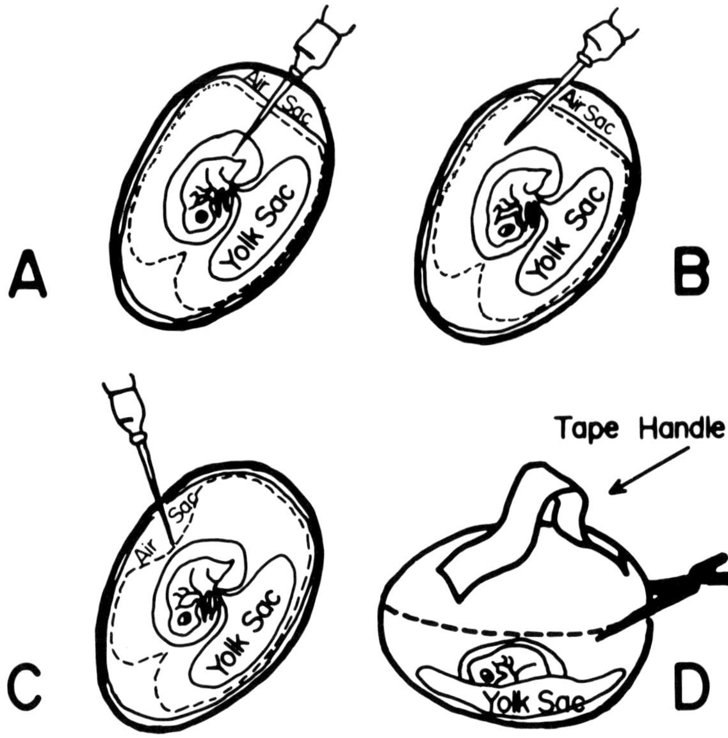

Figure 5.6 Embryonated egg inoculation and harvesting: *(A)* Amniotic cavity inoculation; *(B)* Allantoic cavity inoculation; *(C)* Chorioallantoic membrane inoculation; *(D)* Harvesting chorioallantoic membrane

Amniotic or Allantoic Cavity Inoculation

1. Use 7- to 13-day-old embryonated eggs (the optimum for mumps is 7–8 days; for influenza, 10–13 days).
2. Candle the eggs and make a puncture through the shell over the air sac.
3. Inoculate 0.1–0.2 ml of the specimen into the amniotic sac (figure 5.6*A)* or allantoic cavity (figure 5.6*B)* using a 23-gauge, 1¾″-long needle. Use three or four eggs for each specimen. (Note: These procedures should be done under an egg candler.)
4. Seal the hole in the shell with scotch tape.
5. Incubate eggs at 35°–37°C with the air sac uppermost.
6. Candle inoculated eggs daily. Discard those that die within 24 hours after inoculation.
7. Harvest the amniotic fluids (AMF) and allantoic fluids (ALF) separately, 2–4 days after inoculation for influenza and 5–7 days for mumps, according to the following procedure:
   a. Chill the eggs at 4°C for 2–4 hours.

SECTION I. Immunofluorescent Staining

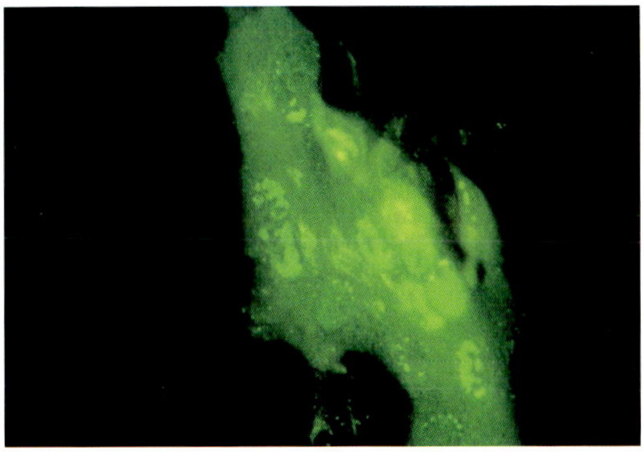

1. HSV-2 in infected HEF monolayer cells stained with antibody to HSV-2 (200×) (Bion). (Compare plate 2.)

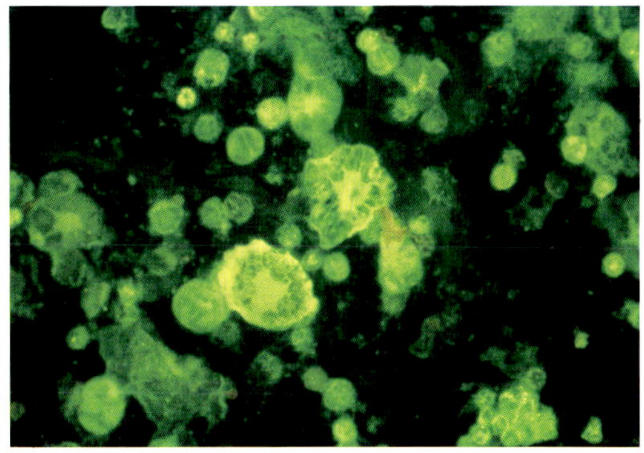

2. HSV-2 in cell spot made from infected A549 monolayer cells stained with monoclonal antibody to HSV-2, showing two multinucleated giant cells in the center of field (200×)

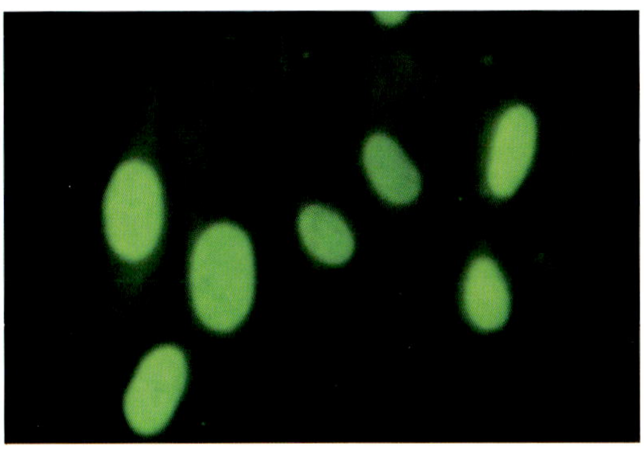

3. Cytomegalovirus-infected MRC-5 monolayer cells in shell vial culture stained with monoclonal antibody to CMV early antigen (400×). (Compare plate 4.)

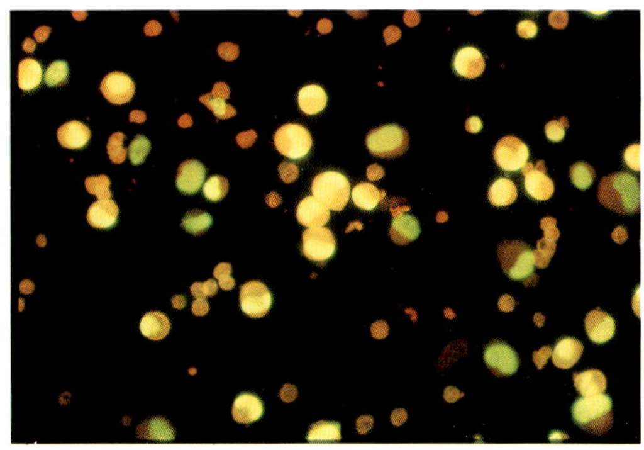

4. Cytomegalovirus in cell spot made from infected MRC-5 monolayer cells stained with monoclonal antibody to CMV early antigen (200×)

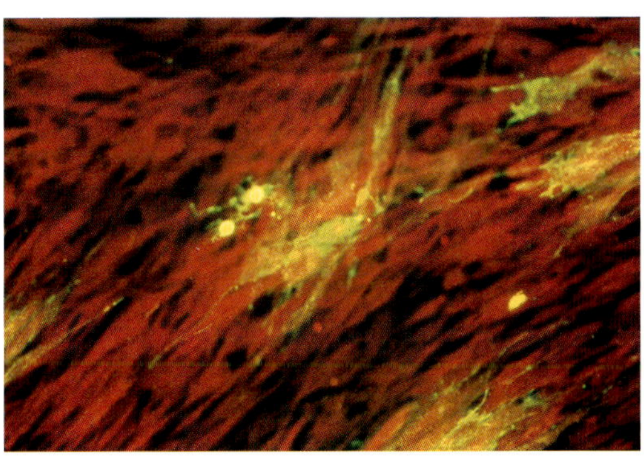

5. Varicella virus–infected MRC-5 monolayer cells in shell vial culture stained with monoclonal antibody to VZV (200×)

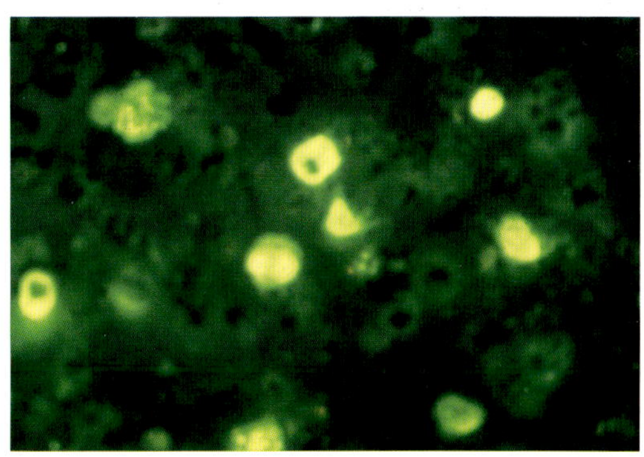

6. Adenovirus in cell spot made from infected A549 monolayer cells stained with monoclonal antibody to hexon (200×)

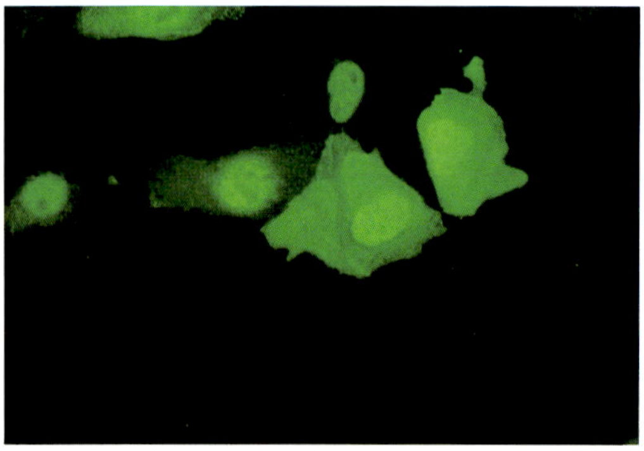

7. Influenza A virus–infected Vero cell monolayer culture stained with antibody to influenza A virus (200×) (Bion). (Compare plate 8.)

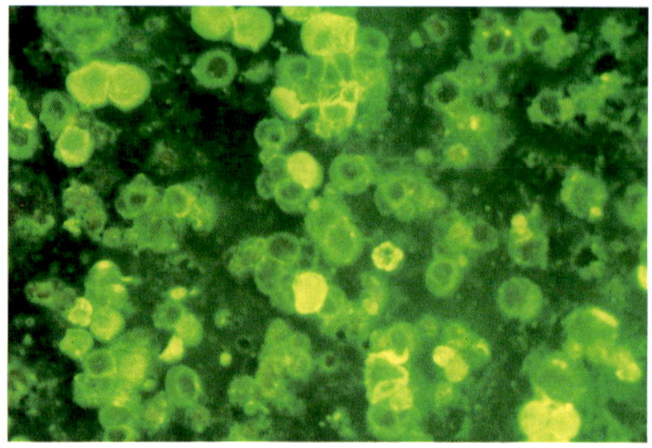

8. Influenza A virus in cell spot made from infected monkey kidney cells stained with monoclonal antibody to influenza A virus (200×)

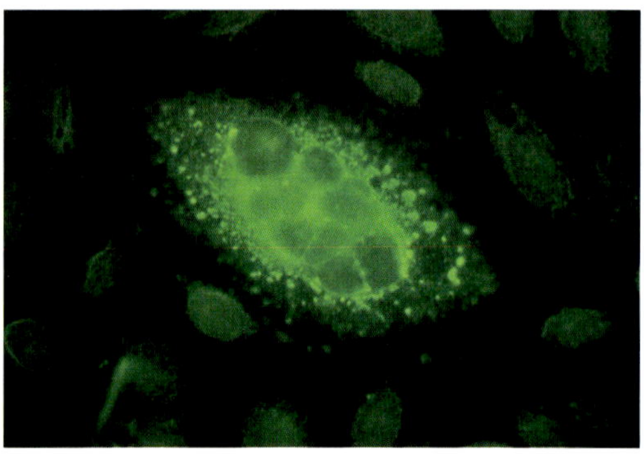

9. Measles virus–infected Hep-2 cell monolayer culture stained with antibody to measles virus. (Note: multinucleated giant cells [200×]) (Bion)

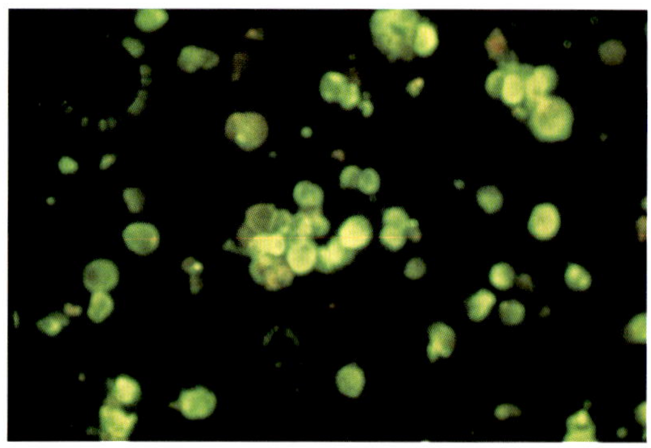

10. Parainfluenza virus 3 in cell spot made from infected monkey kidney cell monolayer culture stained with monoclonal antibody to parainfluenza virus 3 (200×)

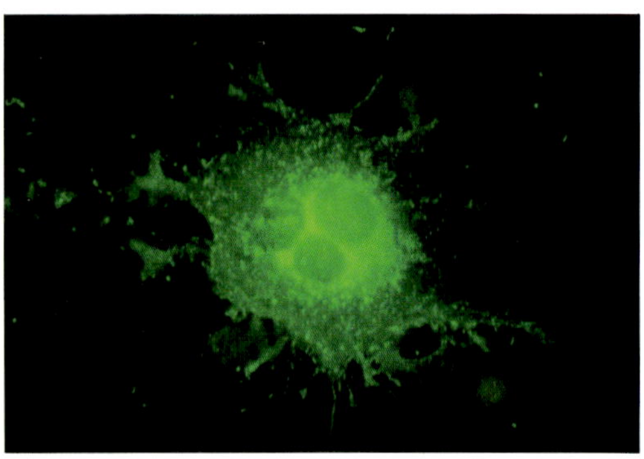

11. Respiratory syncytial virus–infected Hep-2 cell monolayer culture stained with antibody to RSV. (Note: syncytial formation [200×]) (Bion)

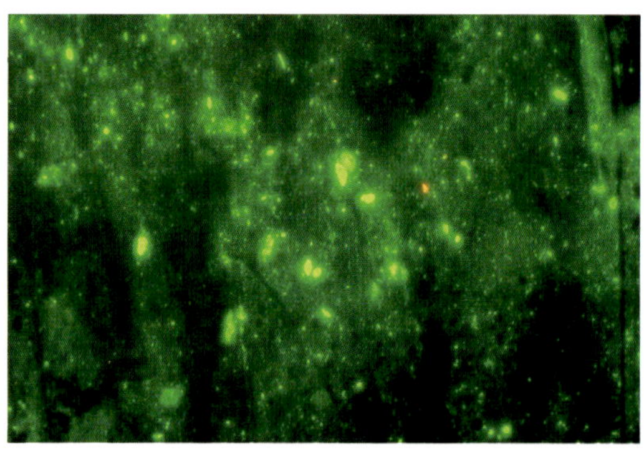

12. Rabies virus nucleocapsid antigen in an impression smear made from a raccoon's brain stained with monoclonal antibody to rabies virus (200×). (Courtesy D. Mayo, State of Connecticut, Divison of Biological Sciences)

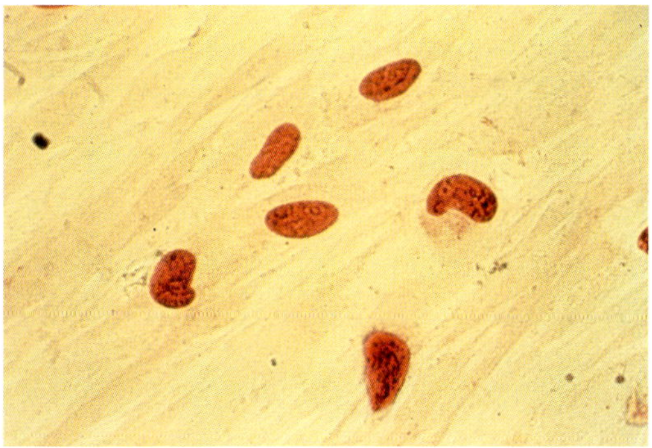

13. Immunoperoxidase staining of CMV-induced early antigen in infected MRC-5 monolayer cells in shell vial culture. (Compare plate 3 [400×])

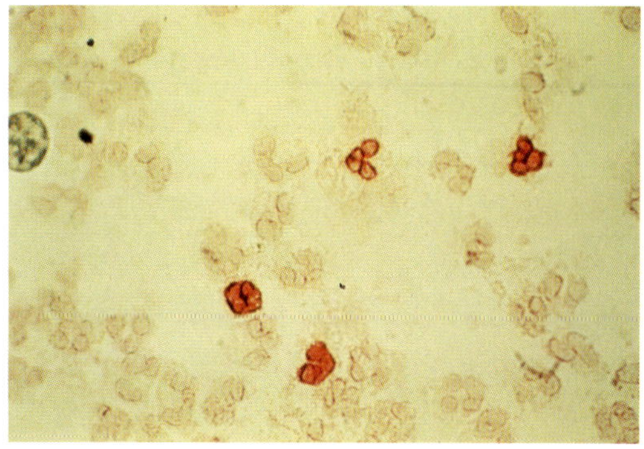

14. Immunoperoxidase staining of CMV-induced early antigens in polymorphonuclear cells in peripheral blood (100×)

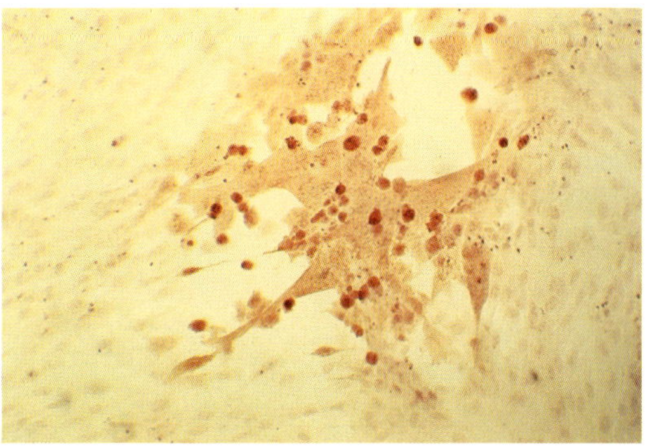

15. Immunoperoxidase staining of HSV-2–infected MRC-5 monolayer cells with polyclonal antibody to HSV-2 (100×)

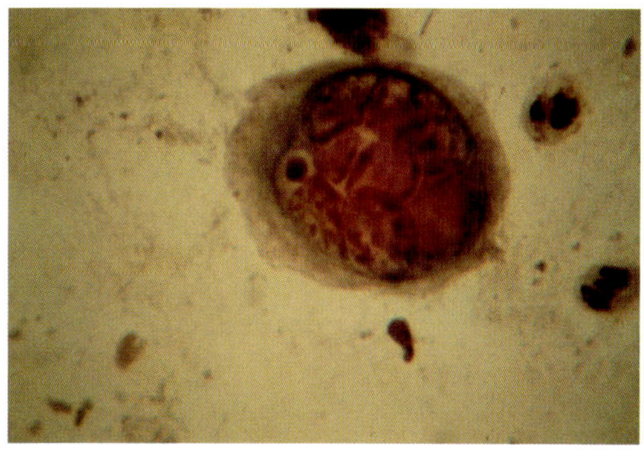

16. A Pap smear showing a multinucleated giant cell prepared from a genital lesion induced by HSV-2 stained with papanicolaou stain (400×)

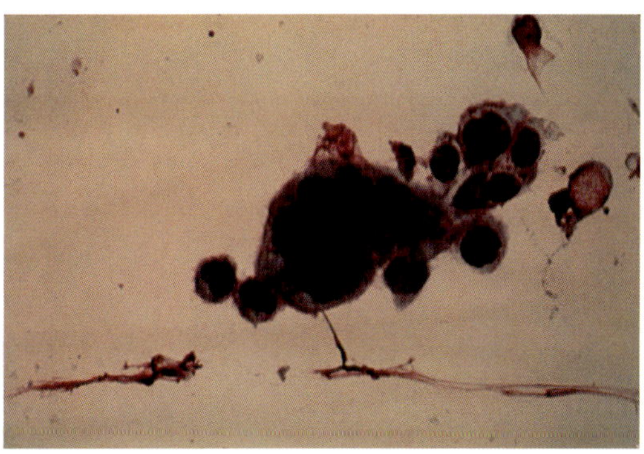

17. A Tzanck smear of an HSV-induced skin lesion stained with Wright's stain showing multinucleated syncytial cells (200×). (Courtesy of L. Milstone, West Haven VA Medical Center)

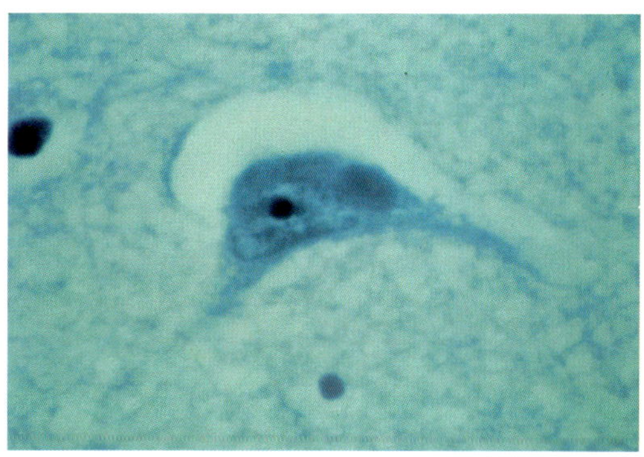

18. A section of hippocampus showing rabies virus–induced Negri bodies in a brain cell (H&E stain, 950×). (Courtesy H. Lucia, University of Mississippi Medical Center)

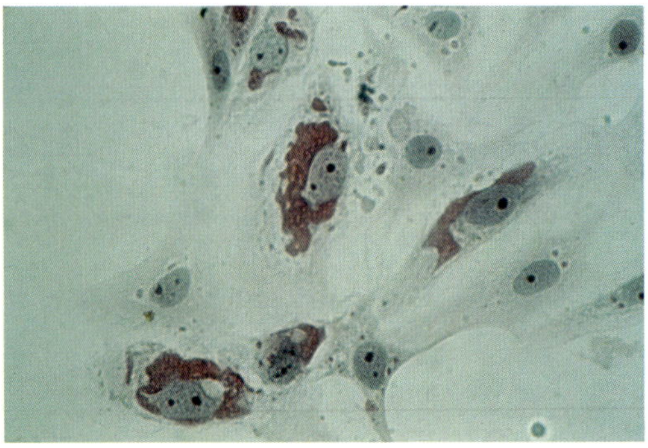

19. Reovirus 1–infected monkey kidney cells showing cytoplasmic eosinophilic inclusions (H&E stain, 400×)

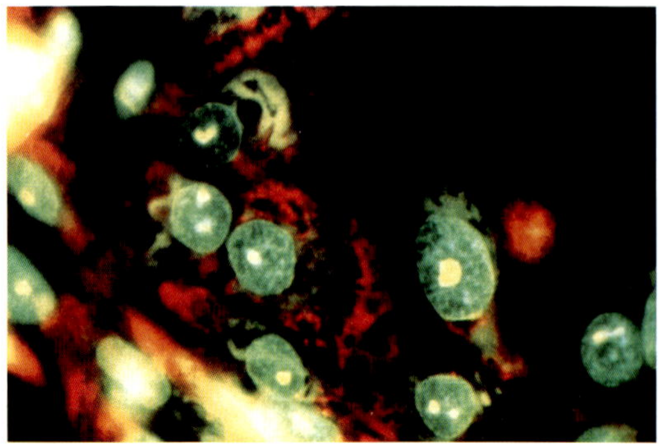

20. Reovirus 1–infected monkey kidney cells showing cytoplasmic inclusions in green color indicating double-stranded RNA (acridine orange stain, 400×)

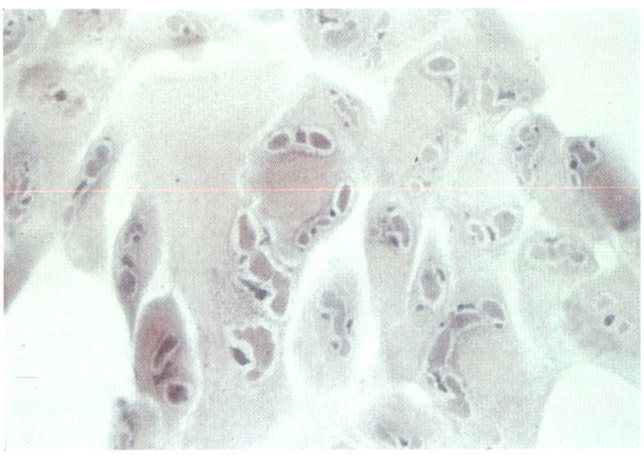

21. Cytomegalovirus-infected MRC-5 cells showing intranuclear inclusions (H&E stain, 400×)

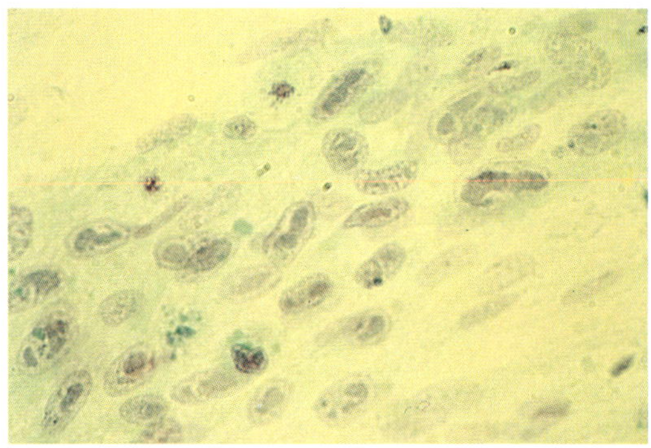

22. Cytomegalovirus-infected MRC-5 cells showing DNA as dark red nuclear inclusions (Feulgen reaction) (400×)

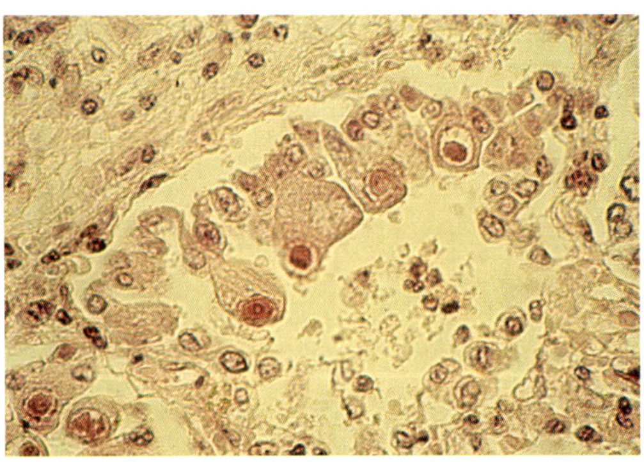

23. Cytomegalovirus in lung tissue section showing nuclear inclusions (H&E stain, 400×)

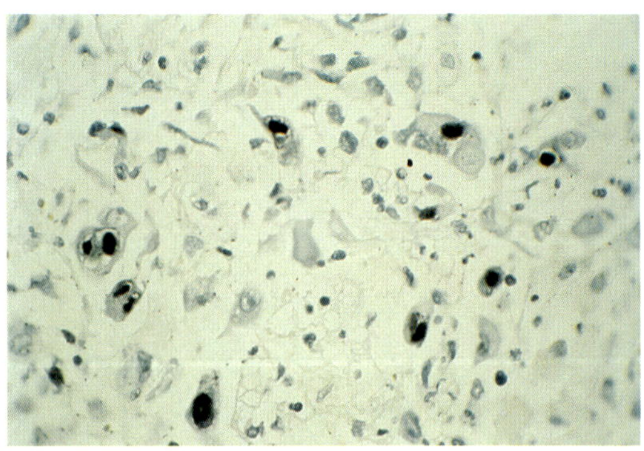

24. Cytomegalovirus in lung tissue section; in situ hybridization using CMV-DNA probe (400×). (Courtesy David Myerson, Fred Hutchinson Cancer Research Center)

b. Open the eggshell over the air sac.
c. Cut with scissors to remove the overlying shell and chorioallantoic membranes.
d. Aspirate the ALF and AMF separately with sterile capillary pipettes.

8. Carry out a spot hemagglutination test by mixing 0.4 ml ALF or AMF and 0.4 ml of a 0.5% suspension of guinea pig RBC. Allow this mixture to stand at room temperature for 30–45 minutes before reading.

Chorioallantoic Membrane Inoculation and Harvesting

1. Candle those eggs containing 9- to 12-day-old developing chicken embryos. Mark an area free from large blood vessels on the side where the embryo is located and near the area over the air sac.
2. Drill two slits in the egg shell, one on the side and the other over the air sac.
3. Puncture the shell membrane under the slits with a sterile needle. Care should be taken not to damage the chorioallantoic membrane (CAM).
4. With a rubber bulb, gently apply suction at the hole over the air sac. If this procedure is carried out while the egg is being candled, one can see the CAM drop and a new, artificial air sac form (figure 5.6*C*).
5. Place 0.05 ml of specimen on the dropped CAM membrane by inserting the needle through the side slit to a depth of about 5 mm. Withdraw the needle very slowly. Seal the opening with scotch tape or wax and incubate at 35°–37°C with the artificial air sac uppermost.
6. Candle the inoculated eggs daily. Discard those that die within 24 hours of inoculation.
7. Harvest the CAM as follows:
   a. Place the inoculated egg on a holder so that the slit through which the inoculum was delivered faces upward.
   b. Make a tape handle over the area of inoculum (figure 5.6*D*).
   c. Cut off the top half of the egg shell, including the infected area, and gently remove the CAM, which is attached to the shell.
   d. Place the infected CAM in a petri dish with a few ml of PBS. Spread the membrane flat against the bottom of the petri dish. Place the dish on a dark surface to facilitate counting the pocks.

6

VIRUS ASSAY, NEUTRALIZATION TEST, AND ANTIVIRAL ASSAY

G. D. HSIUNG

VIRUS ASSAY PROCEDURES

Virus Assay by Cytopathic Effect in Cell Cultures with Liquid Medium

The rate of cellular change and the pattern of CPE induced by different viruses vary greatly depending upon (a) the type of cell culture system used, (b) the concentration of the virus in the specimen, and (c) the properties of each virus strain.

To assay the number of infectious virions in a virus stock, it is necessary to make a serial dilution of the stock virus suspension using separate pipettes.

MAKE SERIAL 10-FOLD DILUTIONS (FIGURE 6.1)

1. Set up a row of six tubes; add 0.9 ml of diluent (PBS or Hanks' BSS) to each tube; discard the pipette.
2. Add 0.1 ml of virus stock to tube 1; discard the pipette.
3. Use a separate sterile pipette, mix the contents in the first tube, and then transfer 0.1 ml of the mixture to tube 2; discard the pipette.
4. Repeat step 3 from tube 2 to tube 6.

INOCULATE CULTURES AND CHECK RESULTS

1. Use the same sterile pipette, starting from the highest dilution—that is, from tube 6—measure 0.1 ml of each dilution, and inoculate into a culture tube or shell vial culture, 2–4 tubes per dilution, until tube 1 is reached.
2. Incubate all inoculated cultures at 37°C.

ISBN 0-300-05845-4.

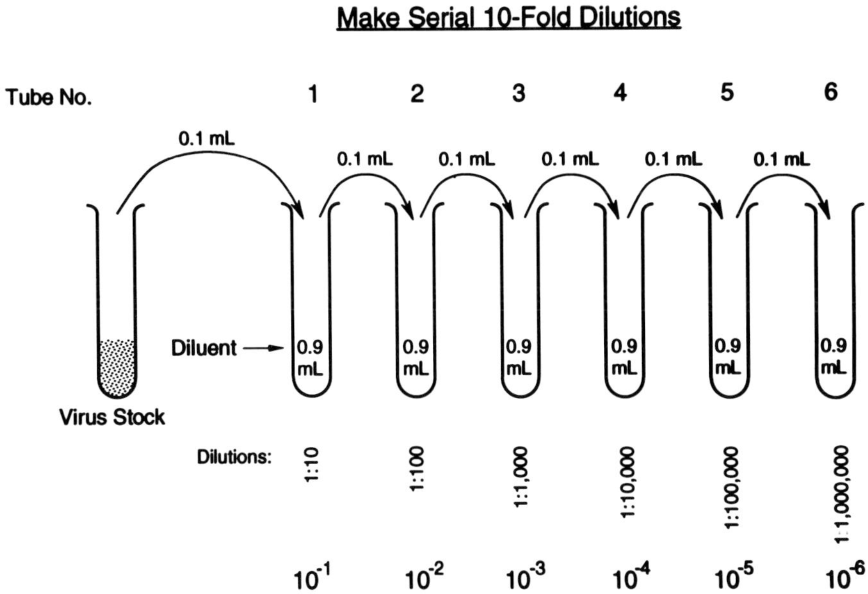

Figure 6.1 Example for making serial 10-fold dilutions. (Note: Pipettes should be changed between each dilution.)

3. Check CPE daily, record results, and calculate virus infectivity titer as outlined below.

CALCULATION OF 50% TISSUE CULTURE INFECTIOUS DOSE ENDPOINT ($TCID_{50}$) EXAMPLE

Dilution of Virus	No. of Cultures Showing CPE vs. No. Inoculated	Cumulative No. Infected	Cumulative No. Not Infected	Calculated Infectivity Ratio	Percent
10^{-3}	4/4	↑ 9	0	9/9	100
10^{-4}	3/4	5	1	5/6	83
10^{-5}	2/4	2	3	2/5	40
10^{-6}	0/4	0	↓ 7	0/7	0

In the example, calculation of the $TCID_{50}$ endpoint using the Reed and Muench formula is as follows:

1. Obtain cumulative numbers of infected cultures that show CPE by adding the numbers in the direction indicated by the arrow. (In this way, one assumes that a culture showing CPE at a given dilution of virus would show CPE at a higher concentration.)
2. Obtain cumulative numbers of noninfected cultures (no CPE): first subtract the total number of inoculated cultures minus the number

showing CPE (at 10^{-4}, 4 less 3 = 1), then add the differences in the direction indicated by the arrow.

3. Calculate the percentages of infectivity by dividing the cumulative number of infected cultures over the sum of the cumulative infected and noninfected cultures (at 10^{-4}, $5 \div 5 + 1 = 0.83 \times 100 = 83\%$).
4. Calculate the proportionate distance between the two dilutions, 10^{-4} and 10^{-5}, where 50% lies:

$$\frac{\text{infectivity above } 50\% - 50\%}{\text{infectivity above } 50\% - \text{infectivity below } 50\%} = \frac{83 - 50}{83 - 40} = 0.7$$

5. Calculate the TCID_{50} endpoint:
   The negative logarithm of the dilution next above 50% is −4.0. The proportionate distance factor (0.7) multiplied by the logarithm of the increment of dilution steps used $(0.7 \times -1) = -0.7$. Thus, the TCID_{50} endpoint $= -4.0 + (-0.7) = -4.7$.
   Therefore the log of TCID_{50} endpoint or titer $= 10^{-4.7}$.
6. Interpretation of TCID_{50} titer: A virus suspension with a TCID_{50} of $10^{-4.7}$ per 0.1 ml represents one TCID_{50}; that is, at such dilution, 50% of the cultures inoculated will become infected (show CPE). A dilution of $10^{-2.7}$ of virus suspension will contain 100 TCID_{50} in a volume of 0.1 ml.

Plaque Formation in Plate Cultures under Methyl Cellulose Overlay Medium

Virus plaques are areas of necrotic cells surrounded by viable cells. Different animal viruses induce plaques of varying sizes and shapes under a solid overlay medium in much the same manner as different bacteria produce characteristic colonies on agar plates. Since one infectious virus unit is theoretically capable of initiating one plaque, this technique can be used both for the accurate quantitative assay of virus infectivity and for the purification of virus strains.

Virus plaques also can be observed in fixed and stained monolayer cultured cells in multiwell plates following incubation under a semisolid overlay medium. Plaques can easily be enumerated, and stained plates can be kept permanently (figure 6.2).

1. Prepare methyl cellulose overlay medium (for formula, see "Media and Solutions" in this chapter).
2. Aspirate or remove the culture fluids from plates containing confluent cell sheets.
3. Make serial 10-fold dilutions of virus suspension using separate pipettes, as stated above, from 10^{-1} to 10^{-6}, and inoculate each

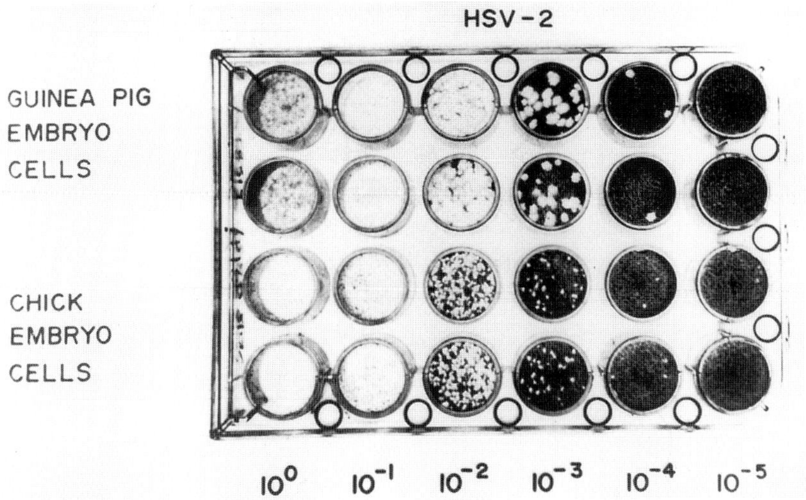

Figure 6.2 Plaque formation of HSV-2 in guinea pig embryo and chick embryo cells in a 24-well panel. (Note: Size and number of plaques are different in the two cell systems.)

dilution into a well—0.1–0.2 ml per well. Tilt the plate to distribute the inoculum evenly over the entire cell sheet.

4. Incubate infected cultures in a 5% CO_2 incubator for 1 hour at 37°C to allow virus adsorption.
5. Add 1% methyl cellulose overlay medium (formula, see below), 1 ml per well for a 24-well plate or 5 ml per well for a 6-well plate.
6. Incubate plates in a 5% CO_2 incubator at 37°C.
7. After 3–4 days (depending upon virus suspected; for example, 3 days for herpes simplex virus, 10 days for cytomegalovirus), aspirate the overlay medium completely.
8. Fix and stain the infected monolayer cultures with a fixative fluid containing 5% formalin and 1.3% crystal violet (for formula, see below) 0.5–2 ml per well for 20–30 minutes. (The entire plate with infected culture can be fixed by soaking in a container with 10% formalin wash and then stained.)
9. Wash the stained plate culture gently but thoroughly under tap water.
10. Count the number of plaques and calculate the virus infectivity titers.

Note that the size of the virus plaques and sensitivity of the cell cultures vary greatly from one type of cell to another. Figure 6.2 shows the size and number of HSV-2 plaques, which are different in the two cell systems.

NEUTRALIZATION TEST

Neutralization Test in Cell Culture

The neutralization test (NT) is considered the most sensitive and accurate method available for the identification of a virus isolate. It is, however, also time consuming and labor-intensive. By this technique, virus infectivity is inhibited by the incubation of the virus isolate with a specific antiserum, followed by inoculation of the virus-antiserum mixture into sensitive cell cultures. The NT remains the major test for the type-specific identification of enteroviruses, rhinoviruses, and adenoviruses since a monoclonal antibody is not yet available commercially against all types.

NEUTRALIZATION TEST (INHIBITION OF CPE OR HEMADSORPTION)

1. Determine virus infectivity titer prior to the test. (For rapid identification, one may select a dilution based on the rapidity with which it induces CPE.)
2. Add equal volumes of a constant virus dilution containing approximately 100 $TCID_{50}$ per 0.1 ml to a known type-specific antiserum at a concentration of 20 units and mix well. (Note: 1 unit is equal to the highest serum dilution that neutralizes 100 $TCID_{50}$ of virus infectivity.)
3. Allow the virus-serum mixture to remain at room temperature for 1 hour. Inoculate 0.2 ml of the virus-serum mixture into each of 2–4 culture tubes.
4. Inoculate 0.1 ml of each serial 10-fold dilution of virus suspension into a set of cultures, 2–4 tubes per dilution as the virus control.
5. Check all tubes for CPE daily for 5–7 days.

Complete inhibition of CPE induced by a virus challenge dose of 100 $TCID_{50}$ in the mixture with a known antiserum type is considered a positive serum neutralization test and indicates the identity of the virus. (It is always advisable to use serum pools if available for a preliminary identification of an isolate. Final identification can then be confirmed by using an individual type-specific antiserum.)

NEUTRALIZATION TEST (PLAQUE REDUCTION)

1. Determine virus infectivity titers by inoculating serial 10-fold dilutions of virus suspensions into 24-well plate cultures; overlay with semisolid overlay medium; and then determine plaque counts.
2. Add equal volumes of a constant virus dilution containing approximately 20–50 PFU (plaque-forming units) per 0.1 ml to a known antiserum of 20 units and mix well in a tube.
3. Allow the virus-serum mixture(s) and the control virus suspensions in serial 10-fold dilutions to remain at room temperature for 1 hour.

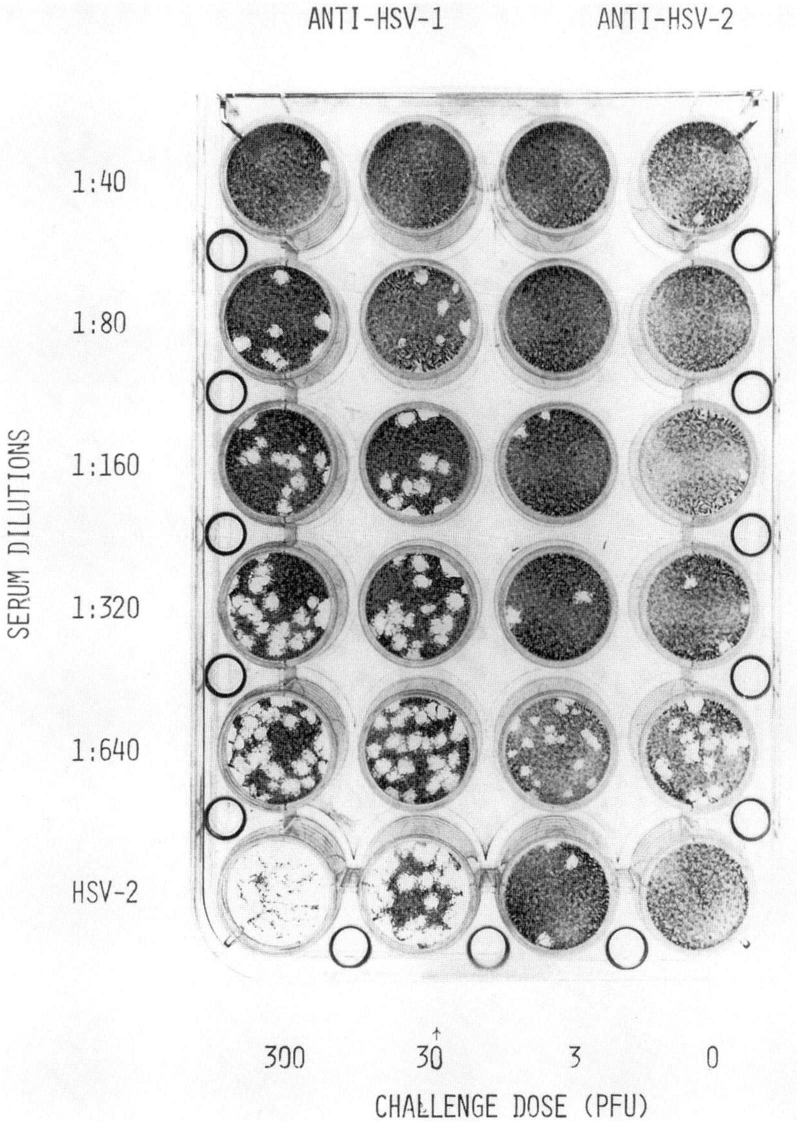

Figure 6.3 Plaque reduction neutralization test of HSV-2 using polyclonal antisera in a 24-well panel. (Note: Anti-HSV-1 cross-reaction inhibited HSV-2 plaques at 1:40 dilution, whereas anti-HSV-2 inhibited HSV-2 plaques at 1:320 dilution.)

4. Inoculate 0.2 ml of each mixture or 0.1 ml of each control virus dilution into duplicate wells in a 24-well plate culture.
5. Incubate all inoculated cultures at 37°C in a CO_2 incubator for 1 hour to allow virus adsorption.
6. Remove plates from the incubator. Overlay with a prewarmed semi-

solid overlay medium, return to a CO_2 incubator, and incubate at 37°C for 2–7 days depending on virus type.

An 80% or greater reduction of plaque counts as compared with virus control is considered a positive serum neutralization test and confirms the identity of the virus (figure 6.3).

ANTIVIRAL ASSAY IN CULTURED CELLS

The effectiveness of a new antiviral agent against a particular virus can be assayed by determining the ability of the antiviral agent to reduce or inhibit viral replication in a cell culture. It should be emphasized that the host cell system used, the virus input multiplicity of infection (MOI), and the virus yield harvesting time can greatly affect the results of an assay. The methods commonly used for the evaluation of antiviral activity in cultured cells are: (1) plaque reduction assay or inhibition of virus-induced CPE, and (2) virus yield reduction assay.

Plaque Reduction Method

Once the infectivity titer of a virus strain to be tested has been determined either by CPE or by plaque formation, the sensitivity of a virus to an antiviral agent can be performed as follows:

1. Determine the challenge dose of the virus to be tested.
2. Make 2- to 4-fold serial dilutions of the antiviral agent.
3. Remove the medium in each well of the cultures and add equal amounts of the virus challenge dose—for example, 50 PFU of virus suspension—to each well in a 6-welled plastic panel monolayer culture (the virus challenge dose will depend on the size of the well). Allow virus adsorption for one hour at 37°C.
4. Add overlay medium containing serial dilutions of antiviral agent to each infected cell monolayer.
5. Incubate all cultures at 37°C in a CO_2 incubator.
6. Check plaque formation 3–7 days after incubation.
7. Aspirate the overlay medium, fix and stain the cultures as indicated above (from step 7 to 10, "Plaque Formation in Plate Cultures under Methyl Cellulose Overlay Medium," page 48).

Figure 6.4 illustrates the plaque reduction of HSV-1 assayed in two cell systems in the presence of increasing concentrations of acyclovir (ACV). In Vero cells, a complete inhibition of HSV-1 was noted at a concentration of 1.56 μM, whereas 25 μM was necessary to inhibit the same number of HSV-1 plaques in guinea pig embryo (GPE) cells. Figure 6.5 illustrates plaque reduc-

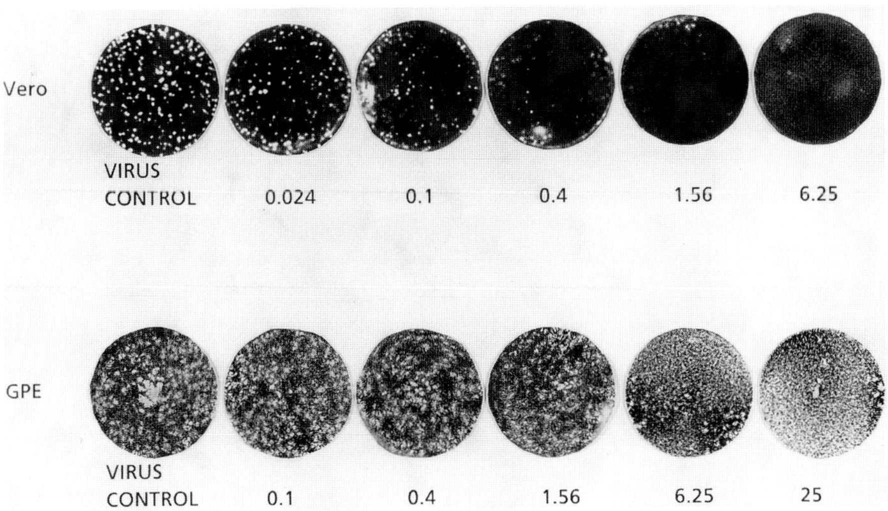

Figure 6.4 Antiviral assay for HSV-1 by plaque reduction method. Comparison of ACV at different concentrations (μM) inhibiting HSV-1 plaque formation in Vero cells *(top)* and GPE cells *(bottom)*. (Reproduced with permission, J. M. Hu and G. D. Hsiung, *Antiviral Res.* 11:219 [1989])

tion of HSV-2 assayed in two cell systems in the presence of ACV. In CE cells, a complete inhibition of HSV-2 was noted at 1.0 μM, whereas 64 μM was necessary to inhibit the same number of HSV-2 plaques in GPE cells.

To test a new virus isolate sensitivity to a given drug, it is necessary to test a standard virus strain with a known antiviral sensitivity endpoint in a given

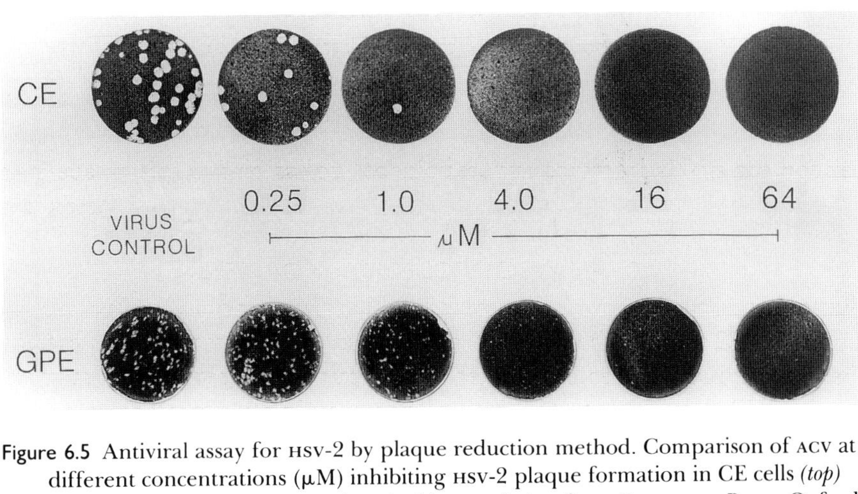

Figure 6.5 Antiviral assay for HSV-2 by plaque reduction method. Comparison of ACV at different concentrations (μM) inhibiting HSV-2 plaque formation in CE cells *(top)* and GPE cells *(bottom)*. (Reproduced with permission from Pergamon Press, Oxford: M. L. Landry; D. Mayo; and G. D. Hsiung, *Pharmacol. Ther.* 40:314 [1989])

cell system at the same time. A resistant virus isolate would require a higher drug concentration to reduce plaque formation. For controls, both known sensitive and resistant strains of viruses should be tested in parallel.

Virus Yield Reduction Method

A virus yield reduction assay for testing antiviral agents is a more stringent and time-consuming method. Depending upon the type of cell culture used, the amount of virus challenge dose applied and the length of time allowed for virus replication will affect test results significantly. In general, the test can be performed as follows:

1. Select an appropriate cell culture system.
2. Remove culture fluid from the flask.
3. Inoculate into each flask a multiplicity of infection of approximately 1.0, that is, 1 PFU per cell from the predetermined virus titer in the virus suspension for challenge dose.
4. Allow virus adsorption for 1 hour at 37°C.
5. Remove the unadsorbed virus by washing three times with PBS.
6. Add medium containing a range of concentrations of antiviral agents to duplicate flasks of virus-infected cultures. (Note: A duplicate control sample should be included consisting of medium *without* any antiviral agent.)
7. Allow the virus to replicate in the presence or absence of the antiviral agents for 2–3 days, or longer for slow-growing viruses, at 37°C.
8. Assay total virus yields in each sample in separate sets of cultures either by plaque or CPE method.

The antiviral concentrations required to reduce 90% (one $\log_{10}$ reduction) or 99% (two $\log_{10}$ reduction) of the total virus yield as compared to the virus control (that is, infected cultures without antiviral agent) are determined from the dose-response curves and are expressed as effective dose 90 (ED_{90}) or ED_{99}, respectively, of the antiviral agent in the virus yield reduction assay. Alternatively, the inhibitory effect of an antiviral agent on the virus yield can also be expressed as $\log_{10}$ reduction at certain concentrations of the antiviral agent as compared to the virus control.

MEDIA AND SOLUTIONS

Methyl Cellulose Semisolid Medium

Methyl Cellulose Solution A

Methyl cellulose powder 4,000 centipoise	0.9 g
Distilled water, demineralized	90 ml

1. Add cold water to methyl cellulose powder, mix well, and autoclave at 15 lb for 15 minutes; refrigerate. The mixture will be lumpy after autoclaving as methocel liquifies at low temperatures and becomes semisolid as the temperature increases.
2. Within the first few hours, constantly shake the flask to dissolve methyl cellulose, which may precipitate a thin layer 12–24 hours after refrigeration. It will dissolve completely, however, after 2–3 days at 4°C if shaken and mixed well.
3. Be sure that all lumps are dissolved before making a complete overlay medium. (Note: This medium should be prepared 2–3 days before use and kept at 4°C in order to ensure a clear solution.)

Nutrient Medium B

Distilled water, demineralized	48 ml
Eagle's medium–Earle's with glutamine 10X	18 ml
Calf serum (heat inactivated)	18 ml
$NaHCO_3$ (7.5%)	6 ml
Antibiotic mixtures	(0.2 ml)
Total	90 ml

Just before use, add solution B to solution A, shaking constantly while pouring.

Fixing and Staining Solutions for Plate Cultures

Crystal Violet Stock Solution (5%)

Crystal violet powder	25 g
Absolute ethanol	475 ml

Filter to remove crystals

Working Staining Solution (1%)

Crystal violet (5% stock)	100 ml
NaCl (0.85%)	375 ml
Formalin (formaldehyde 40%)	25 ml

7

IMMUNOCYTOCHEMICAL ASSAYS: IMMUNOFLUORESCENT AND IMMUNOPEROXIDASE STAINING FOR ANTIGEN DETECTION

G. D. HSIUNG

IMMUNOFLUORESCENT STAINING

The immunofluorescent (IF) staining technique has long been used for the detection of antigens and antibodies in a variety of virus-cell systems. The techniques for fixation and staining may vary according to the specified virus-cell-antibody system under study. IF techniques include direct fluorescent antibody (FA) and indirect fluorescent antibody (IFA) procedures. By the direct fluorescent antibody technique, a specific antibody is tagged with fluorescein isothiocyanate (FITC), allowed to react with the viral antigen in infected cells, and after an incubation period, the antigen-antibody complex is visualized using a microscope with an ultraviolet light source. With indirect IF, the specific antibody is not labeled. Rather, a second antibody, directed against the animal species in which the first (virus-specific) antibody is prepared, is fluorescein-labeled. The direct method is quicker and simpler, while the indirect method requires only one conjugate for the detection of a variety of antigen-antibody reactions, provided that all virus-specific antibodies are made in a single animal species. In the past, only a polyclonal antiserum made from animals was available, and variations in quality were problematic. Now a variety of high-quality monoclonal antibodies produced in mice are available commercially. As a result, the use of IF techniques to identify viruses has become routine in many laboratories. Currently, monoclonal antibodies are available for the rapid identification of a variety of virus isolates, including influenza viruses A and B, parainfluenza viruses 1–4, HSV-1 and HSV-2, VZV, RSV, the measles virus, CMV early antigens, and CMV late antigens, and the adenovirus hexon antigen.

In general, once CPE or hemadsorption is detected and involves 50% or

ISBN 0-300-05845-4.

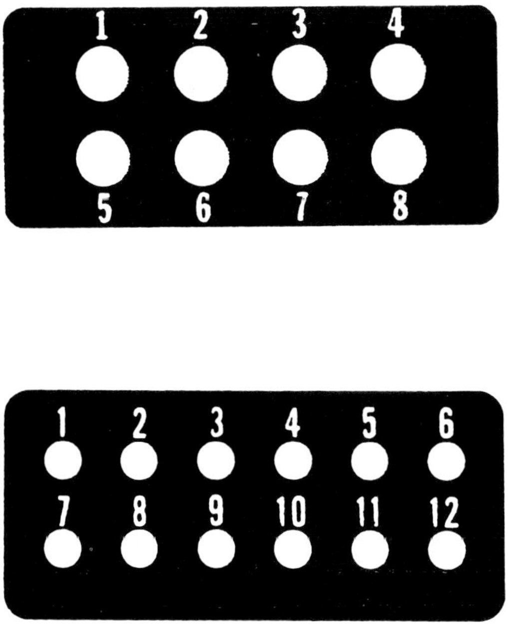

Figure 7.1 Eight- or 12-welled slide used for making cell spots

more of the infected cell monolayer, the cells are removed by trypsin or mechanical scraping, centrifuged, and resuspended in a small volume of buffer. Then a drop of the cell suspension is placed onto welled slides (examples are shown in figure 7.1). The "cell spots" are air-dried, fixed in acetone, and then stained with a specific antibody. The fluorescent pattern seen in infected cells may vary with the virus, the cell type, and the antibody used, but specific identification can be made within hours of the detection of CPE and may even be possible before CPE has appeared.

Preparation of Cell Spots for IF Staining

1. Select a culture of infected cells showing CPE or hemadsorption.
2. Harvest the supernatant fluid into a separate tube for future use.
3. Rinse the cell monolayer with 2 ml PBS.
4. Decant PBS and scrape cells from the glass using a capillary pipette (trypsin can be used for removing cells from the glass and gives better cell morphology).
5. Resuspend cells in 0.2 ml PBS with a capillary pipette.
6. Place a small drop of cells into each well of the 8- or 12-welled (or more) slide (see figure 7.1). The number of cell spots to be prepared depends on the number of antibodies to be tested; each antibody should be tested in duplicate.

7. Air dry cell spots on slides at room temperature (infected cells must be treated with caution).
8. Fix cells on slides in cold acetone for 10 minutes and air dry (at this step, the fixed cells on the welled slide can be stored at −20°C in a tightly covered container for later use).
9. Positive and negative controls must be tested at the same time to ensure specificity of reagents.

Direct Staining Method

1. Overlay cell spots on slide with fluorescein-labeled antibody. Let stand for 30 minutes at 37°C or at room temperature in petri dish containing moist gauze.
2. Wash cell spots with PBS to remove excess labeled antibody (three changes). Dry the slide at room temperature.
3. Mount with buffered glycerine (1 ml PBS + 9 ml glycerine that has been prepared within a three-week period).
4. Cover with a 22 mm × 50 mm (00 thickness) coverslip. When mounting, avoid trapping air bubbles, which could cause a hazy fluorescence.
5. Examine the preparations under a microscope with a dark-field condenser and UV light source.

Indirect Staining Method

1. Overlay cell spots in the welled slides with unlabeled antiserum or monoclonal antibody (most of the monoclonal antibodies are produced in mice) and let stand for 30 minutes in a petri dish containing a piece of moist gauze (as with controls, normal mouse serum can be used instead of an antibody).
2. Wash off excess antibody with PBS (three changes). Dry slide with cell spots at room temperature.
3. Add fluorescein-labeled antiglobulin (in this instance, anti–mouse globulin produced in goats) to the cell spots and let stand to react with the mouse monoclonal antibody that was added in step 1 for 30 minutes in a covered petri dish containing moist gauze.
4. Wash off the excess labeled antiglobulin with PBS (three changes) and dry slide at room temperature.
5. Mount the stained cell spots on the welled slide with buffered glycerine and cover with 22 mm × 50 mm (00 thickness) coverslips. Examine under a dark-field microscope with UV light source.

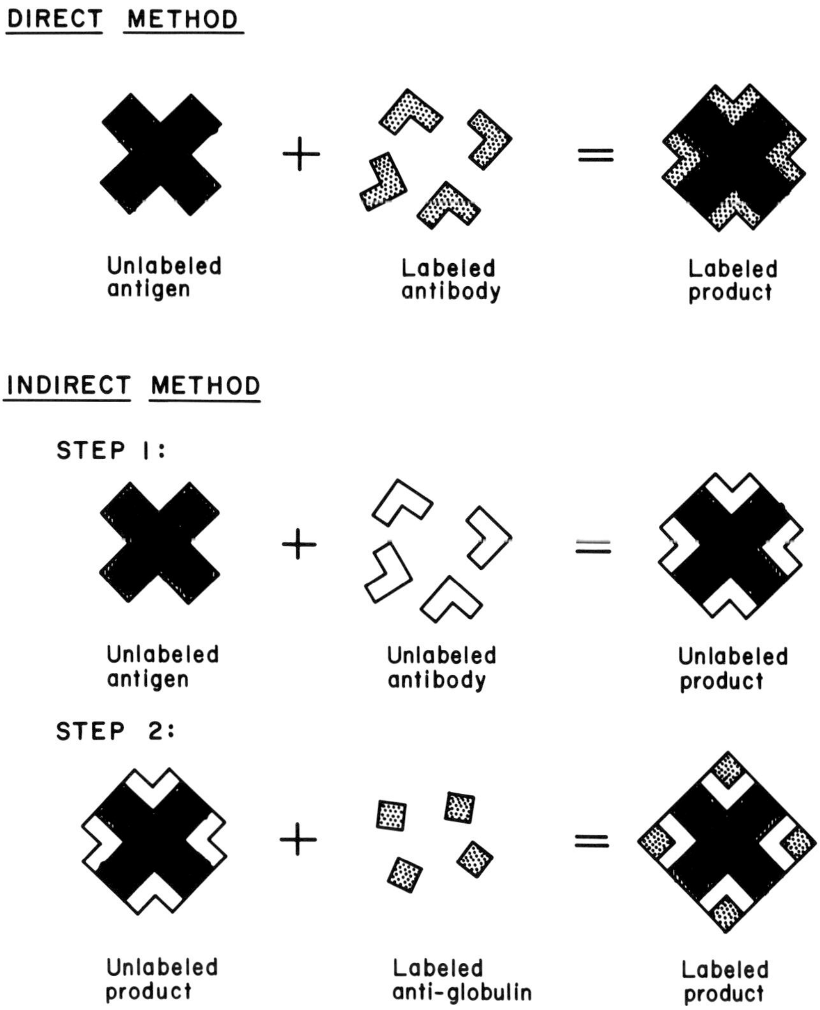

Figure 7.2 Scheme for immunofluorescent staining; direct and indirect methods for antigen detection

Interpretation of Results

The specificity of fluorescent staining of virus-infected cells may be difficult to determine without proper controls. Specificity is, however, indicated by the absence of fluorescent staining in uninfected cells overlaid with a specific antibody plus conjugate; or in infected cells that are overlaid with conjugate alone; or when normal mouse serum and conjugate are used as negative controls. Fluorescent patterns seen in monolayer cell cultures on a coverslip are often better than in cell spots since cell morphology is disrupted during the preparation of the cell spots. (Compare color plates 1 and 2; 3 and 4; 7 and 8.)

Figure 7.2 illustrates the direct and indirect fluorescent labeling reaction. Generally, the direct staining of virus in infected cells using a labeled antibody does not present serious difficulties in interpretation, provided the labeled antibody is of sufficiently high titer, does not cause nonspecific staining, and shows very limited or no cross-reaction with heterologous antigens. A monoclonal antibody shows more specific results than a polyclonal antibody.

The indirect method can be employed to simplify the identification of antigens. For example, if the monoclonal antibody used in the primary reaction is produced in a mouse, a labeled anti–mouse globulin prepared in goats is used to detect the mouse globulin present in the primary reaction. That is, the primary unlabeled antibody plays a dual role, acting as an antibody in the primary reaction and as antigen in the secondary reaction.

Anticomplement Immunofluorescence (ACIF)

To enhance the fluorescence of the antigen-antibody complex, complement is added to the complex followed by the addition of an anticomplement FITC conjugate. Thus, the fluorescent dye is conjugated to an antibody directed at the complement-antigen-antibody complex making the test more sensitive.

IMMUNOPEROXIDASE TEST

Immunoperoxidase (IP) techniques follow the same principles as IF, however, the conjugate is an enzyme, such as horseradish peroxidase. In the direct method, the virus-specific antibody is enzyme-labeled, whereas in the indirect method, an anti–animal species antibody is labeled. The presence of the enzyme-antibody conjugate bound to the virus-infected cells is detected by adding a substrate, either diaminobenzidine or aminoethylcarbazole, then oxidizing it in the presence of hydrogen peroxide, which results in a reddish-brown color. IP staining requires an additional step (the addition of the substrate). IP, however, has several advantages over IF. The reaction is read using only a light microscope or even the naked eye. The slides are permanent, the reagents are more stable, and IP has been more successful than IF on processed tissue sections.

Indirect Staining Method (color plates 13, 14, 15)

1. Wash the infected cells on coverslip with PBS for two or three changes. (Infected cultures must be treated with caution.)
2. Fix cells or tissue sections with cold acetone for 10 minutes at room temperature or with 4% buffered paraformaldehyde for 20 minutes at room temperature.

3. Treat fixed infected cells or tissue sections with H_2O_2-methanol solution (formula below) for 5 minutes or less to remove endogenous peroxidase. (Note: steps 4 and 5 are used for tissues embedded in paraffin. If cells are grown on a coverslip, proceed to step 6.)
4. Use xylene if tissue sections are in paraffin, for 5 minutes, then hydrate in 100% alcohol, followed by 95%, 70%, and 50% alcohol, and finally in water, two changes each. Wash with PBS.
5. Treat with 0.25% trypsin in PBS containing 0.02% $CaCl_2$ for 30 minutes.
6. Overlay with 1:10 or 1:20 normal mouse serum (same host species for the first antibody) for 30 minutes. Remove excess serum. To dilute normal serum, use PBS-BSA (bovine serum albumin) diluent (see formula below).
7. Add virus-specific monoclonal antibody produced in mice to the treated cells and incubate for 60 minutes. Remove excess antibody.
8. Rinse the stained cells with 0.5% Tween 20 in 1% PBS-BSA solution.
9. Add goat anti–mouse immunoglobulin conjugated with peroxidase (1:50 diluted with PBS-BSA) for 30 minutes. Wash with three changes of PBS.
10. Add freshly prepared 0.05% DAB solution (see below for formula) for 5–10 minutes or until brown color appears.
11. Remove substrate. Wash with PBS and water.
12. Counterstain with hematoxylin if needed.
13. Dehydrate in ethanol 50%, 70%, 95%, and 100% and clarify in xylene.
14. Mount the coverslip with stained cells down on microscope slides with mounting fluid (Permount or Harleco synthetic resin).
15. Examine stained preparation with light microscope.

Reagents for Immunoperoxidase Staining

H_2O_2-Methanol Solution

H_2O_2 (30%)	3 ml
Methanol	97 ml

PBS-BSA Diluent

PBS	100 ml
Bovine serum albumin	4 g

DAB Solution (to be prepared just before use)

DAB (3,3′-diaminobenzidine tetrachloride)	5 mg
Distilled water, demineralized	10 ml
H_2O_2 (30%)	0.05 ml

8

ENZYME-LINKED IMMUNOSORBENT ASSAY FOR ANTIGEN DETECTION AND ANTIBODY DETERMINATION

G. D. HSIUNG

Since the introduction of the enzyme-linked immunosorbent assay (ELISA), diagnostic virology has greatly advanced. ELISA is used both to detect virus antigens in clinical specimens and to demonstrate virus-induced antibody in serum samples. ELISA is as sensitive as radioimmunoassay but uses an enzyme instead of a radioisotope as an immunoglobulin marker. Two enzymes, horseradish peroxidase and alkaline phosphatase, are commonly used (table 8.1). When a substrate is added to the enzyme-labeled immunoglobulin, a visible color reaction occurs that can be read visually or quantitated using a spectrophotometer.

For antigen detection, either the antibody sandwich or the competitive assay can be used. In the antibody sandwich method, a specific antibody (most often a monoclonal antibody directed against a single epitope) to the antigen to be detected coats the surface of a solid-phase support, such as polystyrene beads, microtiter plates, or test tubes. Then the test sample (for example, stool, nasopharyngeal aspirate) is added and allowed to react. Detection of the antigen is then determined in two ways: either by the direct, single antibody sandwich assay, or by the indirect, double antibody sandwich assay. For the direct or single antibody sandwich test, the enzyme conjugated to a specific antibody (recognizing a different epitope of the antigen) is then added and allowed to react. For the indirect or double antibody sandwich test, the unlabeled specific antibody is first added, then the enzyme conjugated antiglobulin is added. As a final step, the amount of enzyme bound is detected by the addition of a substrate. The intensity of the resulting color of the reaction product is proportional to the amount of antigen in the test sample and is determined quantitatively by comparing the color obtained to known standards using a spectrophotometer.

ISBN 0-300-05845-4.

Table 8.1 Enzymes Commonly Used in ELISA

Enzyme	Source	Optimum pH	Substrates	Reaction Product Color
Peroxidase	Horseradish	5–7	Benzidine	Blue
			Alphanaphtol	Red
			O-dianisidine	Brown
			3-3-diaminobenzidine	Brown
			3-amino-9-ethylcarbazole	Red-brown
Alkaline phosphatase	Calf intestine	8–10	P-nitrophenyl phosphate	Yellow

Table 8.2 Immunologic Methods Commonly Used for Antibody Determination for Selected Viruses

Virus Group	Virus	ELISA or EIA	HI	CF	NT	IF[b]	LA
RNA	Entero				√		
	Influenza		√	√			
	Measles	√	√	√		√	
	RSV	√		√		√	
	HIV[a]	√				√	
	Rubella	√	√				√
	Rota	√					
DNA	Adeno			√		√	
	CMV	√		√		√	√
	VZV	√		√		√	√
	HSV	√		√		√	
	Hepatitis B	√					

a. Confirmatory test = Western blot
b. For individual cases

The problems encountered in ELISA are similar to those in other immunoassays. The purity, sensitivity, and specificity of the reagents are critical. Careful washing to reduce nonspecific binding, accurate pipetting, and the careful monitoring of assay conditions are all essential. In table 8.2 are several immunologic methods commonly used for the determination of antibodies induced by various viruses. Today, because ELISA kits are made available commercially for most of the viruses listed in table 8.2 and detailed procedures are provided by the manufacturer, they are not described here. The following description serves only to present the general principle of the test.

ELISA METHOD FOR ANTIGEN DETECTION (FIGURE 8.1)

Indirect Method, Using Rotavirus as an Example

1. Precoat round-bottomed polyvinyl microtiter plates with antiserum containing the capture antibody (goat antihuman rotavirus). Plates can be stored at 4°C until used (for at least 24 hours). Wash three times immediately before use.
2. Add specimen to be tested (stool extract, 2%–10%) and incubate at 4°C overnight or at 37°C for two hours, then remove unbound antigen by washing.
3. Add guinea pig antirotavirus serum and incubate for 1 hour at 37°C. Remove unbound antibody by washing.

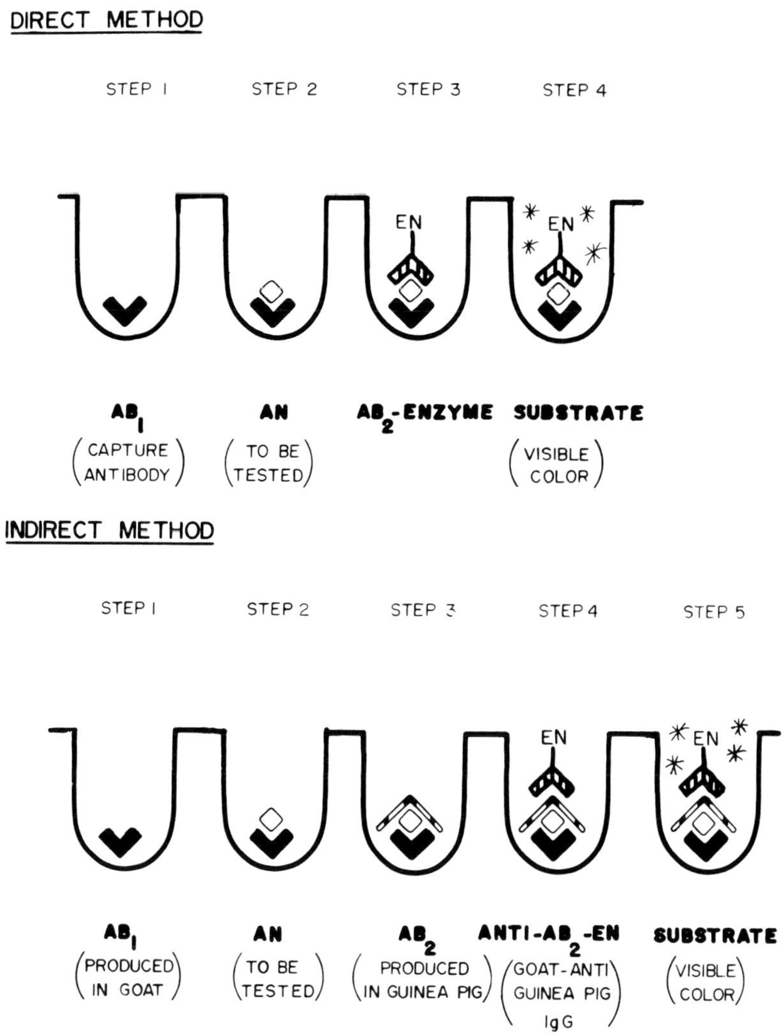

Figure 8.1 Scheme for ELISA: direct and indirect methods for antigen detection

4. Add enzyme conjugate, alkaline phosphatase-labeled goat anti-guinea pig antibody, and incubate at 37°C for 1 hour. The enzyme conjugate binds to the antigen-antibody complex. Excess enzyme conjugate is removed by washing.
5. Add the substrate solution, p-nitrophenyl phosphate disodium, which reacts with the enzyme conjugate until yellow color appears. Stop the reaction by adding 3 M NaOH to each well. Check results with a spectrophotometer. (Compare yellow color with positive standard or dilutions of positive controls). The intensity of the color generated is proportional to the amount of rotavirus antigen in the stool.

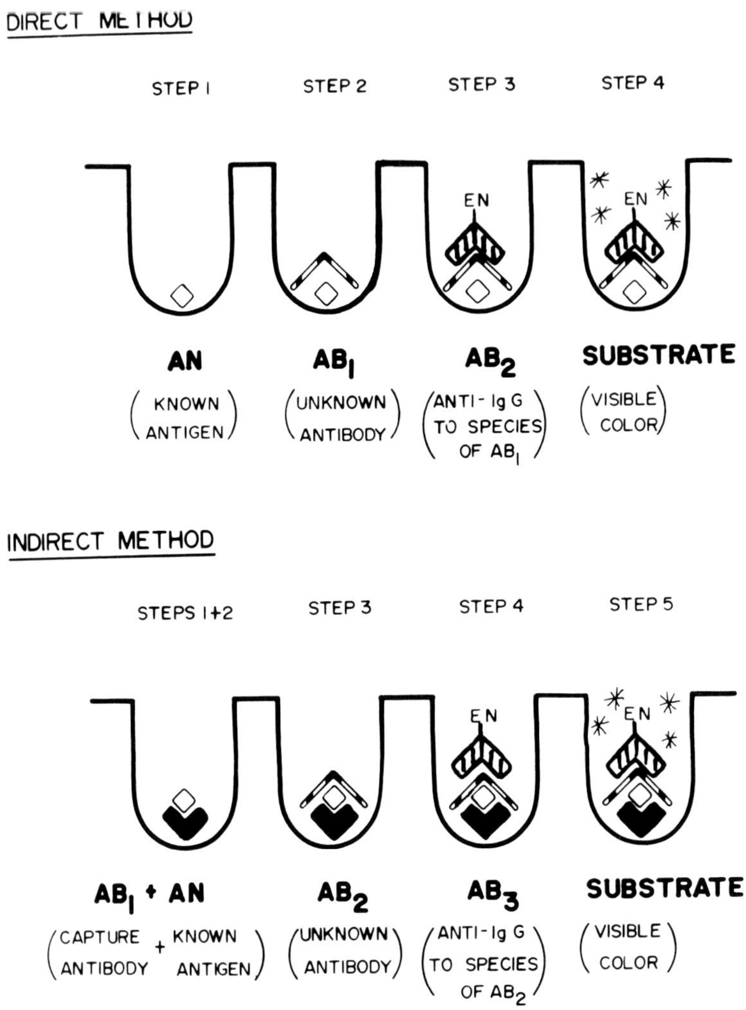

Figure 8.2 Scheme for ELISA: direct and indirect methods for antibody determination

ELISA METHOD FOR ANTIBODY DETECTION (FIGURE 8.2)

Since its introduction, ELISA has been widely applied for the detection of antibodies for a number of viruses. These include antibody detection to CMV, VZV, hepatitis B, HIV, HTLV-1, mumps, measles, RS, rubella, and rotavirus (table 8.2). These are automated tests, and test kits are available commercially. ELISA is suitable for testing large numbers of samples. The test is also more sensitive than the CF test for the detection of low levels of antibody and can be used for the quantitation of antibody titers in order to determine antibody rise.

Indirect Method, Using Rotavirus as an Example

1. Precoat microplate with goat antirotavirus serum (capture antibody) diluted in carbonate buffer and incubate at least overnight at 4°C.
2. Add rotavirus antigen diluted in PBS-Tween 20 and incubate for 2 hours at 37°C or overnight at 4°C.
3. Add test serum (human serum) diluted in PBS-Tween 20 and incubate for 2 hours at 37°C or overnight at 4°C.
4. Wash plate. Add enzyme conjugate, alkaline phosphatase-labeled goat antihuman IgG diluted in PBS-Tween 20 and incubate 1–2 hours (depending on conjugate) at 37°C.
5. Wash, then add prepared substrate solution (p-nitrophenyl phosphate disodium in diethanolamine buffer). Incubate at room temperature for 30 minutes. Add 3 M NaOH to each well to stop the reaction. Compare yellow color with positive controls. The endpoint is recorded as the highest serum dilution having a visible color or the highest serum dilution, indicating an absorbance reading of 0.75 or greater.

REAGENTS

Carbonate-Bicarbonate Buffer (pH 9.6)

Na_2CO_3	1.59 g
$NaHCO_3$	2.93 g
Distilled water, demineralized	1,000 ml

Store at 4°C for not more than two weeks.

PBS-Tween 20 (pH 9.6)

NaCl	8.0 g
KH_2PO_4	0.2 g
Na_2HPO_4	1.15 g
KCl	0.2 g
Demineralized water	1,000 ml
Tween 20	0.50 ml

Store at 4°C.

Diethanolamine Buffer (10%)

Diethanolamine	97 ml
$MgCl_2 \cdot 6H_20$	100 mg
Distilled water, demineralized	800 ml

1. Add HCl 1 M until the pH is 9.8.
2. Make up total volume to 1,000 ml with distilled water. Store at +4°C in the dark.
3. Remove sufficient amount and allow to warm to room temperature 1 to 2 hours before the substrate solution is to be used.

Substrate Solution

P-nitrophenyl phosphate disodium (1 tablet)	5 mg
Diethanolamine buffer (10%)	5 ml

9

HEMAGGLUTINATION AND HEMAGGLUTINATION-INHIBITION TEST

G. D. HSIUNG

Hemagglutination (HA) is based on the ability of certain viruses to agglutinate the red blood cells (RBC) of specific animal species. For example, when influenza virions and RBC are mixed in appropriate ratios, the RBC bind multiple virions and the virions in return bind multiple RBC, forming a latticework. The latticework settles and covers the entire bottom of the container with little or no aggregation of RBC in the center of the tube. (Note that the virions need not be infectious to cause hemagglutination.) Ortho- and paramyxoviruses possess the capacity to agglutinate the RBC of several animal species at low temperatures (4°–22°C), but measles virus agglutinates only certain monkey RBC at 37°C (table 9.1). Several other viruses also possess hemagglutination properties with RBC from specific animal species (for example, adenovirus and rubella) and are discussed in the separate chapters on viruses.

This hemagglutination property of certain viruses can be inhibited by specific antibodies. Thus, the hemagglutination-inhibition (HI) test is useful for both the identification of virus isolates and the measurement of antibody titers of sera. For the latter use of the HI test, first the HA titer of the virus is determined, then a known concentration of the virus is mixed with serial dilutions of the test serum, and the HI titer is determined. Since the virus is bound by the antibody, it can no longer participate in the formation of the latticework of RBC, and hemagglutination will be inhibited. The extent of inhibition of RBC agglutination caused by the highest dilution of the serum is the antibody titer of the serum. In general, prior to the performance of the HI test, sera to be tested should be treated for the removal of nonspecific factors that may inhibit hemagglutination. If the HI test is for antibody

ISBN 0-300-05845-4.

Table 9.1 Hemadsorption and Hemagglutination Characteristics of the Orthomyxoviridae and Paramyxoviridae

Virus Family and Genus	Hemadsorption	Animal Species[a] Erythrocytes Showing Hemagglutination (0.5%)				Optimum Conditions for Erythrocyte Suspension		
		GP	HO	Chick	Mon	pH	Diluent	Temp. (C)
Orthomyxoviridae								
Influenza A,B,C[b]	+	+	+	+[c]	+	7.0	PBS	4°, 22°
Paramyxoviridae								
Parainfluenza 1–3[b,d], mumps, NDV	+	+[e]	+	+	+	7.0	PBS	4°, 22°
Measles	+	–	–	–	+	7.0	Normal saline	37°
Respiratory syncytial virus	–	–	–	–	–			

+ virus-induced agglutination of RBC; – no agglutination of RBC indicated

a. GP guinea pigs
HO human type O
Chick day-old chick
Mon monkey, preferably African green; reactivity variable from animal to animal; must be preselected

b. Elute virus at 37°.

c. Chick cells are not agglutinable by new influenza A isolates or certain parainfluenza-type isolates.

d. With parainfluenza type 4, hemagglutination occurs best at 22°–37°.

e. Tissue culture fluids from cells infected with parainfluenza virus show low titers of hemagglutinin and are not sufficient for the HI test, thus a hemadsorption-inhibition test should be used (see chap. 15).

determination, a known virus antigen is used, or if it is for virus identification, a known type-specific antiserum is used.

PREPARATION OF RBC SUSPENSION

1. Obtain blood from guinea pig or other animal or avian species (by cardiac puncture) and mix with either Alsever's solution or heparin to prevent clotting (for Alsever's solution formula, see chapter 30).
2. Wash freshly obtained RBC three times in PBS and prepare a 10% stock suspension. Store at 4°C for no more than one week until used.
3. Make a 0.5% suspension of the RBC by adding 1 ml of the 10% stock to 19 ml PBS just before use.

DETERMINATION OF HEMAGGLUTININ TITER (FIGURE 9.1)

Culture medium from infected cells that exhibited hemadsorption (HAd) may contain virus hemagglutinin that can be assayed by the following procedure:

1. Prepare 2-fold serial dilutions (1:10 to 1:640) using either infected tissue culture fluid or infected amniotic or allantoic egg fluid pools, in PBS in 0.05-ml amounts in disposable microtiter U or V plates. (The amount can be increased, if plates with large-size wells are used.)
2. Add 0.05 ml of a 0.5% suspension of guinea pig RBC to each dilution of virus suspension and allow the RBC to settle 1–2 hours at 4°C or 22°C. (See table 9.1 for temperature, pH, etc.)
3. Read the virus titer by determining the highest dilution of tissue culture medium (or egg fluid) capable of causing partial or complete hemagglutination. This dilution represents 1 HA unit (see example, figure 9.1).

In the example, the HA titer of the virus is 1:160, which represents 1 HA unit. Thus a dilution of 1:20 would contain 8 HA units in 0.05 ml or 4 HA units in 0.025 ml. (Note: The agglutination of RBC in the presence of an orthomyxovirus is due to the attachment of viral hemagglutinin to glycoprotein receptors present on the surface of erythrocytes.) These attached virions can be eluted from the RBC at 37°C or after prolonged incubation at room temperature as a result of the presence of viral neuraminidase, which destroys the receptors. However, other viruses, for example, the enteroviruses or the adenoviruses, generally do not elute from the RBC after agglutination, because they lack such neuraminidase enzymes on their virions.

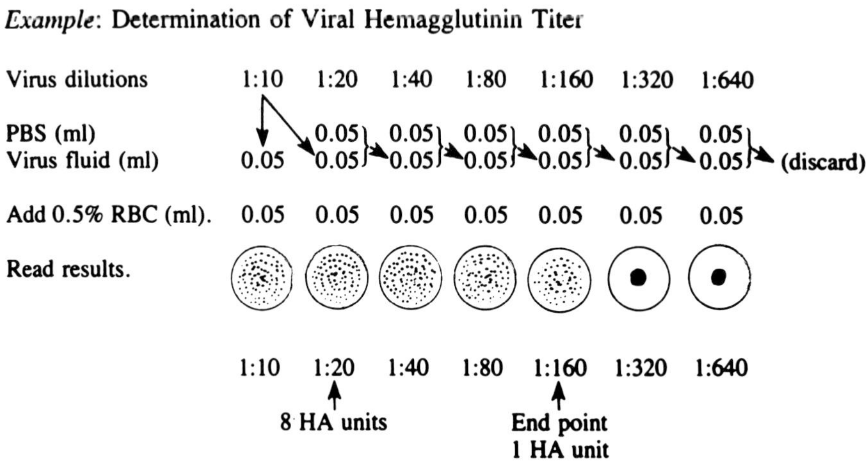

Figure 9.1 An example for the determination of viral hemagglutinin titer. The HA titer of the virus is 1:160, which represents 1 HA unit.

HEMAGGLUTINATION-INHIBITION TEST FOR IDENTIFICATION OF VIRUS ISOLATE (FIGURE 9.2)

The HI test is based on the inhibition of viral agglutination by a specific serum antibody and can be used for virus identification and for antibody assay.

1. Determine the hemagglutinin units of the virus isolate (example described above).
2. Perform the HI test.
   a. Make 2-fold serial dilutions in PBS of type-specific antiserum (pre-

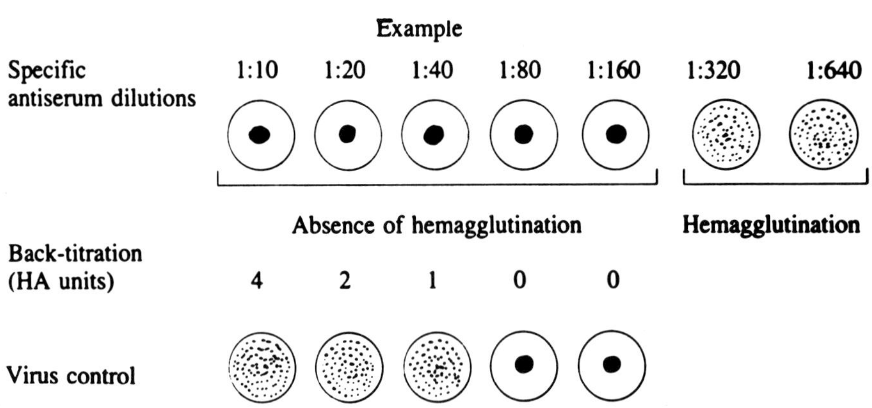

Figure 9.2 An example of hemagglutination-inhibition test for virus identification. Virus hemagglutination is inhibited by a specific antiserum dilution of up to 1:160.

treated for nonspecific inhibitors and RBC agglutinins) from 1:10 to 1:640 in 0.025-ml amounts.

b. Add 0.025 ml of the virus dilution containing 4 HA u/0.025 ml to each serum dilution.
c. Mix well and allow the virus-serum mixtures to remain for 1 hour at room temperature.
d. Add 0.05 ml of a 0.5% suspension of guinea pig RBC to each mixture. Allow to settle for 1–2 hours at 4°C or 22°C.
e. Read the serum titer. The serum titer is the highest dilution of serum inhibiting hemagglutination by the virus and is expressed as the reciprocal of that serum dilution. (Note: To control possible errors in dilution or variations owing to different batches of RBC, it is best to back-titrate the 4-HA-unit dilution as outlined in step 3.)

3. Back-Titration
   a. Prepare five wells, each containing 0.05 ml PBS.
   b. To the first, add 0.05 ml of test virus dilution containing 8 HA units; mix and transfer 0.05 ml to the second; and continue through the fifth well, discarding the last 0.05 ml.
   c. Add 0.05 ml of a 0.5% guinea pig RBC suspension to each well. Mix and allow to settle.
   d. Read for HA. Only the first three wells should exhibit agglutination, indicating 4 HA units were contained in the 0.025 ml used in the test (figure 9.2).
   e. Adjust the test virus concentration, if necessary, by dilution (with PBS if more than three tubes show agglutination) or by the addition of more virus (if fewer than three tubes show agglutination).

An example of a type-specific antiserum inhibition reaction and a back-titration result are shown in figure 9.2. Hemagglutination by virus isolate is inhibited by the type-specific antiserum at a dilution of up to 1:160 and identifies the virus isolate.

HEMAGGLUTINATION-INHIBITION TEST FOR ANTIBODY DETERMINATIONS (FIGURE 9.3)

All serum samples used for the HI test should be heat-inactivated at 56°C for 30 minutes and treated to remove nonspecific inhibitors and also natural agglutinins against erythrocytes of foreign species. Methods for removing nonspecific inhibitors and agglutinins are described below and also under each relevant virus group.

Make serial 2-fold dilutions of acute or convalescent sera in 0.025-ml amounts in PBS using microtiter plates. Add 0.025 ml of the virus suspen-

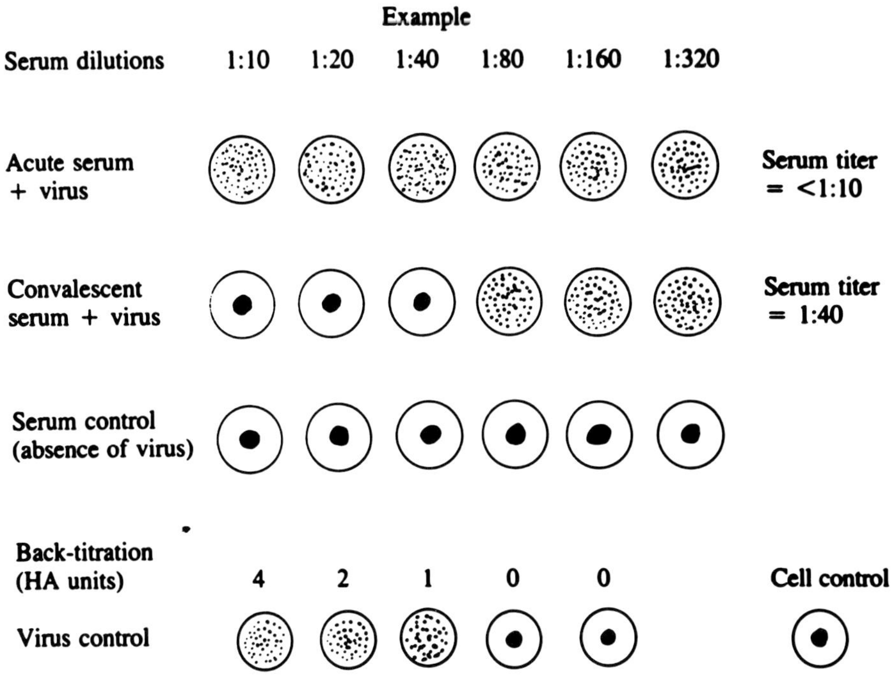

Figure 9.3 An example of a hemagglutination-inhibition test for antibody determinations, and back-titration of the 4-HA-unit antigen. (Note: A greater than 4-fold rise in antibody titer in the convalescent serum [titer = 1:40].)

sion containing 4 HA units to each serum dilution. Allow the virus-serum mixtures to react at room temperature for 1 hour before the addition of RBC. For diagnostic purposes, a 4-fold or greater HI antibody titer rise is suggestive of recent infection by the test virus or a closely related virus, as shown in figure 9.3.

Thus, a greater-than-4-fold rise in antibody titer is observed in the convalescent serum (titer = 1:40).

REMOVAL OF NONSPECIFIC INHIBITORS PRESENT IN SERA

In addition to the procedures for the removal of nonspecific inhibitors listed below, other specific treatments are listed in the specific virus group chapters.

Receptor-destroying Enzyme (RDE) Treatment

1. Add 0.4 ml of RDE (100 units per ml) to 0.1 ml serum and incubate at 37°C overnight.

2. Add 0.3 ml of 2.5% sodium citrate to the above mixture and incubate at 56°C for 30 minutes.
3. Add 0.2 ml PBS to the above mixture to yield a final serum dilution of 1:10.

Kaolin Treatment

1. Prepare a 25% suspension of kaolin (acid-washed) in PBS.
2. Mix serum specimen, diluted 1:5 with PBS, with an equal volume of the kaolin suspension. Shake vigorously and allow to stand at room temperature for 20 minutes, with intermittent shaking.
3. Centrifuge at 2,000 rpm for 30 minutes. The supernatant represents the treated serum at a 1:10 dilution.

Trypsin and Periodate Treatment

1. Mix 0.1 ml of 0.8% trypsin solution in PBS with 0.2 ml of undiluted serum and let stand at 56°C for 30 minutes.
2. Add to this mixture 0.6 ml of a 0.01 M aqueous potassium periodate solution (255 mg per 100 ml distilled water). Allow to stand at room temperature for 15 minutes.
3. Add 0.6 ml of a 1% aqueous glycerol solution and let stand an additional 15 minutes at room temperature.
4. Finally, add 0.5 ml PBS. This final solution gives 2 ml of a 1:10 serum dilution.

REMOVAL OF NATURALLY OCCURRING AGGLUTININS IN SERUM

Certain serum samples contain agglutinins against erythrocytes of foreign species. Therefore, after heat inactivation and nonspecific inhibitor treatment, sera that were diluted 1:10 are adsorbed with RBC. Add 0.1 ml of 50% of the washed, packed RBC, which are to be used for the test of the serum, to 1 ml each of the 1:10-diluted pretreated sera. Place the mixture in the refrigerator for 60 minutes and then centrifuge at 4°C at 1,500 rpm for 10 minutes. Discard the packed cells. The supernatant represents treated serum (1:10) prepared for HI testing.

10

COMPLEMENT-FIXATION TEST

G. D. HSIUNG

The complement-fixation (CF) test is a simple and commonly used serologic method for the determination of antibody titers in large numbers of serum specimens as well as for the identification of virus isolates. The test, however, is limited in that type-specificity within a virus group usually cannot be determined by this method. There are several procedures, each of which requires some level of experience to perform. Both the sensitivity and the specificity of the test are affected by the reagents and by the expertise of the personnel performing the test. Most of the procedures described below are modifications from the manual originally prepared by the Laboratory Branch, Centers for Disease Control, Atlanta, Georgia, which has been designated as the Laboratory Branch complement-fixation test or LBCF.

The CF test depends upon the interaction of antigen, antibody, and complement (C). If either the antigen or antibody is absent, no reaction will occur and the C will not be fixed and will therefore be free to react with the indicator system, that is, sensitized sheep RBC. In the latter, hemolysis of the sheep RBC will occur. The absence or reduction of such hemolysis indicates that the C is fixed by the antigen-antibody complexes in the test system, as illustrated by figure 10.1.

STANDARDIZATION OF CF REAGENTS

All reagents must be standardized prior to the test. Simplified procedures for standardization of the various reagents are described below.

ISBN 0-300-05845-4.

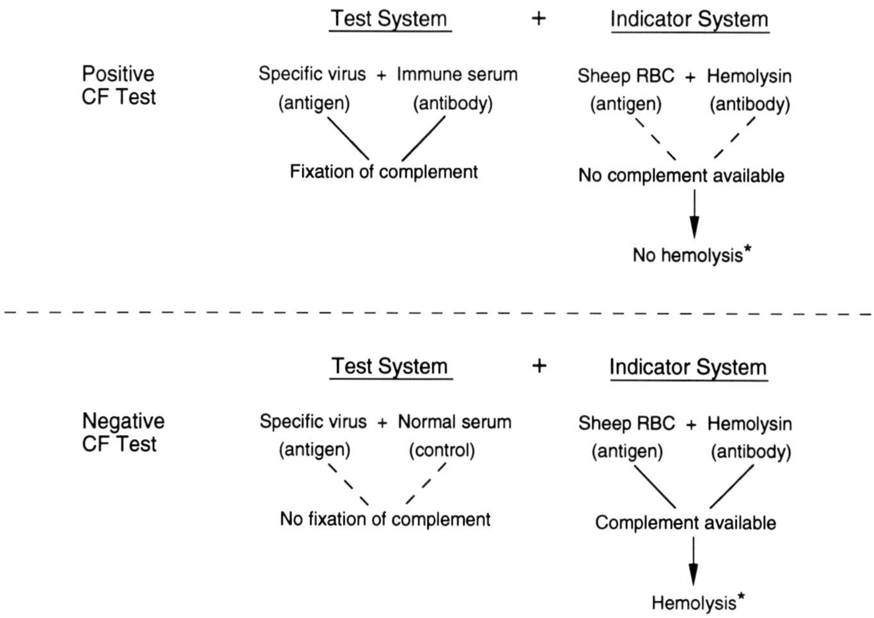

*Final results, either *no hemolysis* or *hemolysis*, indicate positive CF or negative CF

Figure 10.1 Diagram for a CF test

Preparation of Hemolysin Dilutions and Sensitization of Erythrocytes

The incubation of sheep RBC with their homologous antibody is called *sensitization*. The antibody, hemolysin, is prepared from rabbits immunized with sheep RBC. A stock solution of hemolysin can be prepared at a 1:100 dilution in Veronal buffer (for the formula, see the reagents listed below). This stock can be kept at −20°C in small aliquots for months.

1. From the 1:100 stock hemolysin solution make a 1:1,000 dilution. Prepare serial dilutions of the latter using Veronal buffer (VB), as indicated in table 10.1.
2. Add an equal volume of a 2% washed sheep RBC suspension to each dilution of hemolysin, as shown in table 10.1.
3. Incubate the mixture at 37°C for 30 minutes to sensitize the RBC. Sensitized RBC, if not used immediately, should be stored in the refrigerator.

Determination of Complement (C) Units

Commercially prepared C is usually obtained from "normal" guinea pig sera, which may contain antibody to the test antigen. C should therefore be tested

Table 10.1 Preparation of Hemolysin Dilutions and Sensitization of Sheep RBC

Dilutions of Hemolysin Desired	Hemolysin Solution (1 : 1,000) (ml)	+	Veronal Buffer (ml)	Discard Excess (ml)	Hemolysin Final Volume (ml)	+	2% Sheep RBC (ml)
1 : 1,000	2.0		—	—	2.0		2.0
1 : 2,000	1.0		1.0	—	2.0		2.0
1 : 4,000	0.5		1.5	—	2.0		2.0
1 : 6,000	0.4		2.0	0.4	2.0		2.0
1 : 8,000	0.3		2.1	0.4	2.0		2.0
1 : 10,000	0.2		1.8	—	2.0		2.0
1 : 15,000	0.2		2.8	1.0	2.0		2.0
1 : 20,000	0.2		3.8	2.0	2.0		2.0
1 : 30,000	0.2		5.8	4.0	2.0		2.0
1 : 40,000	0.2		7.8	6.0	2.0		2.0

for the presence of specific antibody to the virus under test before use. C can be obtained commercially in a lyophilized preparation. A 1:10 stock solution is prepared in Veronal buffer and kept in an ice bath during the period of testing.

1. Make a dilution of 1:50 from the 1:10 stock solution, followed by serial dilutions of C as shown in table 10.2. Keep in the ice bath.
2. Add 0.2 ml amounts of sensitized RBC to each 0.2 ml of the various C

Table 10.2 Preparations of Complement Dilutions

Final Dilutions of Complement Desired	Complement* (1 : 10) (ml)	Veronal Buffer (ml)
1 : 50	0.5	2.0
1 : 60	0.4	2.0
1 : 70	0.4	2.4
1 : 80	0.3	2.1
1 : 90	0.3	2.4
1 : 100	0.3	2.7
1 : 120	0.2	2.2
1 : 140	0.2	2.6
1 : 160	0.2	3.0
1 : 180	0.2	3.4
1 : 200	0.2	3.8
1 : 220	0.2	4.2
1 : 240	0.2	4.6

*Complement (*C*) obtained commercially is a lyophilized preparation of fresh, normal guinea pig serum and can be reconstituted. A 1 : 10 stock is prepared in VB and kept in ice bath during the period of use. Stock *C* can be stored at −20°C for limited periods without appreciable titer loss.

Table 10.3 An Example for Determination of Complement Units

Hemolysin Dilutions* with Sensitized RBC	Complement Dilutions													Veronal Buffer Only
	50	60	70	80	90	100	120	140	160	180 ↑	200	220	240	
1 : 1,000	0	0	0	0	0	0	0	0	0	0	2	2	4	4
1 : 2,000	0	0	0	0	0	0	0	0	0	0	2	2	4	4
1 : 4,000	0	0	0	0	0	0	0	0	0	0	2	2	4	4
1 : 6,000	0	0	0	0	0	0	2	2	4	4	4	4	4	4
1 : 8,000	0	0	0	0	0	0	2	4	4	4	4	4	4	4
1 : 10,000	0	0	0	1	2	2	4	4	4	4	4	4	4	4
1 : 15,000	0	1	2	2	2	4	4	4	4	4	4	4	4	4
1 : 20,000	4	4	4	4	4	4	4	4	4	4	4	4	4	4
1 : 30,000	4	4	4	4	4	4	4	4	4	4	4	4	4	4
1 : 40,000	4	4	4	4	4	4	4	4	4	4	4	4	4	4

Note. Complete or 100% hemolysis = 0; 75% hemolysis = 1; 50% hemolysis = 2; 25% hemolysis = 3; no hemolysis = 4. In this test, a hemolysin dilution of 1 : 4,000 and a C dilution of 1 : 180 completely hemolyzed the sheep RBC. Therefore, at a 1 : 4,000 hemolysin dilution, 1 : 180 of C = 1 unit, 1 : 90 of C = 2 units, 1 : 45 of C = 4 units. One part of C is added to 89 parts of VB for 2 units of C. In this instance, 1 full unit is considered to give 100% hemolysis. Other laboratories may use 50% as an endpoint, making the test more sensitive.

*Hemolysin dilutions and 2% sheep RBC are combined in equal volumes for sensitization. (See table 10.1.)

dilutions, either with 2 units of hemolysin if known or with a serial dilution of hemolysin starting at 1:1,000, as shown in table 10.3.

3. Add 0.2 ml cold VB to each mixture and incubate for 1 hour at 37°C or overnight in the refrigerator.
4. Read and record the highest dilution of C that gives 100% hemolysis. The highest dilution of C (1:180 in table 10.3) at a hemolysin dilution (1:4,000, representing 1 unit) giving 100% hemolysis is considered to be a full unit of C. For identifying viruses, 2 full units of C should be used in the test. In this case, a 1:90 dilution of C contains 2 units.

CF TEST FOR THE IDENTIFICATION OF AN UNKNOWN VIRUS ISOLATE IN 96-WELL MICROTITER U PLATE (FIGURE 10.2)

1. Preparation of antiserum dilutions using a 0.025-ml micropipette in microtiter plate.
   a. Dispense 0.025 ml VB to wells 2–5 of rows *A–F* and *H* in microtiter plate.
   b. Add 0.025 ml 1:8 antiserum dilution to each well, 1 and 2 of rows *A–F* and *H*.
   c. Make serial 2-fold dilutions of the antiserum from wells 2–5 of row *A*. Repeat this procedure in rows *B–F* and *H* for antiserum controls.
2. Preparation of unknown antigen dilutions in tubes.
   a. Make serial 2-fold dilutions of unknown antigen starting from 1:2 to 1:64.
   b. Transfer 0.025 ml of each antigen dilution to wells 1–5 containing antiserum dilutions and well 7 for antigen controls and (starting from the highest to lowest antigen dilution—row *F* back to row *A*).
   c. Add 0.025 ml VB to row *H* wells (antiserum controls) and column 7 wells (antigen controls).
3. Addition of 2 units of complement
   a. Dilute complement with cold VB to contain 2 units in 0.025 ml.
   b. Add 0.025 ml of the above diluted complement to all wells.
   c. Mix the contents of the wells thoroughly by gently tapping the plate.
4. Complement controls (back-titration).
   a. Add VB to wells 10–12 of rows *A–F*.
   b. Add 0.025 ml containing 2 units of complement to wells of columns 9 and 10 in rows *A–F*.
   c. Make serial 2-fold dilutions of complement in columns 10–12 of

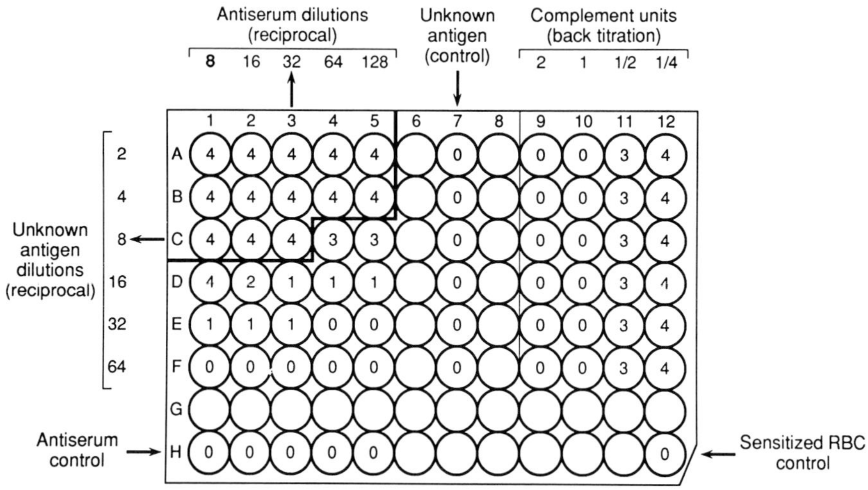

Figure 10.2 An example of a CF test in a microtiter U plate with known antiserum and an unknown virus isolate (○ = empty wells). Degree of fixation: 4 = no hemolysis, complete fixation; 3 = 25% hemolysis; 2 = 50% hemolysis; 1 = 75% hemolysis; 0 = complete hemolysis, no fixation. A complete fixation indicates the identity of the unknown isolate that has a titer of 1:8 when the antiserum diluted at 1:32 is used.

row *A*, to contain 1, ½, and ¼ unit, respectively. Repeat the procedure in rows *B–F*.

d. Add 0.025 ml of each unknown antigen dilution starting from the highest to lowest dilution (1:64 down to 1:2) to wells 9–12 from rows *F* back to *A*.

e. Set up RBC control by adding 0.075 ml VB to well 12 of row *H*.

5. Cover plates with plate sealer and incubate overnight at 4°–6°C.
6. Addition of sensitized sheep RBC.

a. Remove plates from the refrigerator and leave at room temperature for 15–30 minutes while waiting for the sensitization of sheep RBC.

b. Prepare sensitized sheep RBC (add equal amounts of 1.4% sheep RBC + diluted hemolysin to contain 2 units in 0.025 ml as determined above and incubate at 37°C for 30 minutes).

c. Add 0.05 ml of the sensitized sheep RBC suspension to each well.

d. Shake plate gently to mix the contents in each well.

e. Incubate at 37°C for 15–30 minutes or until the antigen controls show clearing (complete hemolysis).

f. Place plate at 4°–6°C to allow the settling of unlysed RBC.

7. Check results.

a. Read for hemolysis:

0 = complete hemolysis

1 = 75% hemolysis
2 = 50% hemolysis
3 = 25% hemolysis
4 = no hemolysis

b. The highest dilution of antigen (1:8) showing 4+ fixation with the highest dilution of antiserum (1:32), as indicated in figure 10.2, gives the final result.
c. The test is not valid for any sample that shows fixation in any of the antigen or antiserum control wells.

INTERPRETATION OF THE TEST

The CF test depends on the interaction of antigen, antibody, and C. If the CF test is used for virus identification (see figure 10.2), complete fixation (no hemolysis at moderate to high dilutions of the unknown antigen) indicates the identity of the virus; complete fixation at low dilutions of the unknown antigen could be just a nonspecific reaction. If the test is used for antibody determination, however, complete fixation at the highest serum dilution is the endpoint and represents the antibody titer of the serum.

REAGENTS

Veronal Buffer Stock Solution (5X)

Solution A

NaCl	83.80 g
$NaHCO_3$	2.52 g
Sodium barbital (sodium 5,5-diethyl barbiturate)	3.00 g
Distilled water, demineralized	1,000 ml

Solution B

Barbital (5,5-diethyl barbituric acid)	4.60 g
$MgCl_2 \cdot 6H_20$	1.00 g
$CaCl_2 \cdot 2H_20$	0.20 g
Hot distilled water, demineralized	500 ml

After cooling, add solution A to solution B and bring to 2,000 ml with distilled water (5X concentrated). Sterilize by filtration if desired. For a working solution for the CF test, to 200 ml of the above mixture add distilled water to 1,000 ml (1:5 dilution used as working solution).

11

HISTOCYTOCHEMICAL STAINING

G. D. HSIUNG

Microscopic examination of fixed and stained infected cells is a technique that should not be overlooked in the diagnostic laboratory. It is good practice and advantageous to include cell cultures on coverslip preparations in shell vials in selected cases. When cultured cells are fixed, stained, and mounted, these preparations are a distinct aid in recognizing and identifying viruses and can be filed for reference as part of the permanent record. In the following sections, three staining procedures that can be used in diagnostic virology are described: (a) hematoxylin-eosin staining for the demonstration of virus-induced inclusions (table 11.1); (b) Feulgen reaction for verification of DNA contents; and (c) acridine orange for verification of RNA contents of virus-infected cells.

HEMATOXYLIN-EOSIN STAINING

The structure of the cell is often altered to a greater extent by Zenker's fixing solution than by formalin since the chromatin often appears as a coarse reticulum and the cytoplasm is somewhat foamy. Zenker's fixing solution, however, is particularly useful for the demonstration of virus-induced intranuclear inclusions and thus preserves sharp structural details of virus-infected cells.

Procedure

1. Wash infected cultures in PBS with three changes (infected cultures must be treated with caution).

ISBN 0-300-05845-4.

Table 11.1 Examples of Virus induced Cytopathology as Revealed by Hematoxylin-Eosin Staining

Virus	Cytopathology		
	Intranuclear Inclusions	Intracytoplasmic Inclusions	Syncytial Formation
RNA viruses			
Reovirus	−	+	−
Parainfluenza	−	+	+
Measles	+	+	+
DNA viruses			
Vaccinia virus	−	+	+
Herpesvirus			
HSV	+	−	+[a]
VZV	+	−	−
CMV	+	+[b]	−
Adenovirus	+	−	−
Polyomavirus	+	−	−[c]

a. Some strains of HSV-2
b. Most seen in tissue sections
c. Vacuolation

2. Fix cells on coverslips in Zenker's fixing solution for 24 hours at room temperature.
3. Wash in cold running water until the water is clear.
4. Dehydrate in 80% ethanol or store in 80% ethanol if samples are not to be stained immediately.
5. Dip in alcohol iodine (0.5% iodine in 95% ethanol) for 1 minute to remove mercury precipitate. Rinse in water.
6. Dip in 0.5% aqueous sodium thiosulfate for 1 minute to remove excess iodine, if any. Rinse in water.
7. Stain in hematoxylin for 1–2 minutes. Rinse in water.
8. Differentiate in 1% ammonium hydroxide solution for 3–5 seconds or until blue color appears. Rinse in water.
9. Counterstain with 0.5% eosin in 95% ethanol for 1–2 minutes.
10. Dehydrate rapidly in two changes of 100% ethanol.
11. Clear in two changes of xylene, 1 minute each.
12. Mount on microscope slides with cell side down using mounting medium (Permount or Harleco synthetic resin).
13. Clean upper surface of coverslip, after mounting medium has set for 24 hours, in order to remove cells or debris that may have adhered to the upper surface of the coverslips.
14. Examine stained preparations with a light microscope (see figure 11.1, *A–H;* color plates 19, 21, and 23).

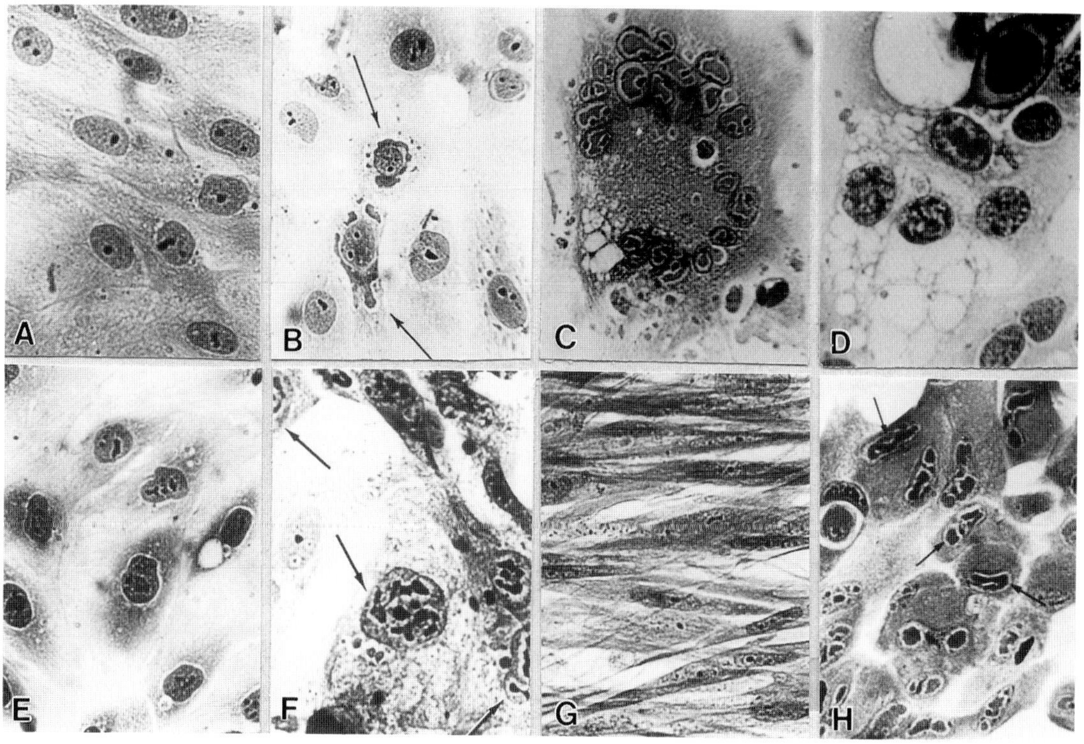

Figure 11.1 Examples of virus-induced intranuclear and intracytoplasmic inclusions (hematoxylin-eosin staining): *(A)* Uninfected RhMK cells (control); *(B)* Reovirus-induced intracytoplasmic inclusions (arrows) in RhMK cells; *(C)* Measles virus–induced intranuclear inclusions in a multinucleated giant cell in GMK cell cultures; *(D)* SV_{40}-induced intranuclear inclusions in patas MK cells with vacuolations; *(E)* Uninfected HEK cells (control); *(F)* Adenovirus-induced intranuclear inclusions (arrows) in HEK cells; *(G)* Uninfected human embryo fibroblasts (control); *(H)* Human CMV-induced intranuclear inclusions (arrows) in HEF cells

Reagents for Hematoxylin-Eosin Staining

Zenker's Fixing Solution

Potassium dichromate	2.5 g
Mercuric bichloride	5.0 g
Distilled water, demineralized	100 ml

The chemicals should be dissolved with the aid of heat. This stock solution keeps very well. Just before using, add glacial acetic acid to 5%.

Harris Hematoxylin

Distilled water	200 ml
Hematoxylin	1 g

Absolute ethanol	10 ml
Ammonium or potassium aluminum sulfate	20.0 g
Mercuric oxide (red)	0.5 g

Dissolve the hematoxylin in the alcohol and the aluminum sulfate in water with the aid of heat. Combine the hematoxylin and the alum solutions and bring to a boil as rapidly as possible. Then add the mercuric oxide. The solution takes on a dark or purple color. Transfer the flask to a cold-water bath. This solution does not need the addition of ripening agents and can be used as soon as it is cold, although better staining results are obtained if it is made a few days in advance. Filter before use and store in a tightly capped bottle. This and other hematoxylin staining solutions may be purchased commercially.

EOSIN

Eosin Y, both water- and alcohol-soluble, gives good results. A 0.5%–1% solution of eosin, either in water or in 95% alcohol, should be used. The solution must be filtered before use.

FEULGEN REACTION

Carnoy's fluid is recommended for cell fixation prior to the use of Feulgen reagent.

Procedure

1. Wash infected cultures in PBS (three changes).
2. Fix cells on coverslips in Carnoy's solution for 20 minutes.
3. Transfer to 95% ethanol for 2 minutes.
4. Transfer to 70% ethanol for 2 minutes.
5. Wash in water for 2 minutes.
6. Transfer to HCl (1N) and leave for 2 minutes.
7. Then transfer to HCl (1N) prewarmed to 60°C and keep at 60° for 10–20 minutes.
8. Transfer to room temperature HCl (1N) for 2 minutes.
9. Wash quickly with water.
10. React with Schiff's reagent for 30 minutes.
11. Wash in running water for 5–15 minutes or until free from unbound reagent.
12. Counterstain with light green (0.5%), if desired, and wash in water.
13. Rapidly dehydrate in 70% ethanol, 2 minutes.
14. Rapidly dehydrate in 95% ethanol (two changes).
15. Rapidly dehydrate in 100% ethanol (two changes).

16. Transfer to xylene.
17. Mount on slides and examine under light microscope.

A Feulgen-positive reaction, detectable by the presence of a red color after reacting with Schiff's reagent, indicates the presence of DNA (see color plate 22).

Reagents for Feulgen Reaction

Carnoy's Fixing Fluid

Absolute ethanol	6 ml
Chloroform	3 ml
Glacial acetic acid	1 ml

Prepare fresh each time. This is one of the best fixatives for Feulgen stain reaction.

SCHIFF'S REAGENT

1. Add 200 ml boiling water to 1 g basic fuchsin.
2. Cool to 50°C and filter.
3. Add 20 ml 1N HCl and cool to room temperature.
4. Add 2 g potassium metabisulfate ($K_2S_2O_5$). Allow to stand overnight.
5. Add a pinch (about 0.5 g) of charcoal if on the following day the reagent is not straw-colored. Shake for 1 minute and then filter.
6. Store in refrigerator. The reagent may be used repeatedly until it turns pink.

ACRIDINE ORANGE STAINING

The acridine orange staining technique can be used for the differentiation of DNA and RNA components of a cell. This method can be adapted for the differentiation of a double-stranded DNA or RNA virus from the single-stranded DNA or RNA virus in an infected cell (table 11.2). The double-stranded DNA or RNA generally shows up as a yellow-green fluorescence; the single-stranded RNA or DNA is orange-red. Better differentiation is obtained if every solution contains 0.002 M $MgSO_4$ including buffers in the following steps.

Procedure

1. Wash infected cultures with PBS, two or three times. (Care should be taken in discarding PBS that contains infectious virus.)

Table 11.2 Acridine Orange Staining Properties of Different Virus-induced Inclusions

Virus Nucleic Acid	Virus	Viral Inclusion	Acridine Orange Staining	Nuclease Susceptibility	
				DNASE	RNASE
Single-stranded RNA	Measles virus	Nucleus and Cytoplasm	Flame red	−	+
Double-stranded RNA	Reovirus	Cytoplasma	Yellow-green	−	+
Single-stranded DNA	Parvovirus	Nucleus	Flame red	+	−
Double-stranded DNA	Adenovirus	Nucleus	Yellow-green	+	−
	Herpesvirus	Nucleus	Yellow-green	+	−
	Poxvirus	Cytoplasm	Yellow-green	+	−

2. Fix infected cells on coverslip in freshly prepared Carnoy's fluid for 2–5 minutes (for formula, see above).
3. Hydrate rapidly by immersing coverslips successively in 95%, 80%, 70%, and 50% ethanol, approximately 1 minute each.
4. Rinse in 0.002 M $MgSO_4$ solution, approximately 1 minute.
5. Place in 1% acetic acid for 1 minute.
6. Rinse in 0.002 M $MgSO_4$ solution for 1 minute.
7. Rinse in phosphate buffer containing 0.067 M Na_2HPO_4 and KH_2PO_4 at pH 6.0 for 3 minutes.
8. Stain in 0.01% acridine orange solution prepared in the same phosphate buffer, pH 6.0, for 3 minutes.
9. Rinse in the same phosphate buffer at pH 6.0 for 2–5 minutes.
10. Differentiate in 0.1 M $CaCl_2$ containing 0.002 M $MgSO_4$ for 2 minutes.
11. Rinse in phosphate buffer, pH 6.0.
12. Mount in the same buffer, cover the stained coverslip with 00-thickness coverslip and seal with nail polish.
13. Examine the preparation in a dark-field microscope with UV light source.

An example of an acridine-orange-stained preparation is shown in color plate 20. Double-stranded reovirus RNA stains green in the cytoplasm, whereas single-stranded cellular RNA stains red in the cytoplasm. Double-stranded DNA stains green in the nucleus (see table 11.2).

Reagents for Acridine Orange Staining

POTASSIUM SODIUM PHOSPHATE BUFFER (0.067 M), pH 6.0

1. Dissolve 9.465 g Na_2HPO_4 in 1,000 ml distilled water.
2. Dissolve 9.072 g KH_2PO_4 in 1,000 ml distilled water.
3. Mix 40 ml Na_2HPO_4 and 230 ml KH_2PO_4 solution.

ACRIDINE ORANGE (AO) SOLUTION

1. Stock solution (0.1%): dissolve 0.1 g of AO powder in 100 ml distilled water.
2. Staining solution (0.01%): dilute 1 ml AO stock solution with 9 ml of 0.067 M phosphate buffer, pH 6.0.

CALCIUM CHLORIDE SOLUTION (0.1 M)

Dissolve 11.099 g $CaCl_2$ in 1,000 ml distilled water.

$MgSO_4$ STOCK SOLUTION (0.2 M)

Dissolve 4.93 g $MgSO_4 \cdot 7H_2O$ in 100 ml distilled water.

To obtain a 0.002 M $MgSO_4$ solution, add 0.1 ml of stock solution to 9.9 ml distilled water.

12

NUCLEIC ACID HYBRIDIZATION TECHNIQUES

M. L. LANDRY

Nucleic acid hybridization techniques have been invaluable tools in the study of viruses and viral pathogenesis. Over the past decade, a variety of hybridization techniques have been applied to the detection of viral nucleic acid directly in clinical specimens and in tissue sections. Yet the use of the hybridization technique has been limited by a lack of clear advantage over immunologic and cell culture techniques and because of the difficulties inherent in transferring a basic research tool to the diagnostic laboratory. Recently developed amplification techniques and advances in the nonradioactive labeling of probes are likely to lead to wider applications in the near future.

Because techniques vary substantially and most clinical laboratories purchase hybridization kits, which come with manufacturer directions, only the general principles of each technique are described here.

PRINCIPLES OF HYBRIDIZATION

Hybridization techniques are based on the specificity and affinity of complementary strands of nucleic acid for each other. Probes complementary to the nucleic acid sequences sought are prepared and labeled. The sample and probe must be single-stranded or denatured by heat or alkali to generate single strands allowed to come in contact with and anneal to each other (figure 12.1). After washing to remove unreacted probes, the annealed probe can be detected by a variety of means depending upon the label used. Hybridization reactions can occur in solution or with samples attached to a solid support.

ISBN 0-300-05845-4.

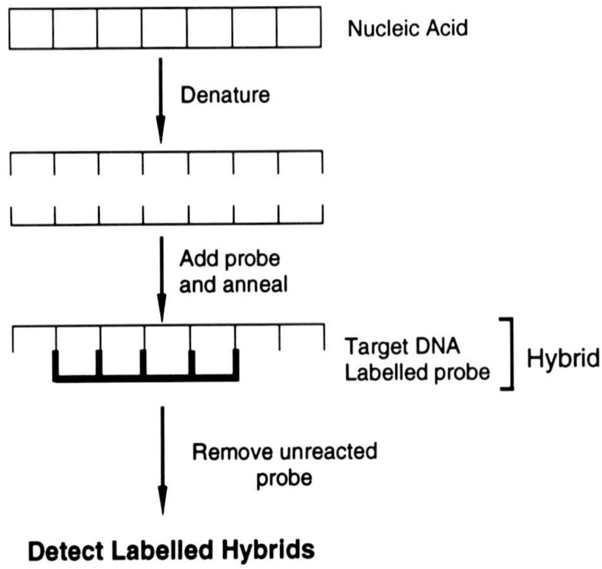

Figure 12.1 Principle of nucleic acid hybridization

PROBE PREPARATION AND LABELING

Currently, there are a number of methods for probe preparation. Restriction fragments of virion nucleic acids can be amplified as DNA inserts in cloning vectors such as bacterial plasmids, then purified and labeled. When cloned into transcription vectors, RNA can be transcribed from the insert with T7 or SP6 polymerase and labeled for use as a probe (riboprobe). If the sequence of the viral nucleic acid is known, probes can be conveniently prepared in an oligonucleotide synthesizer, avoiding lengthy cloning and purification procedures. Polymerase chain reaction can also be used to amplify nucleic acids for use as probes.

The labeling of probes can be accomplished in several ways. By nick translation, DNase I is used to nick the DNA, and DNA polymerase I to repair the strands, incorporating labeled nucleotides in the process. The random primer method uses hexanucleotide segments of DNA and DNA polymerase I to synthesize new stretches of DNA from a template strand, incorporating labeled bases into the new DNA. Alternatively, oligonucleotide probes can be end-labeled with polynucleotide kinase, or enzymes such as horseradish peroxidase can be directly conjugated to probes by linker arms.

For maximum sensitivity, probes have traditionally been labeled with potentially hazardous radioisotopes such as ^{32}P. The short half-life of ^{32}P-labeled probes necessitates frequent resynthesis. Probes labeled with nonradioactive labels such as biotin have a long shelf life (about one year). Protocols using nonradioactive probes have rapid development times and avoid the need for autoradiography, which are additional advantages for the clinical laboratory. Although nonradioactive biotinylated probes have generally been found to be less sensitive than probes labeled with radioisotopes, digoxigenin-labeled probes are highly specific and equivalent to radioisotopes in sensitivity. Digoxigenin-dUTP can be incorporated into DNA probes by random primer labeling; hybrids are detected immunologically using an antibody-enzyme conjugate, anti-digoxigenin-alkaline phosphatase, followed by addition of a substrate to give a color reaction.

Recently, chemiluminescent enzyme substrates have been found to be the most sensitive means for detecting enzyme conjugates. When chemiluminescence has been used to detect digoxigenin-labeled probes, the sensitivity has exceeded radioisotopes with development times that are substantially less.

HYBRIDIZATION TECHNIQUES

Infectious virions are required for virus isolation, whereas nucleic acid hybridization techniques can detect both infectious and noninfectious virions. The steps for various hybridization assays are compared in table 12.1.

Spot or Dot Blot

For spot or dot blot hybridization, nucleic acid can be extracted by simplified procedures, or intact cells can be used. Samples are directly spotted onto nitrocellulose, which avoids restriction enzyme digestion and gel electrophoresis. Speed, simplicity, and ability to assay large numbers of specimens are the obvious advantages of its clinical use. However, only a spot is visualized (see figure 23.3), in contrast to Southern blot, where the restriction fragments of characteristic molecular weights and patterns can be recognized (see figure 22.15). Detecting a nonspecific reaction is therefore more difficult. Careful attention to reaction conditions, probe specificity, and controls is essential.

Southern Blot

The Southern blot was first developed by E. M. Southern and bears his name. For the Southern blot, DNA must first be extracted and purified. It is usually cleaved by restriction enzymes, the fragments separated by gel electrophoresis, and they are then transferred from the gel to nitrocellulose, retaining their electrophoretic position. Before hybridization, the blot is treated to reduce nonspecific binding. The probe, generally labeled with ^{32}P, is then added and hybridization takes place. After hybridization, excess probes are removed by washing, and the radiolabeled hybrids are detected by autoradiography (see figure 22.15).

Table 12.1 Nucleic Acid Hybridization Methods

Hybridization Method	Sample Preparation	Transfer Sample to Solid-Phase Support	Hybridization Process	Final Step
Spot or Dot blot	Extract nucleic acid	Spot onto nitrocellulose	Add probe and hybridize; wash to remove unreacted probe	Detect hybrids as spot (fig. 23.3)
Southern blot	Extract DNA Restriction enzyme digestion Gel electrophoresis	Transfer to nitrocellulose	Add probe and hybridize; wash to remove unreacted probe	Detect hybrids as separate bands by autoradiography (fig. 22.15)
Northern blot	Extract RNA Gel electrophoresis	Transfer to nitrocellulose	Add probe and hybridize; wash to remove unreacted probe	Detect hybrids as separate bands by autoradiography
In situ hybridization	Prepare fixed tissue, cell culture or cell smear	Fix to slide	Add probe and hybridize; wash to remove unreacted probe	Detect hybrids in intact cells (color plate 24)

Northern Blot

The Northern blot is commonly used for RNA. Originally, specially treated paper was used, and nucleic acid was covalently bound to reactive diazo groups. Nitrocellulose is now commonly used, with the transfer performed in a high-salt buffer. Similar to the Southern blot, nucleic acids such as messenger RNA (mRNA) are separated by electrophoresis and transferred onto paper. Radiolabeled single-stranded DNA fragments complementary to the specific mRNA are then hybridized as probes, and hybrids are detected by autoradiography.

In Situ Hybridization

For in situ hybridization, intact cells—in the form of fixed or frozen tissue, cell smears, or a cell culture—are probed for viral sequences (color plate 24). This method is appropriate for the pathology laboratory because histology can also be evaluated and the cell type infected can be determined. In situ hybridization can also be applied to an infected cell culture in the diagnostic virology laboratory for rapid detection and the identification of virus isolates.

AMPLIFICATION TECHNIQUE

Despite the many technical improvements in hybridization, probe preparation, and labeling, the limit of detection for the most sensitive nucleic acid probes remains around 10^3 to 10^4 infectious virions. This level of sensitivity is often insufficient for direct detection of virus in clinical specimens. The development of amplification techniques is directed at enhancing the detection signal, thus providing greatly increased levels of sensitivity, approaching that of culture techniques.

Amplification techniques are varied and are undergoing continuing modifications and enhancements. The polymerase chain reaction (PCR) has had the widest application and uses repeated cycles of DNA synthesis with heat-stable Taq DNA polymerase to carry out in vitro replication of target nucleic acid sequences (figure 12.2). For RNA viruses, reverse transcriptase is used as the first step to generate a DNA copy of the RNA genome, which is then amplified with DNA polymerase. The amplified sequences can be detected by gel electrophoresis and visualized with ethidium bromide staining, or by blotting to nitrocellulose (Southern blot), or by dot blot followed by hybridization with a ^{32}P-labeled probe, or more recently, by using a digoxigenin-labeled probe. The careful selection of primers is essential for good results.

The sensitivity of this technique has led to the problem of distinguishing how many organisms have clinical significance because the presence of small but detectable amounts of organisms may be normal. The development of quantitative assays may improve clinical utility. Another problem for clinical

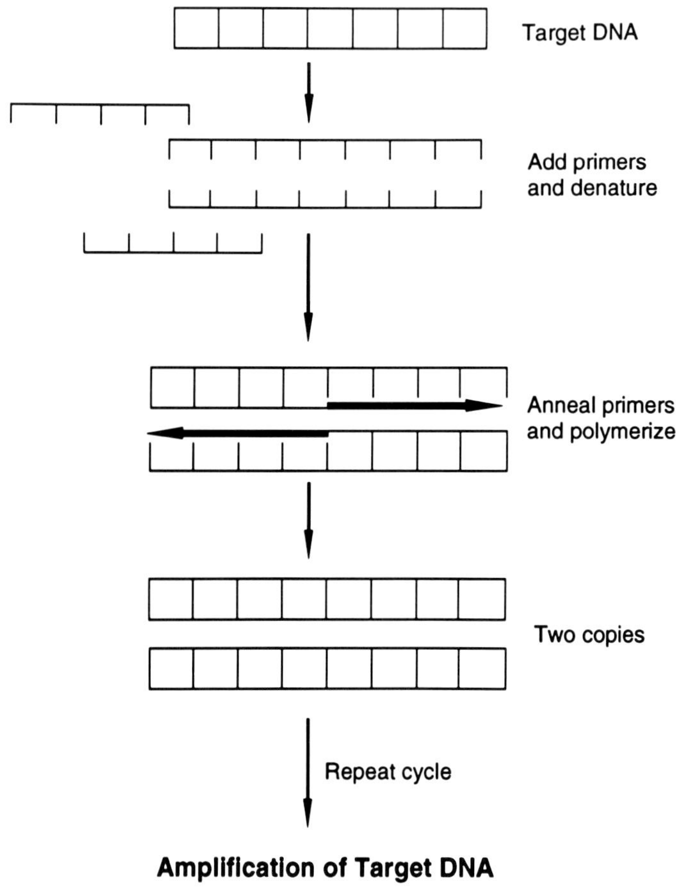

Figure 12.2 Principle of the polymerase chain reaction (amplification of target DNA)

diagnosis is the contamination of specimens with amplified fragments, which may contaminate glassware, buffers, and air systems, causing false-positive results.

USE IN THE VIRAL DIAGNOSTIC LABORATORY

Virus Identification in Cell Culture

DNA probes can be used to identify virus isolates in cell culture and to detect viral nucleic acid in centrifugation culture but in general offer no advantage over immunologic techniques. An exception is the enteroviruses, which have many antigenic types and no group antigen. The 5′ noncoding region of the genome is highly conserved between enterovirus types, allowing one probe to identify a number of antigenic types. PCR has now been used for the rapid identification of enteroviruses, differentiation of HTLV-I from HTLV-II, and differentiation of RSV subgroups.

Direct Detection of Virus in Clinical Specimens

DNA probes can be used to detect a virus directly in clinical specimens. Yet for viruses that can be routinely cultivated, the comparative sensitivity of hybridization techniques has been insufficient. For human papillomaviruses that are not cultivable and have many antigenic types, the hybridization technique is the diagnostic method of choice (see chapter 23). The diagnosis of EBV-associated lymphomas is also best established by detection of the viral genome. Hybridization techniques have been useful for hepatitis B virus, parvovirus B19, and human immunodeficiency virus in blood, and for the Norwalk virus, rotavirus, adenovirus, and enterovirus in stools. For maximum sensitivity, however, the best approach is first to amplify sequences directly in clinical specimens by PCR, then detect the amplified sequences by hybridization. Inhibitors of PCR must be recognized and avoided—such as heparin in blood samples (ethylene diaminetetraacetic acid [EDTA] and acid citrate dextrose are recommended instead), or nonspecific inhibitors in stools. When PCR is employed, the detection of HIV is as sensitive as a culture method.

Direct Detection of Virus in Pathologic Specimens

In situ hybridization kits using nonradioactive probes are commercially available for many viruses. In tissue sections, viral diagnosis based on inclusions may be insensitive and occasionally inaccurate (color plate 23). The application of viral probes can give specific a identification of the infecting virus and can detect a virus infection in cells without inclusions (color plate 24). In addition, if fixation is too long or suboptimal, viral antigens may be destroyed, resulting in a false-negative result. Since viral DNA is more stable, it may be detectable by hybridization even after a sample has been stored for a long time.

Antiviral Susceptibility Testing

With the increasing use of antiviral therapies, drug resistance is becoming a reality, which has led to a demand for antiviral sensitivity testing. Hybridization techniques have been increasingly used to quantitate the inhibition of nucleic acid replication in the presence of antiviral drugs. A virus is first inoculated into a cell culture in the presence of increasing concentrations of an antiviral agent. After an appropriate incubation period for virus replication, the infected cells are lysed and the lysates are adsorbed or spotted onto membranes. To assess the amount of viral replication that has occurred in the presence of a drug, the viral nucleic acid in the infected cell lysate is hybridized with a radiolabeled probe and the hybrids are subsequently quantitated using a scintillation counter or a densitometer. Thus, the hybridization technique has been used as a reproducible, quantitative, and time-saving alternative to a standard plaque-reduction assay. Before it can be widely applied in a clinical laboratory setting, however, conversion to a nonradiometric assay is essential.

Unfortunately, all assays that rely on quantitating viral replication to assess susceptibility of an isolate to an antiviral drug will be inherently slow for viruses with long replication cycles, such as CMV. In the future, direct testing of virus isolates by PCR for the presence of defined mutations in viral genes known to be associated with resistance to antiviral drugs will greatly improve the speed of testing.

13

ELECTRON MICROSCOPY AND IMMUNOELECTRON MICROSCOPY

C. K. Y. FONG

Electron microscopy (EM) has been used for many years in the recognition of viruses in clinical specimens. Historically, EM has been used successfully to detect smallpox and varicella-zoster viruses in vesicular fluid or pustular materials—an approach that makes a rapid and accurate diagnosis possible.

EM is particularly useful in the discovery and diagnosis of noncultivable viruses. Such viruses include human papilloma viruses (HPV), the etiology of common warts, molluscum contagiosum virus causing skin tumor, and rotavirus (RV), the etiologic agent of nonbacterial gastroenteritis. Subsequently, several other noncultivable viruses associated with gastroenteritis, such as the Norwalk virus, enteric adenoviruses, and calicivirus were first discovered by EM, which is still the most reliable method for their recognition.

The identification of viruses by EM is based on virus morphology and the characteristic ultrastructure of virus-cell interactions. It is not limited by the specificity of the test reagents used by most immunodiagnostic methods. Over the years, EM has been used for the rapid identification of unknown or known viruses with unusual biological properties and of endogenous and contaminant agents in cell cultures. Clinical specimens in which EM is used as the method of laboratory diagnosis are given in table 13.1. Two major EM techniques routinely used for virologic examination are negative staining and thin sectioning.

NEGATIVE STAINING TECHNIQUE

The negative staining technique for the visualization of virions was introduced in 1959 by R. W. Horne and P. Wildy. Heavy metal salts are used to

ISBN 0-300-05845-4.

Table 13.1 Use of Electron Microscopy as a Diagnostic Tool for Selected Viral Infections

Disease Associated	Virus Suspected	Specimen Collected
Gastroenteritis	Rotavirus Adenovirus Norwalk agent	Feces
Exanthematous disease	Poxvirus* Varicella-zoster virus* Herpes simplex virus*	Vesicular fluid
Warts	Papillomavirus	Tissue biopsy
Skin tumor	Molluscum contagiosum	Tissue biopsy
Progressive multifocal leukoencephalopathy	Polyomavirus	Brain biopsy
Subacute sclerosing panencephalitis	Measles virus	Brain biopsy
Identification of unknown virus isolates or endogenous virus	Not specified	Not specified

*Conventional culture method is available for isolation of these viruses. (Modified from C. K. Y. Fong, *VA Practitioner* 6:65–75 [1989].)

enhance the contrast between background and the virion's image. The procedure is simple and needs no special equipment. To be able to detect viruses in clinical specimens by EM, the virus concentration in a suspension should be at least 10^6 virions per ml. If the virus concentration is too low, ultracentrifugation is necessary. A grossly contaminated specimen should first be subjected to a low-speed centrifugation to remove debris. The clarified supernatant is then centrifuged at a higher speed (15,000 $\times$ g for 1 hour) in order to obtain a concentrated virus suspension. After removing the supernatant completely, the pellet is resuspended in a small amount of distilled water. As a rule, distilled water is used as a diluent as it helps to lyse cellular structures but leaves the virus intact. Clinical specimens should be handled with care, and safety precautions followed—namely, working in a biological safety cabinet. Contaminated instruments and glassware must be properly disinfected.

Specimen Source and Processing

VESICLE FLUID

Vesicle fluid aspirated from an intact vesicle is the best specimen. Puncture the vesicle using a sterile needle and collect the fluid using a sterile Pasteur pipette that contains a small amount of distilled water. The mixture of vesicle fluid and water is placed on an EM grid and stained with 2% phosphotungstic acid (PTA) as described below.

FECES

Make a 10%–20% suspension of stool in distilled water and clarify by centrifugation at 3,000 × *g* for 30 minutes. The supernatant may be placed directly onto an EM grid for staining and examination or can be centrifuged at 15,000 × *g* for 1 hour to concentrate the virus in the pellet, which is then resuspended in distilled water for EM examination.

URINE

If cloudy, it should be clarified first by centrifugation at 3,000 × *g* for 15 minutes to remove the gross contaminants. The supernatant is then centrifuged at 15,000 × *g* for 1 hour. The pellet is treated as above.

SERUM

Mix equal volumes of serum with distilled water, then centrifuge at 15,000 × *g* for 1 hour. Discard the supernatant and resuspend the pellet with distilled water to its original volume. Recentrifuge as before; the pellet is then treated as above.

TISSUE BIOPSY

Make a 10% tissue suspension in distilled water with a glass homogenizer. Clarify by centrifugation at 3,000 × *g* for 10 minutes, then centrifuge at 15,000 × *g* for 1 hour. The final pellet is processed as above.

Methods for Negative Staining Procedure

CONVENTIONAL METHOD

1. Deposit a drop of concentrated virus suspension or clinical specimen at the bottom of a petri dish.
2. Place a Formvar-coated EM grid 300 or 400 mesh on the virus drop, Formvar membrane side down, for about 1 minute or longer.
3. Pick the grid up with a fine forceps. Remove the excess fluid by touching the grid to the edge of a piece of torn filter paper.
4. Hold the grid with a forceps and add a drop of 2% PTA salt solution pH 7.0 for 1 minute.
5. Remove the excess PTA by touching the grid with a piece of torn filter paper.
6. Place the stained grid in a petri dish and irradiate with UV light for 5 minutes on each side to inactivate infectious virus.
7. Examine the grid under EM.

The virions are surrounded by heavy metal atoms and are revealed against a dark background, the so-called negative stain, since the electron beam can pass through the low-electron density of the virus but not through the metallic compound. Yet most clinical specimens contain a small number of virions. Pseudoreplica and agar-diffusion methods have been developed for concentrating virus suspensions and are often used in diagnostic laboratories. Only the latter method is described here.

AGAR-DIFFUSION METHOD

This method was first described by N. Anderson and F. W. Doane. It is similar to the pseudoreplica method originally described by D. G. Sharp but is simpler to perform. The procedure is as follows:

1. Place a 1% agar block on a microscope slide. Add a small drop of virus suspension to the surface of the agar block.
2. Place 1–2 EM grids (Carbon-Formvar coated, 300 mesh) over the drop of virus suspension (Carbon-Formvar side down) and allow the aqueous solution and salts to diffuse into the agar until the suspension disappears, thus leaving the concentrated virions on the agar surface, where they have adhered to the coated grids.
3. Pick up the grids with the adhered virions and stain with a 2% PTA solution for 1 minute. Remove excess PTA solution by blotting with a piece of torn filter paper.
4. Inactivate infectious virus with UV irradiation, 5 minutes on each side.
5. Examine the grids under an EM.

IMMUNOELECTRON MICROSCOPY

Immunoelectron microscopy (IEM) has been used for the rapid identification of viruses using a homologous antiserum or for enhancing the concentration of a virus in order to facilitate the recognition of small number of virions in the specimen. IEM has been widely used for recognition and identification of small-sized viruses, such as the Norwalk virus and the hepatitis A virus in stool specimens and to detect the hepatitis B surface antigen in serum. The procedure is as follows:

1. Mix virus suspension with homologous antiserum at optimal dilution.
2. Incubate the mixture at 37°C for 1 hour or overnight at 4°C to permit formation of virus-antibody complexes.
3. Centrifuge the virus-antibody mixture at 15,000 × *g* for 30 minutes.
4. Resuspend the pellet in a small amount of distilled water.
5. Apply the suspension of virus-antibody complexes to a Formvar-coated EM grid and remove the excess by blotting with a piece of torn filter paper. Immediately stain the grid with 2% PTA for 1 minute (avoid drying the preparation).
6. Examine the grid under EM.

Alternatively, after step 1, the virus-antiserum mixture can be placed directly on an agar block and processed by the agar-diffusion method as described above, without ultracentrifugation. In addition, the Airfuge, a table model ultracentrifuge, has become available for centrifuging minute amounts of virus suspension directly onto EM grids. Following centrifugation, the grids can be stained with PTA and then examined by EM.

THIN SECTIONING TECHNIQUE

This method is used to prepare fixed and embedded tissues or cells for EM. It is more time-consuming than other methods and requires trained personnel. It does offer advantages because it allows the direct observation of the well-preserved virus-cell interaction, which in turn reveals the site of virus replication and maturation in the host cells, thus aiding in the recognition of unknown viruses. The choice of reagents for fixing and embedding varies from laboratory to laboratory and usually depends on each one's preference. For detailed procedures concerning the use of various buffer solutions, fixatives, and embedding media, the reader is referred to specific articles cited at the end of part II. For rapid diagnosis, the fixation and embedding procedures can be shortened from several days to several hours. The following procedures are routinely used for fixation and embedding and have proved satisfactory in the laboratory.*

Primary Fixation

1. For monolayer culture, fix cultured cells in situ with 2% glutaraldehyde in 0.1 M Sorensen's phosphate buffer, pH 7.2, for 1 hour at 4°C, scrape, and then pellet at 700 × *g* for 15 minutes.
2. For tissue fragments (1 mm^3), fix with 3% glutaraldehyde solution. All fixed specimens are washed in three changes of 0.1 M Sorensen's phosphate buffer and stored at 4°C.

Postfixation

Remove the buffer, add 1.33% osmium-collidine fixative, pH 7.4, to the above fixed tissue fragments or cell pellets for 1 hour at 0°–4°C (for formula, see below).

En Bloc Staining

Rinse the specimens with triple-distilled water three times; stain en bloc with 0.5% uranyl acetate for 4 hours at 4°C to enhance contrast of virions.

*Virology Reference Laboratory, the Veterans Affairs Medical Center, West Haven, Connecticut

Dehydration

Use the following concentrations of ethanol: 50%, 70%, 95%, and 100% (three changes), 14 minutes each.

Penetration and Infiltration

Place specimens in a 1:1 mixture of ethanol 100% and Spurr embedding medium for 60–90 minutes, then two changes in Spurr, and leave in the refrigerator overnight.

Embedding

Embed the tissue block in fresh Spurr in a Beem capsule. Put the capsule into a 60°C oven for 2 days for polymerization. Many embedding media are available for use. Low-viscosity Spurr medium is easy to use, especially when specimens contain a small quantity of single cells that require a centrifugation step to embed the medium.

Trimming

Cut off excess plastic at the tips of the blocks in order to reach the specimen easily.

Sectioning

Thin sectioning is the most tedious step and requires considerable experience. The two ultramicrotomes most widely used are the Sorvall and the LKB. Microtomes from other sources are also available. Thin sections are collected from the water surface in a trough. The thickness of the sections can be accurately judged from the interference colors, ranging from bright colors for fairly thick sections to dull grays for the very thin ones (gray: less than 600Å; silver: 600–900Å). Sections are placed on grids for staining and examination.

Staining

A double staining of uranyl and lead salts is employed:

1. Place a drop of uranyl acetate staining solution (0.5%) in a wax-lined petri dish. Place the grid with sections down on the staining solution for 10–12 minutes and rinse three times in distilled water. (The length of time required for staining varies depending on tissue specimen and embedding media.)
2. Place a drop of lead citrate staining solution in a wax-lined petri dish

with a small weighing dish containing a few pellets of NaOH to absorb CO_2 in the air and to prevent precipitation of the lead salt solution during staining. Place the grids with sections down over the staining solution for 10–12 minutes, rinse first in 0.02 N NaOH, then in distilled water. The double-stained sections are then ready for EM examination.

Figures 13.1 and 13.2 illustrate the major groups of viruses that infect humans, as revealed in negatively stained preparations and in thin-sectioned infected cells together with their diagrams.

REAGENTS

Sorensen's Phosphate Buffer

Solution A: 0.2 M NaH_2PO_4

$NaH_2PO_4 \cdot H_20$	5.52 g
Distilled water	200 ml

Solution B: 0.2 M Na_2HPO_4

Na_2HPO_4 (anhydrous)	14.2 g
Distilled water	500 ml

To make 0.1 M phosphate buffer, pH 7.2, mix:

Solution A	70 ml
Solution B	180 ml
Distilled water	250 ml

2% Glutaraldehyde in 0.1 M Phosphate Buffer, pH 7.2

Solution A	70 ml
Solution B	180 ml
50% glutaraldehyde	20 ml
Distilled water	230 ml

3% Glutaraldehyde in 0.1 M Phosphate Buffer, pH 7.2

Solution A	70 ml
Solution B	180 ml
50% glutaraldehyde	30 ml
Distilled water	220 ml

Osmium-Collidine Fixative, pH 7.4

2% OsO_4	6.7 ml
0.2 M s-collidine buffer, pH 7.4 (2,4,6-trimethylpyridine)	3.3 ml

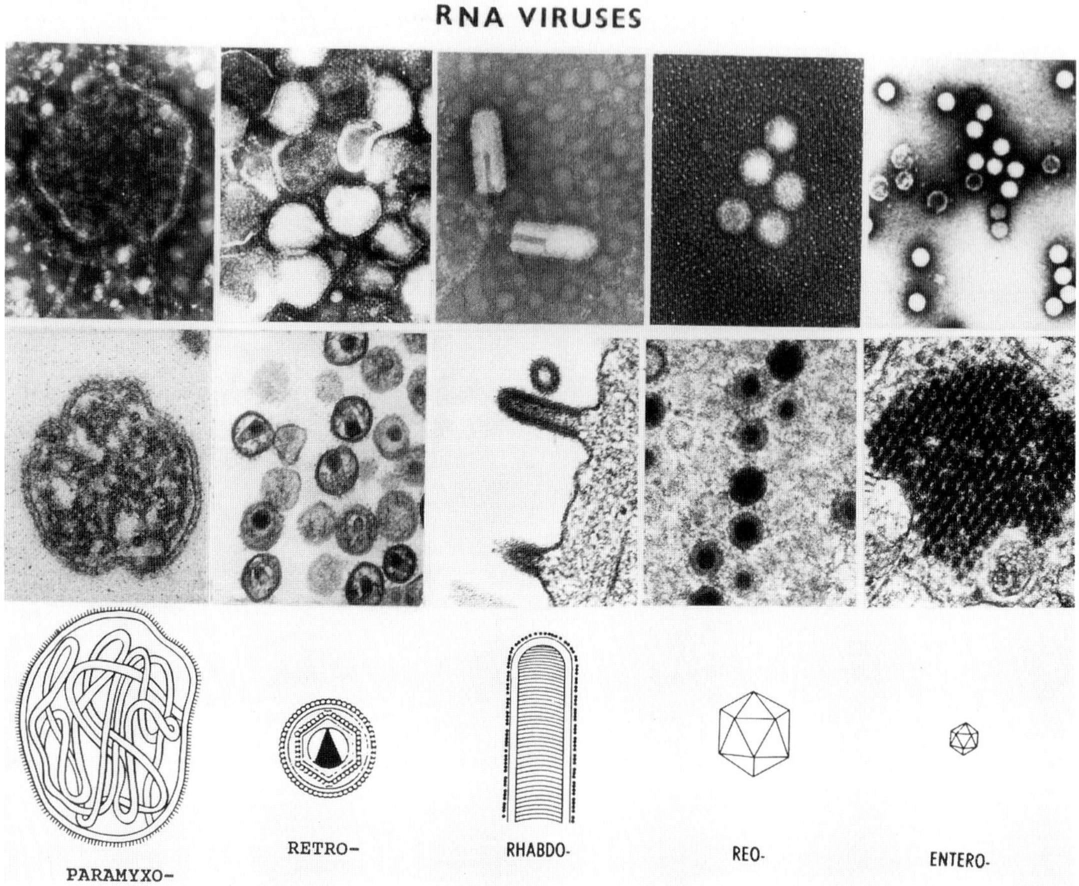

Figure 13.1 Relative size and shape of RNA viruses as revealed in negatively stained preparations *(top row)*, in thin-sectioned cells *(middle row)*, and as compared with diagrams *(bottom row)*. (Modified G. D. Hsiung et al., *Progr. Med. Virol.* 25:155 [1979])

Spurr Embedding Medium

Vinyl cyclohexane dioxide (VCD)	10.0 g
Diglycidyl ether of polypropylene glycol (DER 736)	7.0 g
Nonenyl succinic anhydride (NSA)	26.0 g
Dimethylaminoethanol (DMAE)	0.4 g

The embedding medium should be freshly prepared, but it may be kept for convenience in measured aliquots. These are best stored in sealed containers at −20°C or lower.

Uranyl Acetate Solution (10%)

Uranyl acetate	5 g
25% ethanol	50 ml

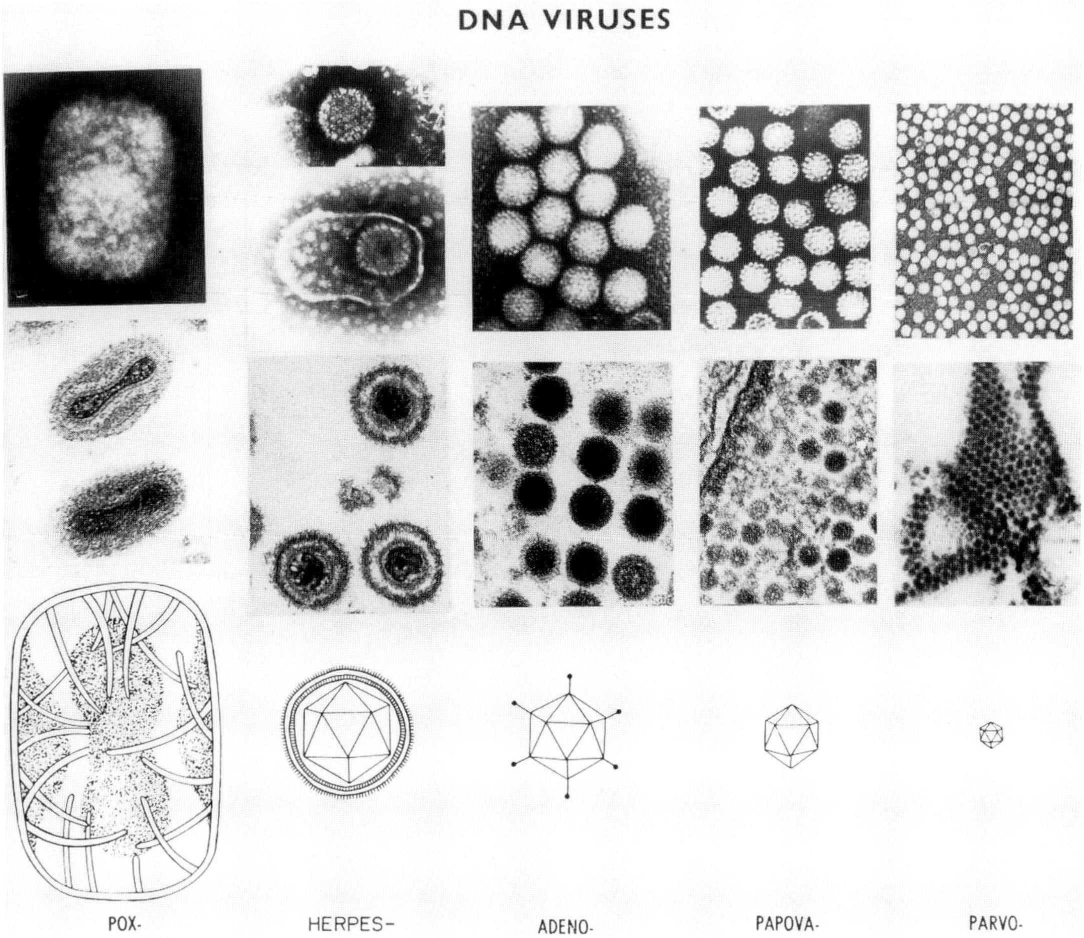

Figure 13.2 Relative size and shape of DNA viruses as revealed in negatively stained preparations *(top row)*, in thin-sectioned cells *(middle row)*, and as compared with diagrams *(bottom row)* (G. D. Hsiung et al., *Progr. Med. Virol.* 25:154 [1979])

The solution is filtered through Millipore membrane (0.2 um) and stored in a tightly capped tube at room temperature protected from light.

Lead Citrate Solution

Lead citrate	30–35 mg
Distilled water	10 ml

Shake immediately for 30 minutes, add drops of 5 N NaOH and shake vigorously until the mixture is clear. Centrifuge for 15 minutes before use. This solution should be stored in a tightly stoppered tube and clarified by centrifugation each time before use.

Chapter 5: Virus Isolation and Identification Methods

Beckwith, D. G.; Halstead, D. C.; and Beckwith, C. H. Comparison of the Ortho Cultureset (Vero cells) and Difco Cellmatics system (mink lung and primary rabbit kidney cells) for the detection of herpes simplex virus from clinical specimens. *Diagn. Microbiol. Infect. Dis.* 6:145–49 (1987).

Enders, J. F.; Weller, T. H.; and Robbins, F. C. Cultivation of the Lansing strain of poliomyelitis virus in cultures of various human embryonic tissues. *Science* 109:85–87 (1949).

Espy, M. J.; Hierholzer, J. C.; and Smith, T. F. The effect of centrifugation on the rapid detection of adenovirus in shell vials. *Am. J. Clin. Path.* 88:358–60 (1987).

Espy, M. J.; Smith, T. F.; Harmon, M. W.; and Kendal, A. P. Rapid detection of influenza virus by shell vial assay with monoclonal antibodies. *J. Clin. Microbiol.* 24:677–79 (1986).

Espy, M. J.; Wold, A. D.; Jespersen, D. J.; Jones, M. F.; and Smith, T. F. Comparison of shell vials and conventional tubes seeded with rhabdomyosarcoma and MRC-5 cells for the rapid detection of herpes simplex virus. *J. Clin. Microbiol.* 29:2701–3 (1991).

Gleaves, C. A.; Smith, T. F.; Shuster, E. A.; and Pearson, G. R. Rapid detection of cytomegalovirus in MRC-5 cells inoculated with urine specimens by using low-speed centrifugation and monoclonal antibody to an early antigen. *J. Clin. Microbiol.* 19:917–19 (1984).

Hsiung, G. D. Further studies on characterization and grouping of ECHO viruses. *Ann. N.Y. Acad. Sci.* 101:413–22 (1962).

———. The impact of cell culture sensitivity on rapid viral diagnosis: A historical perspective. *Yale J. Biol. Med.* 62:79–88 (1989).

Hsiung, G. D., and Melnick, J. L. Comparative susceptibility of kidney cells from different monkey species to enteric viruses (poliomyelitis, coxsackie, and ECHO groups). *J. Immun.* 78:137–46 (1957).

James, J., and Bell, G. T. Human monoclonal antibody production. *J. Immun. Meth.* 100:5–40 (1987).

Johnston, S. L. G., and Siegel, C. S. Presumptive identification of enteroviruses with RD, HEP-2 and RMK cell lines. *J. Clin. Microbiol.* 28:1049–50 (1990).

Krisher, K. K., and Menegus, M. A. Evaluation of three types of cell culture for recovery of adenovirus from clinical specimens. *J. Clin. Microbiol.* 25:1323–24 (1987).

Landry, M. L.; Madore, H. P.; Fong, C. K. Y.; and Hsiung, G. D. Use of guinea pig embryo cell cultures for isolation and propagation of group A coxsackieviruses. *J. Clin. Microbiol.* 13:588–93 (1981).

Landry, M. L.; Mayo, D.; and Hsiung, G. D. Comparison of guinea pig embryo cells, rabbit kidney cells, and human embryonic lung fibroblast cell strains for isolation of herpes simplex virus. *J. Clin. Micro.* 15:842–47 (1982).

Langenberg, A.; Zbanyszek, R.; Dragavon, J.; Ashley, R.; and Corey, L. Comparison of diploid fibroblast microtiter plate system for the isolation of herpes simplex virus. *J. Clin. Microbiol.* 26:1772–74 (1988).

Paya, C. V.; Wold, A. D.; and Smith, T. F. Detection of cytomegalovirus infections in specimens other than urine by shell vial assay and conventional tube cell cultures. *J. Clin. Microbiol.* 25:755–57 (1987).

Rubin, S. J., and Rogers, S. Comparison of cultureset and primary kidney cell culture for the detection of herpes simplex virus. *J. Clin. Microbiol.* 19:920–22 (1984).

Salmon, V. C.; Kenyon, B. R.; and Overall, J. C., Jr. Cross contamination of viral specimen related to shell vial caps. *J. Clin. Microbiol.* 28:2820 (1990).

Salmon, V. C.; Stanberry, L. R.; and Overall, J. C., Jr. More rapid isolation of herpes simplex virus in a continuous line of mink lung cells than in Vero or human fibroblast cells. *Diagn. Microbiol. Infect. Dis.* 2:317–24 (1984).

Smith, M. C.; Creutz, C.; and Huang, Y. T. Detection of respiratory syncytial virus in nasopharyngeal secretions by shell vial technique. *J. Clin. Microbiol.* 29:463–65 (1991).

Chapter 6: Virus Assay, Neutralization Test, and Antiviral Assay

Anderson, L. J.; Hatch, M. H.; Flemister, M. R.; and Marchetti, G. E., Jr. Detection of enterovirus 70 with monoclonal antibodies. *J. Clin. Microbiol.* 20:405–8 (1984).

Bartova, L.; Rajcani, J.; and Pogady, J. Herpes simplex virus antibodies in the cerebrospinal fluid of schizophrenic patients. *Acta Virolog.* 31:443–46 (1987).

Cao, Y. Z.; Friedman-Kien, A. E.; Mirabile, M.; Li, X. L.; Alan, M.; Dieterich, D.; and Ho, D. D. HIV-1 neutralizing antibodies in urine from seropositive individuals. *J. Acq. Imm. Def. Syndr.* 3:195–99 (1990).

Hu, J. M., and Hsiung, G. D. Evaluation of new antiviral agents. 1, In vitro perspectives. *Antiviral Res.* 11:217–32 (1989).

Hudson, R. W.; Herrmann, J. E.; and Blacklow, N. R. Plaque quantitation and virus neutralization assays for human astroviruses. *Arch. Virol.* 108:33–38 (1989).

Matsumoto, K.; Kobayashi, T.; and Kimura, Y. Isolation and preliminary characterization of antigenic variant of echovirus type 11. *J. Med. Virol.* 31:253–58 (1990).

Peigue-Lafeuille, H.; Fuchs, F.; Gharabaghi, F.; Chambon, M.; and Aymard, M. Impact on routine diagnosis of echovirus infections of intratypic differentiation and antigenic variation in echovirus 25 studied by using monoclonal antibodies. *J. Clin. Microbiol.* 28:2291–96 (1990).

Prichard, M. N.; Turk, S. R.; Coleman, L. A.; Engelhardt, S. L.; Shipman, C.; and Drach, J. C. A microtiter virus yield reduction assay against human cytomegalovirus and herpes simplex virus. *J. Virolog. Meth.* 28:101–6 (1990).

Reed, L. J., and Muench, H. A simple method for estimating fifty percent endpoints. *Am. J. Hyg.* 27:493–97 (1938).

Sidwell, R. W. Determination of antiviral activity. *Drugs Pharm. Sci.* 27 *(Mod. Anal. Antibio.):* 433–80 (1986).

Taguchi, F.; Hara, K.; and Kajioka, J. Rapid diagnostic procedure for virus neutralization test. *Kitasato Arch. Exptl. Med.* 57:279–83 (1984).

Watson, D. G.; Patterson, L. J.; Rinehart, J. E.; Hughes, J. H.; and Hamparian, V. V. Detection of serotypic variants in viral preparations: Sensitivity of a procedure for selection. *J. Virolog. Meth.* 23:333–38 (1989).

Chapter 7: Immunocytochemical Assays: Immunofluorescent and Immunoperoxidase Staining for Antigen Detection

Avrameas, S., and Ternynck, T. Peroxidase labeled antibody and Fab conjugates with enhanced intracellular penetration. *Immunochemistry* 8:1175–79 (1971).

Balkovic, E. S., and Hsiung, G. D. Comparison of immunofluorescence with commercial monoclonal antibodies and biochemical and biological techniques for typing clinical herpes simplex isolates. *J. Clin. Microbiol.* 22:870–72 (1985).

Gallo, D.; Diggs, J. L.; Shell, G. R.; Daily, P. J.; Hoffman, M. N.; and Riggs, J. L. Comparison of detection of antibody to the acquired immune deficiency syndrome virus by enzyme immunoassay, immunofluorescence, and Western blot methods. *J. Clin. Microbiol.* 23:1049–51 (1986).

Gardner, P. S.; Grandien, M.; and McQuillin, J. Comparison of immunofluorescence and immunoperoxidase methods for viral diagnosis at a distance. A WHO Collaborative Study. *Bull. WHO* 56:105–10 (1978).

Gardner, P. S., and McQuillin, J. *Rapid Virus Diagnosis: Application of Immunofluorescence.* Toronto: Butterworth, 1974.

Gleaves, C. A.; Smith, T. F.; Shuster, E. A.; and Pearson, G. R. Rapid detection of cytomegalovirus in MRC-5 cells inoculated with urine specimens by using low-speed centrifugation and monoclonal antibody to an early antigen. *J. Clin. Microbiol.* 19:917–19 (1984).

Herrmann, J. E.; Perron-Henry, D. M.; and Blacklow, N. R. Antigen detection with monoclonal antibodies for the diagnosis of adenovirus gastroenteritis. *J. Infect. Dis.* 155:1167–71 (1987).

Kadi, Z.; Dali, S.; Bakouri, S.; and Bougermouh, A. Rapid diagnosis of respiratory syncytial virus infection by antigen immunofluorescence detection with monoclonal antibodies and immunoglobulin M immunofluorescence test. *J. Clin. Microbiol.* 24:1038–40 (1986).

Minnich, L. L.; Smith, T. F.; Ray, C. G.; and Specter, S. Rapid detection of viruses by immunofluorescence. In *Cumitech,* 24:1–13. Washington, D.C.: American Society for Microbiology, 1988.

Nakane, P., and Kawaoi, A. Peroxidase-labeled antibody. A new method of conjugation. *J. Histochem. Cytochem.* 22:1084–91 (1974).

Ozeki, T.; Mizuno, S.; Sanefuzi, H.; Tanaka, N.; Fujiwara, H.; Masuda, H.; and Funakoshi, K. Localization of hepatitis B core antigens in chronic active hepatitis using immunoperoxidase and chromotrope aniline blue staining. *Br. J. Exptl. Pathol.* 68:605–12 (1987).

Rossier, E.; Miller, H. R.; and Phipps, P. H. Rapid viral diagnosis by immunofluorescence: An atlas and practical guide. Ottawa: University of Ottawa Press, 1989.

Sandstrom, E. G.; Schooley, R. T.; Ho, D. D.; Byington, R.; Sarngadharan, M. G.; MacLane, M. E.; Essex, M.; Gallo, R. C.; and Hirsch, M. S. Detection of human anti-HTLV-III antibodies by indirect immunofluorescence using fixed cells. *Transfusion* 25:308–12 (1985).

Smaron, M. R.; Saxon, E.; Wood, L.; McCarthy, C.; and Morello, J. A. Diagnosis of measles by fluorescent antibody and culture of nasopharyngeal secretions. *J. Virolog. Meth.* 33:223–29 (1991).

Swenson, P. D., and Kaplan, M. H. Rapid detection of cytomegalovirus in cell culture by indirect immunoperoxidase staining with monoclonal antibody to an early nuclear antigen. *J. Clin. Microbiol.* 21:669–73 (1985).

Waner, J. L.; Whitehurst, N. J.; Downs, T.; and Graves, D. G. Production of monoclonal antibodies against parainfluenza 3 virus and their use in diagnosis by immunofluorescence. *J. Clin. Microbiol.* 22:535–38 (1985).

Zhao, L.; Landry, M. L.; Balkovic, E. S.; and Hsiung, G. D. Impact of cell culture sensitivity and virus concentration on rapid detection of herpes simplex by cytopathic effects and immunoperoxidase staining. *J. Clin. Microbiol.* 25:1401–5 (1987).

Chapter 8: Enzyme-linked Immunosorbent Assay

Engvall, E., and Perlmann, P. Enzyme-linked immunosorbent assay (ELISA). Quantitative assay of immunoglobulin G. *Immunochemistry* 8:871–74 (1971).

Gerna, G.; Sarasini, A.; Di Matteo, A.; Parea, M.; Torsellini, M.; and Battaglia, M. Rapid detection of human rotavirus strains in stools by single-sandwich enzyme-linked immunosorbent assay systems using monoclonal antibodies. *J. Virolog. Meth.* 24:43–56 (1989).

Gonik, B.; Seibel, M.; Berkowitz, A.; Woodin, M. B.; and Mills, K. Comparison of two enzyme-linked immunosorbent assays for detection of herpes simplex virus antigen. *J. Clin. Microbiol.* 29.436–38 (1991).

Grandien, M.; Peterson, C. A.; Gardner, P. S.; Linde, A.; and Stanton, A. Rapid viral diagnosis of acute respiratory infections: Comparison of enzyme-linked immunosorbent assay and the immunofluorescence technique for detection of viral antigens in nasopharyngeal secretions. *J. Clin. Microbiol.* 22:757–60 (1985).

Harmon, M. W.; Drake, S.; and Kasel, J. A. Detection of adenovirus by enzyme-linked immunosorbent assay. *J. Clin. Microbiol.* 9:342–46 (1979).

Needham, C. A., and Hurlbert, P. Evaluation of an enzyme-linked immunoassay employing a covalently bound capture antibody for direct detection of herpes simplex virus. *J. Clin. Microbiol.* 30:531–32 (1992).

Patel, H.; Frenkel, L. D.; Greenhalgh, M.; Howell, R.; and Patel, S. Rapid culture confirmation of herpes simplex virus by a monoclonal antibody-based enzyme immunoassay. *J. Clin. Microbiol.* 29:410–12 (1991).

Peigue-Lafeuille, H.; Bailly, J. L.; Aymard, M.; and Fuchs, F. Use of non-neutralizing monoclonal antibodies in an ELISA for intratypic differentiation of twenty-eight echovirus type 25 clinical isolates. *J. Virolog. Meth.* 36:91–99 (1992).

Smith, K. O.; Ludwig, M. J.; Stigall, B. W.; and Boswell, N. Enzyme-linked immunosorbent assay (ELISA) for HIV antibody by a glass slide technique. *J. Immun. Meth.* 136:239–46 (1991).

Thomas, E. E., and Book, L. E. Comparison of two rapid methods for detection of respiratory syncytial virus (RSV) (Testpack RSV and Ortho RSA ELISA) with direct immunofluorescence and virus isolation for the diagnosis of pediatric RSV infection. *J. Clin. Microbiol.* 29:632–35 (1991).

Thompson, K. D., and Van Enk, R. A. Changing technologies in immunodiagnosis: Enzyme immunoassays versus immunofluorescence assays. *Clin. Microbiol. Newsltr.* 13:73–76 (1991).

Verano, L., and Michalski, F. J. Herpes simplex virus antigen direct detection in standard virus transport medium by Du Pont Herpchek enzyme-linked immunosorbent assay. *J. Clin. Microbiol.* 28:2555–58 (1990).

Voller, A.; Bartlett, A.; Bidwell, D.; Clark, M.; and Adams, A. The detection of viruses by enzyme-linked immunosorbent assay. *J. Gen. Virol.* 33:165–67 (1976).

Voller, B. A., and Bidwell, D. E. Bacteria microplate enzyme immuno-assay system for measurement of cytomegalovirus. *J. Clin. Microbiol.* 11:546–51 (1980).

———. Enzyme-linked immunosorbent assay. In Rose, N. R.; Friedman, H.; and Fahey, J. L., eds. *Manual of Clinical Immunology,* 17:99–109. Washington, D.C.: American Society for Microbiology, 1986.

Yolken, R. H. Enzyme immunoassays for the detection of microbial agents and prospects for improved assays. *Yale J. Biol. Med.* 59:25–31 (1986).

———. Enzyme-linked immunosorbent assay (ELISA): A practical tool for rapid diagnosis of viruses and other infectious agents. *Yale J. Biol. Med.* 53:85–92 (1980).

Yolken, R. H.; Kim, H. W.; Clem, T.; Wyatt, R. G.; Kalica, A. R.; Chanock, R. M.; and Kapikian, A. Z. Enzyme-linked immunosorbent assay (ELISA) for detection of human reovirus-like agent of infantile gastroenteritis. *Lancet* 2:263–67 (1977).

Chapter 10: Complement-Fixation Test

Arsenakis, M., and May, J. T. Confirmation of herpes simplex virus type 2 infections in herpes-like genital lesions by a simple complement-fixation test. *Br. J. Venereal Dis.* 58:48–51 (1982).

Dzierzanowska, D.; Kolodziejczyk, J.; and Lourie, I. Determination of cytomegalovirus antibody titers in sera of volunteer blood donors by use of complement fixation and two immunoenzymatic tests. *Arch. Immun. Ther. Exp.* (Warsaw) 34:397–402 (1986).

Ferrante, P.; Achilli, G.; Gerna, G.; and Bergamini, F. Subacute sclerosing panencephalitis: Detection of measles antibody in serum and cerebrospinal fluid by enzyme-linked immunosorbent assay, complement fixation, and hemagglutination inhibition. *Microbiologica* 10:111–18 (1987).

Inouye, S.; Matsuno, S.; and Kono, R. Difference in antibody reactivity between complement fixation and immune adherence hemagglutination tests with virus antigens. *J. Clin. Microbiol.* 14:241–46 (1981).

Julkunen, I.; Kleemola, M.; and Hovi, T. Serological diagnosis of influenza A and B infections by enzyme immunoassay. Comparison with the complement fixation test. *J. Virol. Meth.* 9:7–14 (1984).

Midthun, K.; Pand, L. I.; Flores, J.; and Kapikian, A. Z. Comparison of immunoglobulin A (IgA), IgG, and IgM enzyme-linked immunosorbent assays, plaque reduction neutralization assay, and complement fixation in detecting seroresponses to rotavirus vaccine candidates. *J. Clin. Microbiol.* 27:2799–2804 (1989).

Miller, H.; McCulloch, B.; Landini, M. P.; and Rossier, E. Comparison of immunoblotting with other serological methods and virus isolation for the early detection of primary cytomegalovirus infection in allograft recipients. *J. Clin. Microbiol.* 27:2672–77 (1989).

Ossewaarde, J. M., and deBooij, A. D. Development of a monoclonal antibody for use as an amboceptor in complement fixation tests. *J. Virol. Meth.* 25:13–20 (1989).

Pauri, P.; Bagnarelli, P.; and Clementi, M. Complement-fixation test for rotavirus detection: Comparison and analysis of different methods to reduce anticomplementary activity of some specimens. *J. Virolog. Meth.* 3:329–35 (1982).

Sato, S.; Ochiai, H.; and Niwayama, S. Application of the single radial complement fixation test for serodiagnosis of influenza, respiratory syncytial, mumps, adeno type 3, and herpes simplex type 1 virus infections. *J. Med. Virol.* 24:395–404 (1988).

Schmidt, N. J., and Emmons, R. W. General principles of laboratory diagnostic methods for viral, rickettsial and chlamydial infections. In Schmidt, N. J., and Emmons, R. W., eds., *Diagnostic Procedures for Viral, Rickettsial and Chlamydial Infections,* 6th ed., pp. 21–28. Washington, D.C.: American Public Health Association, 1989.

Sethi, J.; Pei, D.; and Hirshaut, Y. Choice and specificity of complement in complement fixation assay. *J. Clin. Microbiol.* 13:888–90 (1981).

Swack, N. S.; Gahan, T. F.; and Hausler, W. J., Jr. The present status of the complement fixation test in viral diagnosis. *Infect. Agents Dis.* 1:219–24 (1992).

Taguchi, F. New complement fixation test with peroxidase-labeled complement C1q for direct and quantitative determination of antibodies to herpes simplex virus. *Microbiol. Immun.* 32:1167–73 (1988).

Chapter 11: Histocytochemical Staining

Gluck, L., and Kulovich, M. V. Histochemical studies of the distribution of RNA in tissues of the developing chick embryo. *Yale J. Biol. Med.* 36:379–87 (1964).

Malherbe, H. H., and Strickland-Cholmley, M. Viral cytopathology; Boca Raton: CRC Press, 1980.

Mayor, H. D., and Melnick, J. L. Intracellular and extracellular reactions of viruses with vital dyes. *Yale J. Biol. Med.* 34:340–58 (1961).

Springer, S.; Hackman, R. C.; Holle, R.; Greenberg, G. M.; Weems, C. E.; Myerson, D.; Meyers, J. D.; and Thomas, E. D. Use of bronchoalveolar lavage to diagnose acute diffuse pneumonia in the immunocompromised host. *J. Infect. Dis.* 154:604–10 (1986).

Traystman, M. D.; Gupta, P. K.; Shah, K. V.; Reissig, M.; Cowles, L. T.; Hillis, W.; and Frost, J. K. Identification of viruses in the urine of renal transplant recipients by cytomorphology. *Acta Cytol.* 24:501–10 (1980).

Vernon, S. E. Cytologic features of nonfatal herpes virus tracheobronchitis. *Acta Cytol.* 26:237–42 (1982).

Chapter 12: Nucleic Acid Hybridization Techniques

Abraham, R.; Chonmaitree, T.; McCombs, J.; Prabhakar, B.; Lo Verde, P. T.; and Ogra, P. L. Rapid detection of poliovirus by reverse transcription and polymerase chain amplification: Application for differentiation between poliovirus and nonpoliovirus enteroviruses. *J. Clin. Microbiol.* 31:395–99 (1993).

Andiman, W.; Gradoville, L.; Heston, L.; Neydorff, R.; Savage, M. E.; Kitchingman, G.; Shedd, D.; and Miller, G. Use of cloned probes to detect Epstein-Barr viral DNA in tissues of patients with neoplastic and lymphoproliferative diseases. *J. Infect. Dis.* 148:967–77 (1983).

Brigati, D. J.; Myerson, D.; Leary, J. J.; Spalholz, B.; Travis, S. Z.; Fong, C. K. Y.; Hsiung, G. D.; and Ward, D. C. Detection of viral genomes in cultured cells and paraffin-embedded tissue section using biotin-labelled hybridization probes. *Virology* 126:32–50 (1983).

Buchbinder, A.; Josephs, S. F.; Ablashi, D.; Salahuddin, S. Z.; Klotman, M. E.; Manak, M.; Krueger, G. R. F.; Wong-Staal, T.; and Gallo, R. Polymerase chain reaction amplification and in situ hybridization for the detection of human B-lymphotropic virus. *J. Virol. Meth.* 21:191–97 (1988).

Chou, S., and Merigan, T. C. Rapid detection and quantitation of human cytomegalovirus in urine through DNA hybridization. *N. Engl. J. Med.* 308:921–25 (1983).

De Leon, R.; Matsui, S. M.; Baric, R. S.; Herrmann, J. E.; Blacklow, N. R.; Greenberg, H. B.; and Sobsey, M. D. Detection of Norwalk virus in stool specimens by reverse transcriptase-polymerase chain reaction and nonradioactive oligoprobes. *J. Clin. Microbiol.* 30:3151–57 (1992).

Demmler, G. J.; Buffone, G. J.; Schimbor, C. M.; and May, R. A. Detection of cytomegalovirus in urine from newborns by using polymerase chain reaction DNA amplification. *J. Infect. Dis.* 158:1177–84 (1988).

Dickover, R. W.; Donovan, R. M.; Goldstein, E.; Danerkar, S.; Bush, C. E.; and Carlson, J. R. Quantitation of human immunodeficiency virus DNA by using the polymerase chain reaction. *J. Clin. Microbiol.* 28:2130–33 (1990).

Englund, J. A.; Zimmerman, M. E.; Swierkosz, E. M.; Goodman, J. L.; Scholl, D. R.; and Balfour, H. H. Herpes simplex virus resistant to acyclovir. *Ann. Intern. Med.* 112:416–22 (1990).

Espy, M. J., and Smith, T. F. Detection of herpes simplex virus in conventional tube cell cultures and in shell vials with a DNA probe kit and monoclonal antibodies. *J. Clin. Microbiol.* 26:22–24 (1988).

Forghani, B.; Dupuis, K. W.; and Schmidt, N. J. Rapid detection of herpes simplex virus DNA in human brain tissue by in situ hybridization. *J. Clin. Microbiol.* 22:656–58 (1985).

Fung, J. C.; Shanley, J.; and Tilton, R. C. Comparison of the detection of herpes simplex virus in direct clinical specimens with herpes simplex virus-specific DNA probes and monoclonal antibodies. *J. Clin. Microbiol.* 22:748–53 (1985).

Gadler, H.; Larsson, A.; and Solver, E. Nucleic acid hybridization: A method to determine effects of antiviral compounds on herpes simplex virus type I DNA synthesis. *Antiviral Res.* 4:63–70 (1984).

Gleaves, C. A.; Hursh, D. A.; Rice, D. H.; and Meyers, J. D. Detection of cytomegalovirus from clinical specimens in centrifugation culture by in situ DNA hybridization and monoclonal antibody staining. *J. Clin. Microbiol.* 27:21–23 (1989).

Gleaves, C. A.; Myerson, D.; Bowden, R. A.; Hackman, R. C.; and Meyers, J. D. Direct detection of cytomegalovirus from broncho-alveolar lavage samples by using rapid in situ DNA hybridization assay. *J. Clin. Microbiol.* 27:2429–32 (1989).

Heneine, W.; Khabbaz, R. E.; Lal, R. B.; and Kaplan, J. E. Sensitive and specific polymerase chain reaction assays for diagnosis of human T-cell lymphotropic virus type I (HTLV-I) and HTLV-II infections in HTLV-I/II seropositive individuals. *J. Clin. Microbiol.* 30:1605–7 (1992).

Ho, M.; Miller, G.; Atchison, R. W.; Breinig, M. K.; Dummer, J. S.; Andiman, W.; Starzl, T. E.; Eastman, R.; Griffith, B. P.; Hardesty, R. L.; Bahnson, H. T.; Hakala, T. R.; and Rosenthal, J. T. Epstein-Barr virus infections and DNA hybridization studies in posttransplantational lymphoma and lymphoproliferative lesions: The role of primary infection. *J. Infect. Dis.* 5:876–86 (1985).

Holodniy, M.; Kim, S.; Katzenstein, D.; Konrad, M.; Groves, E.; and Merigan, T. C. Inhibition of human immunodeficiency virus gene amplification by heparin. *J. Clin. Microbiol.* 29:676–79 (1991).

Hsia, K.; Spector, D. H.; Lawrie, J.; and Spector, S. A. Enzymatic amplification of human cytomegalovirus sequences by polymerase chain reaction. *J. Clin. Microbiol.* 27:1802–9 (1989).

Jacobson, M. A.; Berger, T. G.; Fikrig, S.; Becherer, P.; Moohr, J. W.; Stanat, S. C.; and Biron, K. K. Acyclovir-resistant varicella zoster virus infection after chronic acyclovir therapy in patients with the acquired immunodeficiency syndrome (AIDS). *Ann. Intern. Med.* 112:187–91 (1990).

Kejian, G., and Bowden, D. S. Digoxigenin-labeled probes for the detection of hepatitis B virus DNA in serum. *J. Clin. Microbiol.* 29:506–9 (1991).

Landry, M. L. Nucleic acid hybridization in viral diagnosis. *Clin. Biochem.* 23:267–77 (1990).

Landry, M. L.; Zibello, T. A.; and Hsiung, G. D. Comparison of in situ hybridization and immunologic staining with cytopathology for detection and identification of herpes simplex virus infection in cultured cells. *J. Clin. Microbiol.* 24:968–71 (1986).

Morrison, E. A. B.; Goldberg, G. L.; Kadish, A. S.; and Burk, R. D. Polymerase chain reaction detection of human papillomavirus: Quantitation may improve clinical utility. *J. Clin. Microbiol.* 30:2539–43 (1992).

Musiani, M.; Zerbini, M.; Gibellini, D.; Gentilomi, G.; Venturoli, S.; Gallinella, G.; Ferri, E.; and Girotti, S. Chemiluminescence dot blot hybridization assay for detection of B19 parvovirus DNA in human sera. *J. Clin. Microbiol.* 29:2047–50 (1991).

Myerson, D.; Hackman, R. C.; and Meyers, J. D. Diagnosis of cytomegaloviral pneumonia by in situ hybridization. *J. Infect. Dis.* 150:272–77 (1984).

Persing, D. H., and Landry, M. L. In vitro amplification techniques for the detection of nucleic acids: New tools for the diagnostic laboratory. *Yale J. Biol. Med.* 62:159–71 (1989).

Peterson, E. M.; Aarnaes, S. L.; Bryan, R. N.; Ruth, J. L.; and de la Maza, L. M. Typing of herpes simplex virus with synthetic DNA probes. *J. Infect. Dis.* 153:757–62 (1986).

Redfield, D. C.; Richman, D. D.; Albanil, S.; Oxman, M. N.; and Wahl, G. M. Detection of herpes simplex virus in clinical specimens by DNA hybridization. *Diagn. Microbiol. Infect. Dis.* 1:117–28 (1983).

Rotbart, H. A. Enzymatic RNA amplification of the enteroviruses. *J. Clin. Microbiol.* 28:438–42 (1990).

Scott-Taylor, T. H., and Hammond, G. W. Conserved sequences of the adenovirus genome for detection of all human adenovirus types by hybridization. *J. Clin. Microbiol.* 30:1703–10 (1992).

Seidlin, M.; Takiff, H. E.; Smith, H. A.; Hay, J.; and Straus, S. E. Detection of varicella-zoster virus by dot-blot hybridization using a molecularly cloned viral probe. *J. Med. Virol.* 13:53–61 (1984).

Sorbello, A. F.; Elmendorf, S. L.; McSharry, J. J.; Venezia, R. A.; and Echols, R. M. Rapid detection of cytomegalovirus by fluorescent monoclonal antibody staining and in situ DNA hybridization in a dram vial cell culture system. *J. Clin. Microbiol.* 26:1111–14 (1988).

Sullender, W. M., and Wertz, G. W. Synthetic oligonucleotide probes differentiate respiratory syncytial virus subgroups in a nucleic acid hybridization assay. *J. Clin. Microbiol.* 29:1255–57 (1991).

Swierkosz, E. M.; Scholl, D. R.; Brown, J. L.; Jollick, J. D.; and Gleaves, C. A. Improved DNA hybridization method for detection of acyclovir-resistant herpes simplex virus. *Antimicrob. Agents Chemother.* 31:1465–69 (1987).

Zipeto, D.; Revello, M. G.; Silini, E.; Parea, M.; Percivalle, E.; Zavattoni, M.; Milanesi, G.; and Gerna, G. Development and clinical significance of a diagnostic assay based on the polymerase chain reaction for detection of human cytomegalovirus DNA in blood samples from immunocompromised patients. *J. Clin. Microbiol.* 30:527–30 (1992).

Chapter 13: Electron Microscopy and Immunoelectron Microscopy

Almeida, J. D. Practical aspects of diagnostic electron microscopy. *Yale J. Biol. Med.* 53:5–18 (1980).

Anderson, N., and Doane, F. W. Agar diffusion method for negative staining of microbial suspensions in salt solutions. *Appl. Microbiol.* 24:295 (1972).

Chernesky, M. A. The role of electron microscopy in diagnostic virology. In Lennette, D.; Spector, S.; and Thompson, K., eds. *Diagnosis of Viral Infections: The Role of Clinical Laboratory*, pp. 125–42; Baltimore: University Park Press, 1979.

Doane, R. W., and Anderson, N. *Electron Microscopy in Diagnostic Virology: A Practical Guide and Atlas.* New York: Cambridge University Press, 1987.

Flewett, T. H.; Bryden, A. S.; and Davies, H. Virus particles in gastroenteritis. *Lancet* 2:1497 (1973).

Fong, C. K. Y. Electron microscopy in the rapid, accurate diagnosis of virus infections. *V.A. Practitioner* 6:65–75 (1989).

Fong, C. K. Y., and Landry, M. L. An adventitious viral contaminant in commercially supplied A549 cells: Identification of infectious bovine rhinotracheitis virus and its impact on diagnosis of infection in clinical specimens. *J. Clin. Microbiol.* 30:1611–13 (1992).

Gary, G. W., Jr.; Hierholzer, J. C.; and Black, R. E. Characteristics of noncultivable adenoviruses associated with diarrhea in infants: A new subgroup of human adenoviruses. *J. Clin. Microbiol.* 10:96–103 (1979).

Hammond, G. W.; Hazelton, P. R.; Chuang, I.; and Klisko, B. Improved detection of viruses by electron microscopy after direct ultracentrifugation preparation of specimens. *J. Clin. Microbiol.* 14:210–21 (1981).

Hayat, M. A. *Principles and Techniques of Electron Microscopy: Biological Application*, vol. 1. Baltimore: University Park Press, 1981.

Horne, R. W., and Wildy, P. Virus structure revealed by negative staining. *Adv. Virus Res.* 10:101 (1963).

Hsiung, G. D.; Fong, C. K. Y.; and August, M. J. The use of electron microscopy in diagnosis of virus infection. *Prog. Med. Virol.* 25:133–59 (1979).

Kapikian, A. Z.; Wyatt, R. G.; Dolin, R.; Thornhill, T. S.; Kalica, A. R.; and Chanock, R. M. Visualization by immune electron microscopy of 27-nm particles associated with acute infectious nonbacterial gastroenteritis. *J. Virol.* 10:1075–81 (1972).

Nagler, F. P. O., and Rake, G. The use of electron microscope in diagnosis of variola, vaccinia, and varicella. *J. Bact.* 55:45–51 (1948).

Palmer, E. L., and Martin, M. L. Electron Microscopy in Viral Diagnosis. Boca Raton: CRC Press, 1988.

Spur, A. R. A low-viscosity epoxy resin embedding medium for electron microscopy. *J. Ultrastructure Research* 26:31–43 (1969).

PART THREE

RNA VIRUSES: RECOGNITION AND CHARACTERIZATION OF VIRUSES INFECTING HUMANS

VIRUS MORPHOLOGY AND VIRUS-INDUCED CELLULAR CHANGES

14

PICORNAVIRIDAE

G. D. HSIUNG

Picornaviridae includes the enteroviruses and the rhinoviruses (table 14.1). The name was adopted by the International Committee on Taxonomy of Viruses as a derivation of *pico*, meaning very small, and *RNA*, the nucleic acid type present in all members of the group.

ENTEROVIRUSES

The enteroviruses include 3 types of poliovirus, 23 types of coxsackievirus A, 6 types of coxsackievirus B, and some 31 types of the echovirus or enteric cytopathogenic human orphan virus. Starting with enterovirus type 68, all newly isolated enterovirus types are designated numerically. Enterovirus type 72 is hepatitis A virus (see chapter 26). The human alimentary tract is the natural habitat for these viruses. They can be recovered from patients with a variety of illnesses, including poliomyelitis, aseptic meningitis, myocarditis, and pericarditis. In addition to their pathogenicity in humans, the polioviruses produce paralysis in monkeys.

Morphologically, all enteroviruses, whether in infected cultured cells or tissues, are similar (figures 14.1, 14.2). They are small, 25–30 nm, have icosahedral symmetry with 32 capsomeres, and contain a single-stranded RNA genome. Viruses in this group are relatively stable at pH 3.0 and are ether-resistant. Storage at −20°C is satisfactory for several years.

ISBN 0-300-05845-4.

Table 14.1 Levels of Host Cell Sensitivity to and Pathogenicity of Human Enteroviruses and Rhinoviruses[a]

Viruses	Serotypes	Primary Cell Cultures			Passaged Cell Lines				Newborn Mice
		HEK	RhMK or GMK	GPE	HEF	Hep-2	BGMK	RD	
Enterovirus									
Poliovirus	1–3	++	+++	–	++	++	++	++	–
Coxsackievirus B	1–6	++	+++	–	++	++	++	–	++
Coxsackievirus A[b]	1–22,24	+/–	+/–	+/–	–	–	–	++/–	++
Echovirus[c]	1–7,9,11–27,29–34	+	++	–	+	–	–	+++	–
Enterovirus[d]	68–71	+	++	e	+	e	e	e	e
Rhinovirus	1–100	+/–	+/–	–	++	+/–	–	e	–

+++ highly sensitive; ++ moderately sensitive; + less sensitive; – not sensitive

a. Can be differentiated by acid stability (pH 3.0) and also genomic amplification (Olive et al., *J. Gen. Vir.* 71:2141 [1990]).

b. Coxsackievirus A23 deleted since it cross-reacts with echovirus 9.

c. Echovirus 8 deleted since it cross-reacts with echovirus 1; echovirus 10 reclassified as reovirus type 1. Echovirus 28 reclassified as rhinovirus type 1A.

d. Enteroviruses starting with type 68 designated numerically; enterovirus type 72 is classified as hepatitis A.

e. Test not done

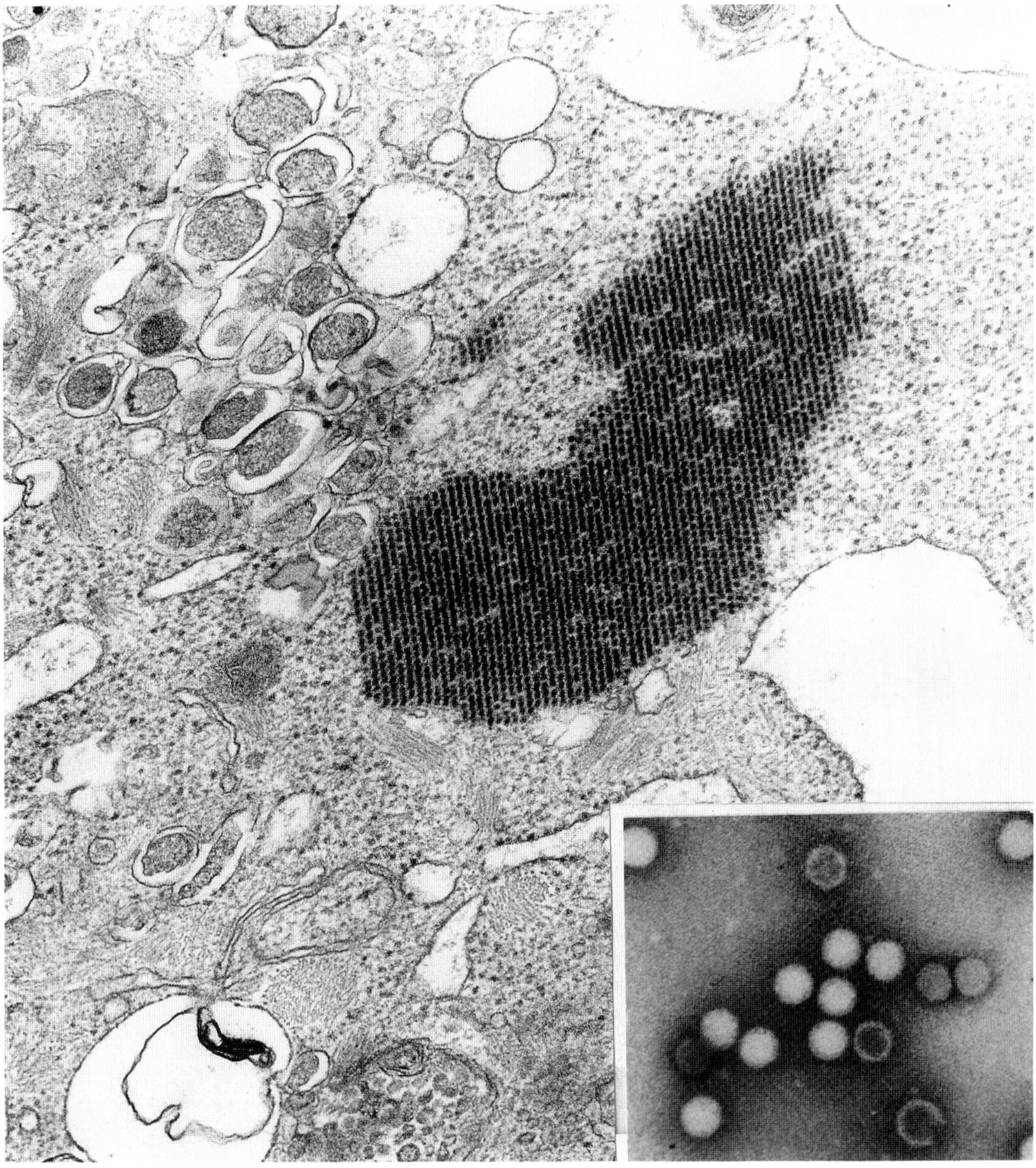

Figure 14.1 Electron micrograph of poliovirus 1 in an infected Hep-2 cell. A large aggregate of virions in crystalline arrays is present in the cytoplasm (48,000×). Inset: Poliovirus particles with PTA (168,000×).

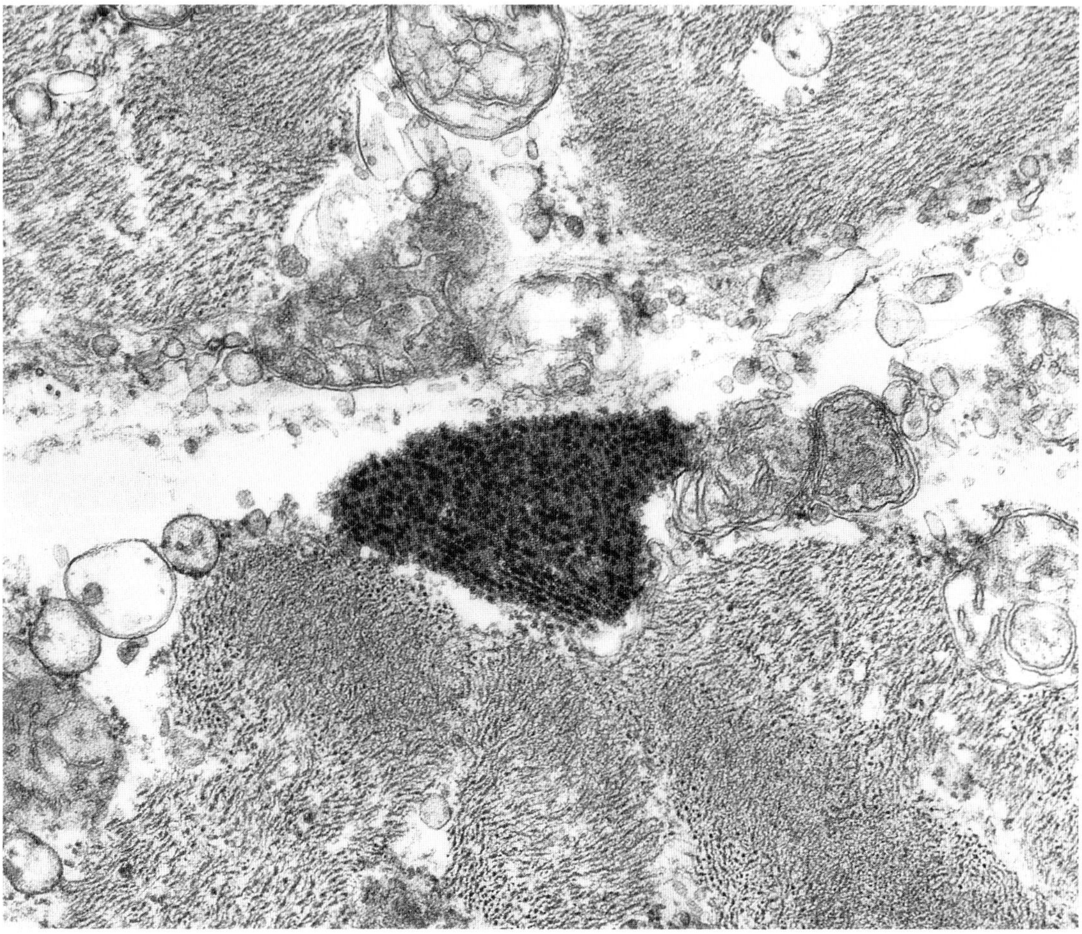

Figure 14.2 Electron micrograph of an aggregate of coxsackievirus A-10 virions in an infected mouse muscle cell (38,400×)

Host Range

The discovery by J. F. Enders and co-workers in 1949 that poliovirus can be propagated in a cell culture of nonneural tissue origin greatly facilitated the cultivation of viruses and the ultimate control of poliomyelitis. Most enteroviruses propagate readily only in cell cultures of primate origin, although certain types of coxsackie A virus can replicate in GPE cell cultures. Furthermore, the susceptibility of various species of monkey kidney cell cultures to these viruses has been found to differ. Of the primate cell cultures, rhesus, cynomolgus, and African green monkey kidney cells are highly sensitive to polio, coxsackie B, and most echoviruses. Cell cultures of other African monkeys, for example, the patas monkey, are susceptible only to certain enterovirus types. New World monkey kidney cells are generally resistant. Human kidney culture is a most suitable system for the isolation of enteroviruses, but the supply is limited. Nonprimate cell cultures are generally resistant to infection by human enteroviruses.

In recent years, the supply of monkeys has become scarce. Several passaged cell lines, including human embryo lung fibroblast cells, buffalo green monkey kidney (BGMK), and several cell lines derived from human carcinoma cells (A549, Hep-2, and RD), are also used for enterovirus isolation. The differential susceptibility of cell cultures has been used for presumptive grouping of polio-, coxsackie A and B, and echoviruses (table 14.1). Note that some difficulties may be encountered when cell lines are used since susceptibility patterns may be altered with the passage of cell sublines in different laboratories. Newborn mice provide the most uniform and sensitive system for the isolation of coxsackieviruses, especially group A, of which only selected types can be propagated in cell cultures.

Specimen Source and Inoculation

Stool or rectal swab, throat swab, and cerebrospinal fluid are common specimens for the isolation of enteroviruses. Suitable cell culture types for inoculation of specimens include primary monkey kidney cells and human cell lines as listed in the section above. The usage of one type of cell culture over another is dictated by availability and cell susceptibility. If, however, group A coxsackievirus is suspected, the specimen may be inoculated into either newborn mice or GPE or RD cells.

Virus Isolation and Propagation

CYTOPATHIC EFFECT IN MONOLAYER CELL CULTURES

Observation for cytopathic effect induced by enteroviruses in cell cultures is the most sensitive and rapid method for the recognition of this group of viruses (table 14.1). Examples of typical CPE induced by poliovirus 1 in RhMK and Hep-2 cells are shown in figure 14.3. Extensive cell destruction usually occurs 24–48 hours after inoculation in either cell system. Sensititity of echoviruses, however, is selectively dependent upon the type of cell culture used. Differences in the selectivity of some cell culture systems to enterovirus infection are illustrated in figures 14.4, 14.5, and 14.6. Group B coxsackieviruses induce extensive CPE in BGMK cells but no CPE in RD cells, whereas echoviruses induce extensive CPE in RD cells but no CPE in BGMK cells (figure 14.4). Both group B coxsackievirus and echovirus induce distinct CPE in RhMK cell cultures, but only the former induced CPE in A549 cells and only uncertain CPE is noted in A549 cells infected with echoviruses (figure 14.5). Certain strains of echovirus 11, however, induce extensive CPE in a human kidney cell culture within a few hours of inoculation, but only small areas of degeneration are noted in RhMK cultures two to three days after

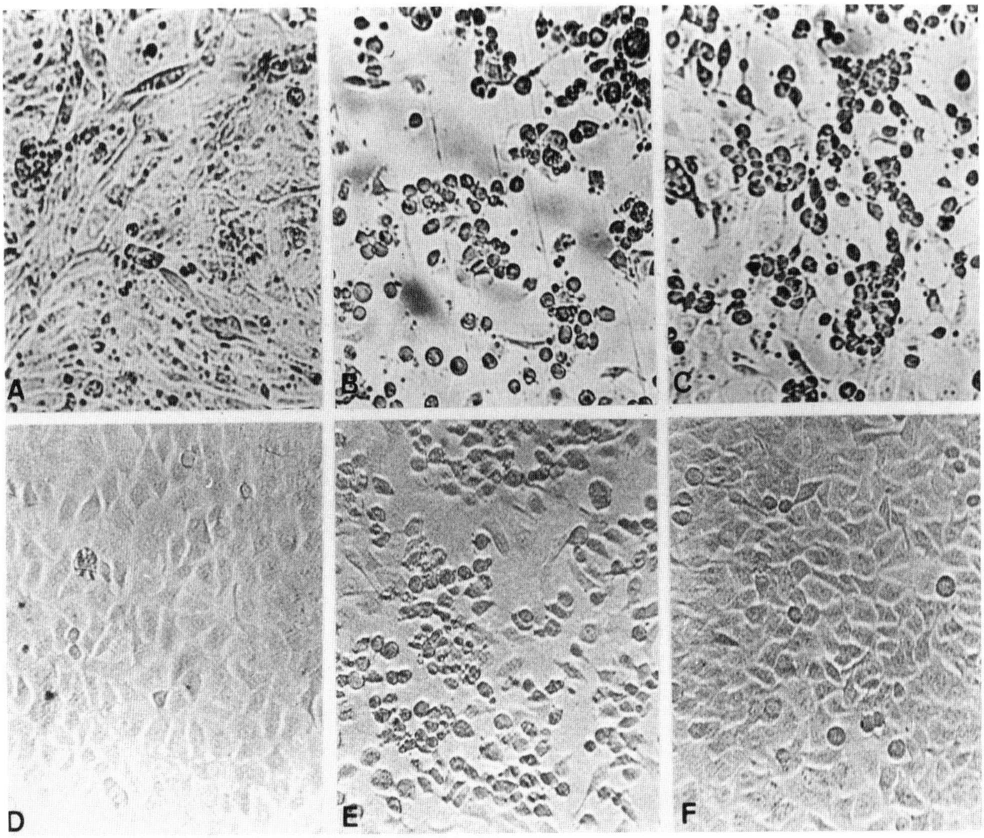

Figure 14.3 Cytopathic effect induced by poliovirus 1 and echovirus 11 in RhMK and Hep-2 cultures (100×): *(A)* Uninoculated RhMK culture; *(B)* CPE induced by poliovirus 1 in RhMK culture, one day after inoculation; *(C)* CPE induced by echovirus 11 in RhMK culture, four days after inoculation; *(D)* Uninoculated Hep-2 culture; *(E)* CPE induced by poliovirus 1 in Hep-2 culture, one day after inoculation; *(F)* Absence of CPE in Hep-2 culture infected with echovirus 11, four days after inoculation

inoculation (figure 14.6), and no CPE is noted in HEP-2 cells. Certain group A coxsackieviruses induce distinctive CPE in GPE cells (figure 14.7), a useful additional diagnostic tool when GPE cells are available. Thus, the use of multiple appropriate cell cultures increases the yield of enterovirus isolations and enhances the rapidity of presumptive diagnosis.

MOUSE PATHOGENICITY

Newborn mice infected with both group A and B coxsackieviruses show illness four to seven days after inoculation. Hindlimb paralysis is usually observed in mice inoculated with coxsackieviruses; death occurs in five to seven days. The coxsackieviruses can be subdivided into groups A and B,

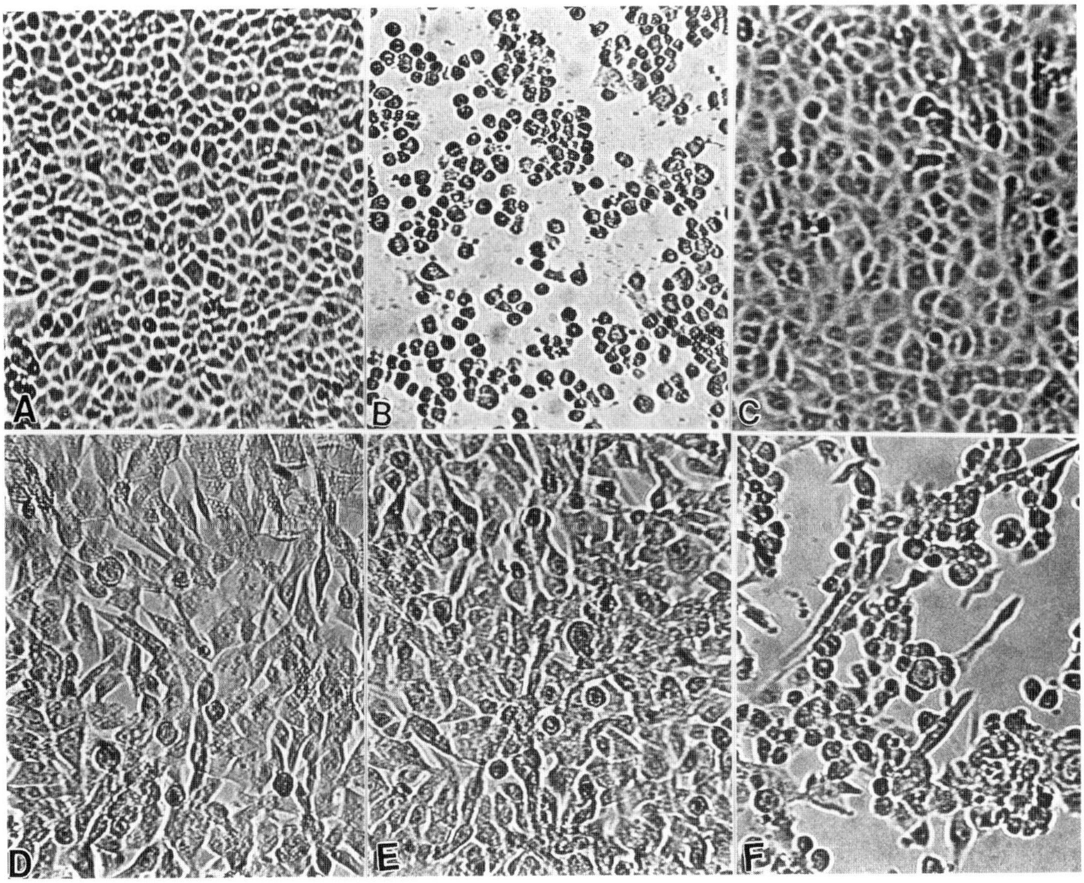

Figure 14.4 Cytopathic effect induced by coxsackievirus B2 and echovirus 30 in BGMK and RD cell cultures (100×): *(A)* Uninfected BGMK cell culture; *(B)* CPE induced by coxsackievirus B2 in BGMK cells one day after inoculation; *(C)* Absence of CPE in BGMK cell culture infected with echovirus 30; *(D)* Uninfected RD cell culture; *(E)* Absence of CPE in RD cell culture infected with coxsackievirus B2; *(F)* CPE induced by echovirus 30 in RD cells one day after inoculation

as determined by histological lesions produced in mice; group A coxsackieviruses cause generalized myositis, whereas group B coxsackieviruses cause focal myositis, especially in heart muscle.

Identification of Isolates

CELL CULTURE SELECTIVITY

Owing to the very large number of serotypes in the enterovirus group and a limited source of antisera to the enteroviruses, a presumptive identification tool is urgently needed for rapid typing within the enterovirus group.

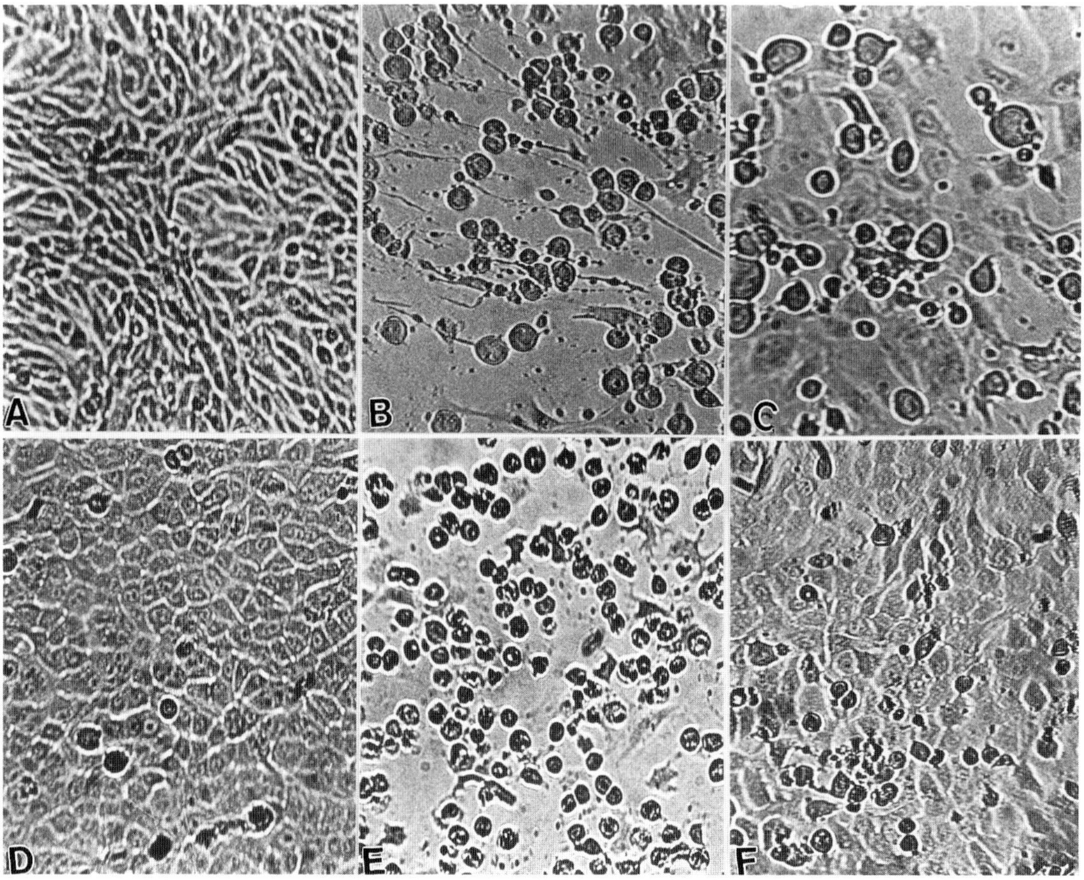

Figure 14.5 Cytopathic effect induced by coxsackievirus B2 and echovirus 30 in RhMK and A549 cell cultures (100×): *(A)* Uninfected RhMK cell culture; *(B)* CPE induced by coxsackievirus B2 in RhMK culture; *(C)* CPE induced by echovirus 30 in RhMK culture; *(D)* Uninfected A549 cells; *(E)* CPE induced by coxsackievirus B2 in A549 cell culture; *(F)* Uncertain CPE in A549 cell culture infected with echovirus 30

Advantage has been taken of the differences in susceptibility of host cell systems, the rate of appearance of enterovirus CPE, and mouse pathogenicity for presumptive identification of these viruses. For example, group B coxsackievirus induces CPE rapidly in BGMK cells but not in RD cells, and the reverse is true for echoviruses, that is, the latter induce rapid CPE in RD cells but not in BGMK cells (figure 14.4). A presumptive identification can be made as soon as typical CPE appears in certain cell culture systems (table 14.1). Thus, many serum pools can be eliminated, and serologic testing can be simplified. Final identification, however, of a virus type within the group can be made only by a neutralization test using type-specific antiserum or by an immunofluorescent test using available monoclonal antibodies. These

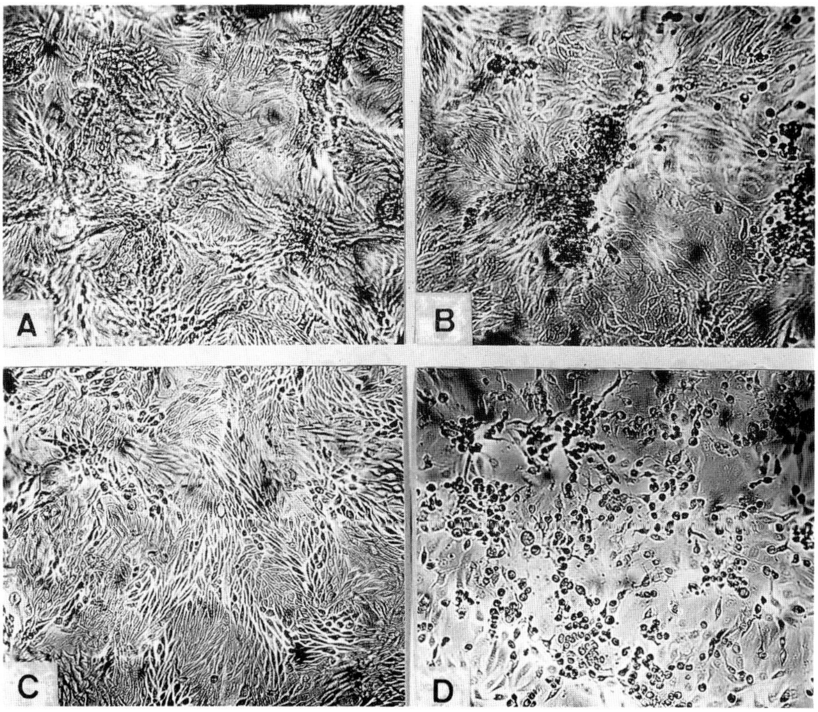

Figure 14.6 Cytopathic effect induced by echovirus 11, WB strain, in RhMK and HEK cultures (100×) (G. D. Hsiung, *Yale J. Biol. Med.* 33:359 [1961]). *(A)* Uninoculated RhMK culture; *(B)* CPE induced by echovirus 11 in RhMK culture, two days after infection, showing small foci of degeneration; *(C)* Uninoculated HEK culture; *(D)* CPE induced by echovirus 11, eight hours after infection, showing extensive cellular degeneration

include coxsackie A9 and A24, coxsackie B1–6, several echoviruses, and a few enterovirus types.

NEUTRALIZATION TEST

Identifying enteroviruses by serological means is a laborious task because of the many virus types in this group. A neutralization test in tissue culture (see chapter 6) using serum pools is one means of facilitating the typing of enteroviruses. The intersecting serum scheme originally described by K. A. Lim and M. Benyesh-Melnick involves the use of serum pools containing permuted combinations of antisera, each serum being represented in 1-, 2-, or 3-class pools. The identity of the virus type is indicated by the serum type neutralizing the agent present in either the pool of 1-class or 2- or 3-class pools sharing a common component (tables 14.2, 14.3, and figure 14.8). The use of eight serum pools has proved to be a relatively rapid and satisfactory technique for identifying the 42 types of enteroviruses that exist, especially

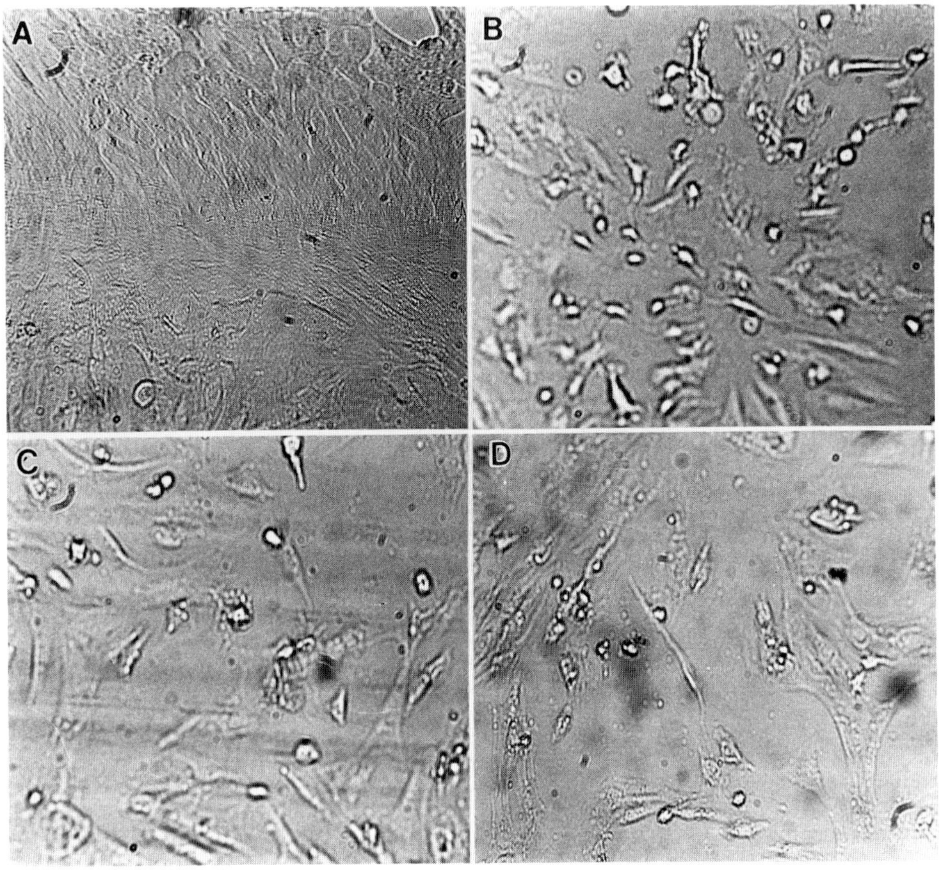

Figure 14.7 Cytopathic effect induced by group A coxsackieviruses in GPE cell cultures (100×).(Modified from M. L. Landry et al., *J. Clin. Microbiol.* 13:588 [1981]) *(A)* Uninoculated GPE culture; *(B)* Coxsackievirus A-10, two days after inoculation; *(C)* Coxsackievirus A-2, three days after inoculation; *(D)* Coxsackievirus A-5, three days after inoculation

when the new Lim and Benyesh-Melnick (LBM) antiserum pools are made available.*

ELECTRON MICROSCOPY

Electron microscopy is not generally used to diagnose enterovirus infections. Rapid identification of an isolate as an enterovirus, however, can be made by the negative staining of a suspension of the virus isolate grown in a cell culture. Enterovirus virions can also be identified by size and shape in thin sections of virus-infected cells. This is illustrated by the presence of virus

*Information on how to obtain these reagents is available from the Virus Diseases Unit, World Health Organization, 1211 Geneva 27, Switzerland (see Melnick and Wimberly 1985).

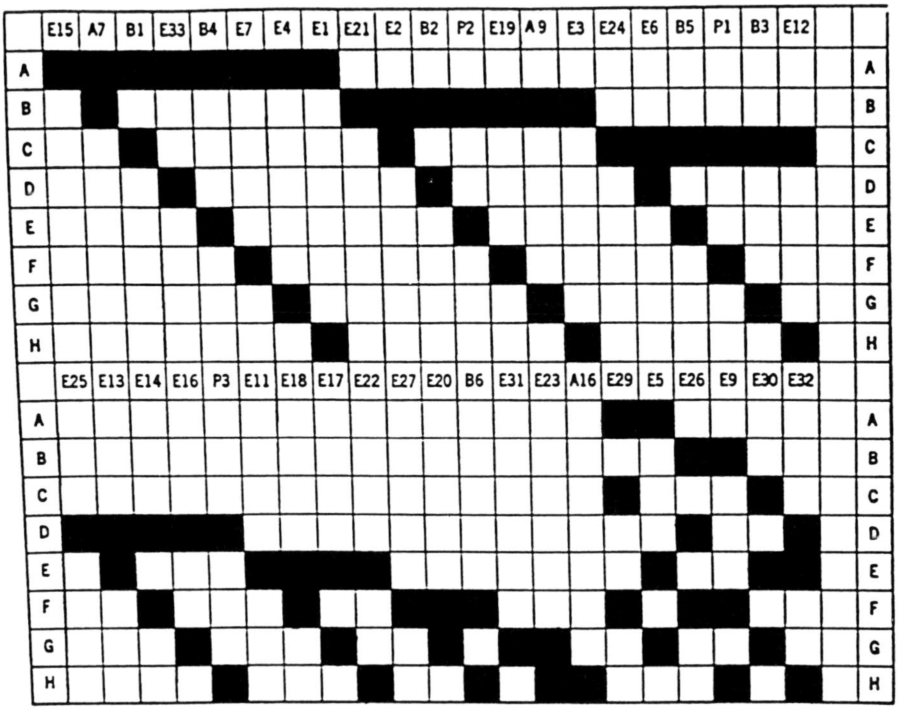

Figure 14.8 Diagram of the eight enterovirus antiserum pools *(A–H)*. Multiple antisera are incorporated into different pools. A virus isolate can be typed against the eight antiserum pools and identified among the 42 enteroviruses. (Reproduced with permission from *Diagnostic Procedures for Viral, Rickettsial and Chlamydial Infections,* 5th ed., E. H. Lennette and N. J. Schmidt, eds., Washington, D.C.: American Public Health Association, 1979, 512)

aggregates forming characteristic crystalline array in the cytoplasm of infected cells (figures 14.1, 14.2). For the group A coxsackieviruses that do not grow readily in cell culture, visualization by EM of typical virions in tissue specimens, such as the skeletal muscle of infected mice (figure 14.2), is a means to detect and identify this group of viruses. In addition, immunoelectron microscopy is used to recognize and identify certain enteroviruses that have not been grown in cultured cells, such as enterovirus type 72 (hepatitis A virus), as discussed in chapter 26.

MOLECULAR BIOLOGY TECHNIQUE

A polymerase chain reaction assay has been used to detect and differentiate enteroviruses and rhinoviruses in clinical specimens. Using primers homologous to the 5′ noncoding and viral protein 2 regions of the genomes of enterovirus and rhinovirus, the two can be identified (see Olive et al., *J. Gen. Virol.* 71:2141 [1990]). The PCR assay can detect as little as a single PFU using ethidium bromide-stained gels.

Table 14.2 Enterovirus Antiserum Constituents Included in Each Pool

Pool	Constituent Sera
A	CA7, CB1, CB4, E1, E4, E5, E7, E15, E29, E33
B	P2, CA7, CA9, CB2, E2, E3, E9, E19, E21, E26
C	P1, CB1, CB3, CB5, E2, E6, E12, E24, E29, E30
D	P3, CB2, E6, E13, E14, E16, E25, E26, E32, E33
E	P2, CB4, CB5, E5, E11, E13, E17, E18, E22, E30, E32
F	P1, CB6, E7, E14, E18, E19, E20, E26, E27, E29
G	CA9, CB3, E4, E5, E16, E17, E20, E23, E30, E31
H	P3, CA16, CB6, E1, E3, E9, E12, E22, E23, E32

Reproduced with permission from *Diagnostic Procedures for Viral, Rickettsial and Chlamydial Infections,* 5th ed., E. H. Lennette and N. J. Schmidt, eds., Washington, D.C.: American Public Health Association, 1979, 514.

CA coxsackie A virus E echovirus
CB coxsackie B virus P poliovirus

Table 14.3 Codes for Reading Results on Neutralization Test Using Enterovirus Antiserum Pools

If neutralized by pool(s):	Virus is:	If neutralized by pool(s):	Virus is:
A	E15	CD	E6
B	E21	CE	CB5
C	E24	CF	P1
D	E25	CG	CB3
E	E11	CH	E12
F	E27	DE	E13
G	E31	DF	E14
H	CA16	DG	E16
AB	CA7	DH	P3
AC	CB1	EF	E18
AD	E33	EG	E17
AE	CB4	EH	E22
AF	E7		
AG	E4	FG	E20
AH	E1	FH	CB6
BC	E2	GH	E23
BD	CB2		
BE	P2	ACF	E29
BF	E19	AEG	E5
BG	CA9	BDF	E26
BH	E3	BFH	E9
		CEG	E30
		DEH	E32

Reproduced with permission from *Diagnostic Procedures for Viral, Rickettsial and Chlamydial infections,* 5th ed., E. H. Lennette and N. J. Schmidt, eds., Washington, D.C.: American Public Health Association, 1979, 515.

CA coxsackie A virus E echovirus
CB coxsackie B virus P poliovirus

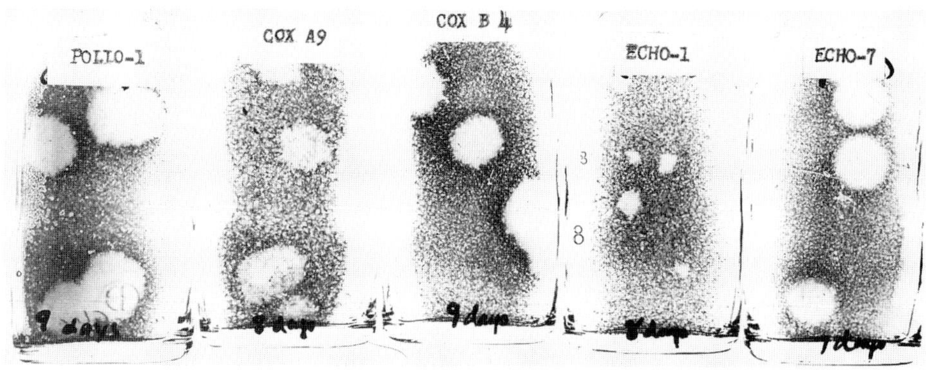

Figure 14.9 Plaque morphology of different enteroviruses in RhMK cultures. Left to right: poliovirus 1, coxsackievirus A-9, coxsackievirus B-4, echovirus 1, and echovirus 7. (G. D. Hsiung, *Yale J. Biol. Med.* 33:359 [1961])

PLAQUE FORMATION UNDER AGAR OVERLAY MEDIUM

Virus plaques are observable not only under a semisolid overlay (see chapter 6) but also under a solid overlay medium. Although this method is not commonly used in a clinical laboratory, the distinct plaque morphology exhibited by members of the enterovirus group under an agar overlay medium can be used for differentiating each virus type within the group (figure 14.9). Selective differences in cell susceptibility are also illustrated under agar overlay. For example, poliovirus 1 induces large clear plaques in both rhesus and patas monkey kidney cells, whereas echovirus 1 produces plaques in the RhMK but not in the patas cell culture (figure 14.10). Thus, the plaque method in adjunct with selective cell susceptibility is a useful tool for the presumptive rapid diagnosis of poliovirus or echovirus infection, especially in the recognition and separation of *mixed* virus infections. In the following, detailed procedures are set forth.

1. Prepare agar overlay medium (for formula, see below).
2. Drain culture fluids from 25-cm^2 culture flasks with confluent monolayer cells.
3. Make serial 10-fold dilutions of virus suspension from 10^{-1} to 10^{-6} and inoculate each dilution into a flask culture, 0.1–0.5 ml per flask, and tilt the flask to distribute the inoculum evenly over the entire sheet.
4. Incubate infected cultures at 37°C for 1 hour to allow virus adsorption. (Note: The flask culture can be incubated in a regular incubator without CO_2.)
5. Add prewarmed agar overlay medium containing neutral red, approximately 5 ml per 25-cm^2 flask, and lay flat to cover the entire cell sheet.
6. Avoid exposure of overlaid cultures to light since photosensitization

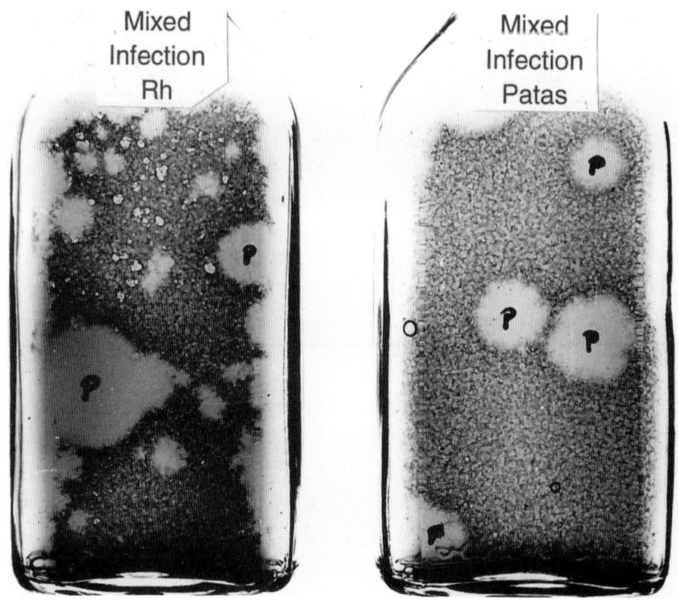

Figure 14.10 Mixed infection with two enteroviruses: Monkey kidney cultures inoculated with a fecal specimen containing two different enteroviruses. *Left,* Rhesus monkey kidney culture showing both poliovirus 1 (large circular plaques) and echovirus 1 (small irregular plaques); *right,* Patas monkey kidney culture showing only poliovirus plaques (G. D. Hsiung and D. M. Horstmann, *Conn. Med.* 25:403 [1961])

may injure cultured cells as a result of the presence of neutral red in the overlay medium.

7. Invert culture bottles after agar has solidified and incubate at 37°C in the dark.
8. Check overlaid cultures daily for 2–3 weeks for the appearance of plaques induced by the virus.
9. Pick plaques, if present, by scraping the areas surrounding the degenerated cells under the agar and within the plaque area with the bent tip of a capillary pipette. Transfer the material from the pipette into a fresh culture tube or a flask culture to obtain purified virus stock.

Serodiagnosis

Several methods have been used in the serologic diagnosis of enterovirus infection. Diagnosis of enterovirus infections without the isolation of a specific virus type, however, is not routinely attempted except in selected cases. In this instance, the neutralization test in cell cultures is usually applied with acute and convalescent sera for detecting the antibody rise to the virus type involved.

RHINOVIRUSES

To date, the rhinovirus group consists of one hundred established serotypes. They are known principally as the causative agents of the common cold. Human beings are their natural hosts, and no laboratory animal has yet been found susceptible to infection. Rhinoviruses can be isolated from the nose and throat but not from the feces of patients with acute febrile upper-respiratory disease.

The physical and chemical as well as morphological properties of the rhinoviruses are similar to those of the enteroviruses, except that the rhinoviruses are unstable at acid pH (pH 3.0). The growth of rhinoviruses is generally facilitated by incubation at 33°C in rotating culture tubes that contain media at neutral pH, conditions similar to those of the mucous membrane of the nasal cavity.

Tissue cultures represent the only known system for the propagation of the rhinoviruses. Early studies showed that the rhinoviruses could be subdivided into the H strain and the M strain. The H viruses propagate best in human cells, both primary cultures and human embryo fibroblasts (HEF) such as MRC-5 and others. The M viruses produce CPE in primary rhesus monkey as well as in human cells, including some of the continuous cell lines such as those derived from the carcinoma of human nasopharynx (KB) and of human cervix (HeLa). Since the ability of rhinoviruses to replicate in monkey cells is variable, however, this method is not commonly used for differentiation.

Specimen Source and Inoculation

Nasal washings or nasopharyngeal swabs are the best sources for the isolation of rhinoviruses. These specimens should be inoculated directly into cultures immediately after collection, without freezing, if possible. Human embryo fibroblast cells, MRC-5 or WI-38, are satisfactory, but variations in susceptibility occur. All cultures should be maintained at a neutral pH and incubated at 33°C in a roller drum, if available.

Virus Isolation, Propagation, and Assay

CYTOPATHIC EFFECT IN MONOLAYER CELL CULTURES

The CPE induced by rhinovirus in human embryo fibroblast and in HeLa cell cultures is shown in figures 14.11 and 14.12. These cellular changes resemble those of an enterovirus but develop rather slowly upon primary isolation. Under favorable conditions—that is, in culture tubes maintained at pH 6.8–7.0 and incubated at 33°C in a roller drum—rhinovirus-induced CPE generally appears within two weeks.

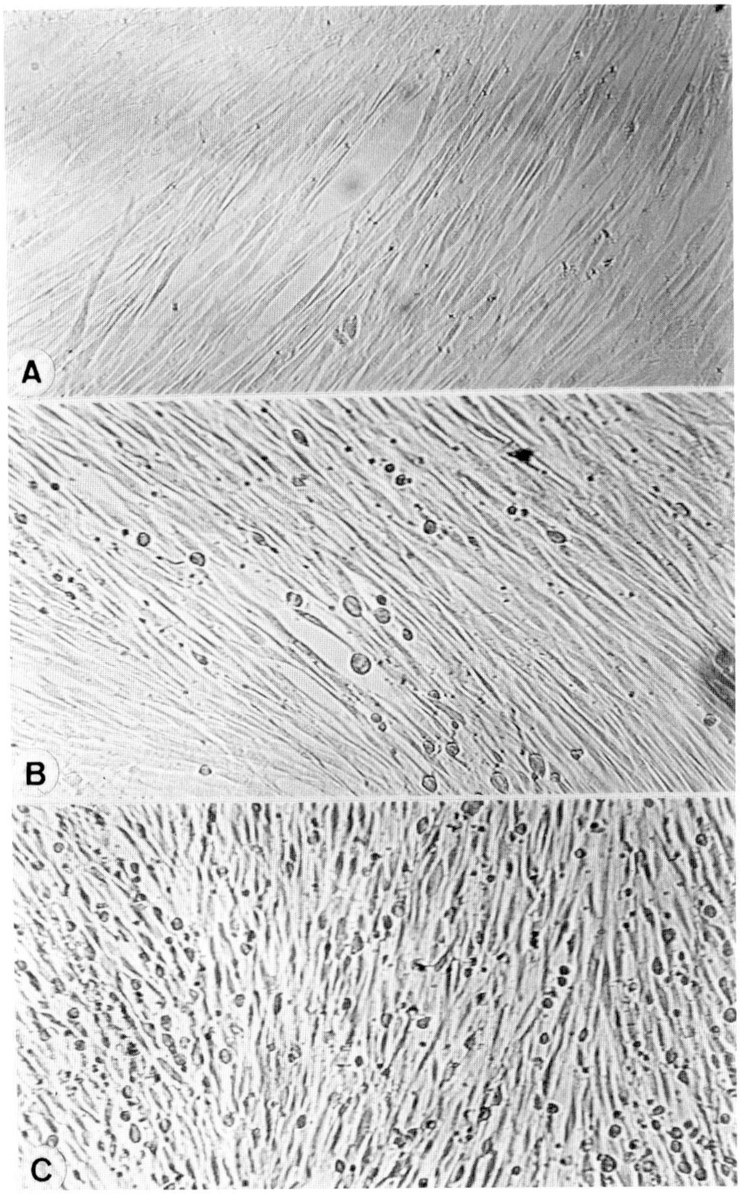

Figure 14.11 Cytopathic effect induced by rhinovirus 14 in human embryo fibroblast (HEF) cell culture (100×): *(A)* Uninoculated HEF cell culture; *(B)* Early CPE, 24 hours after infection; *(C)* Late CPE, 72 hours after infection

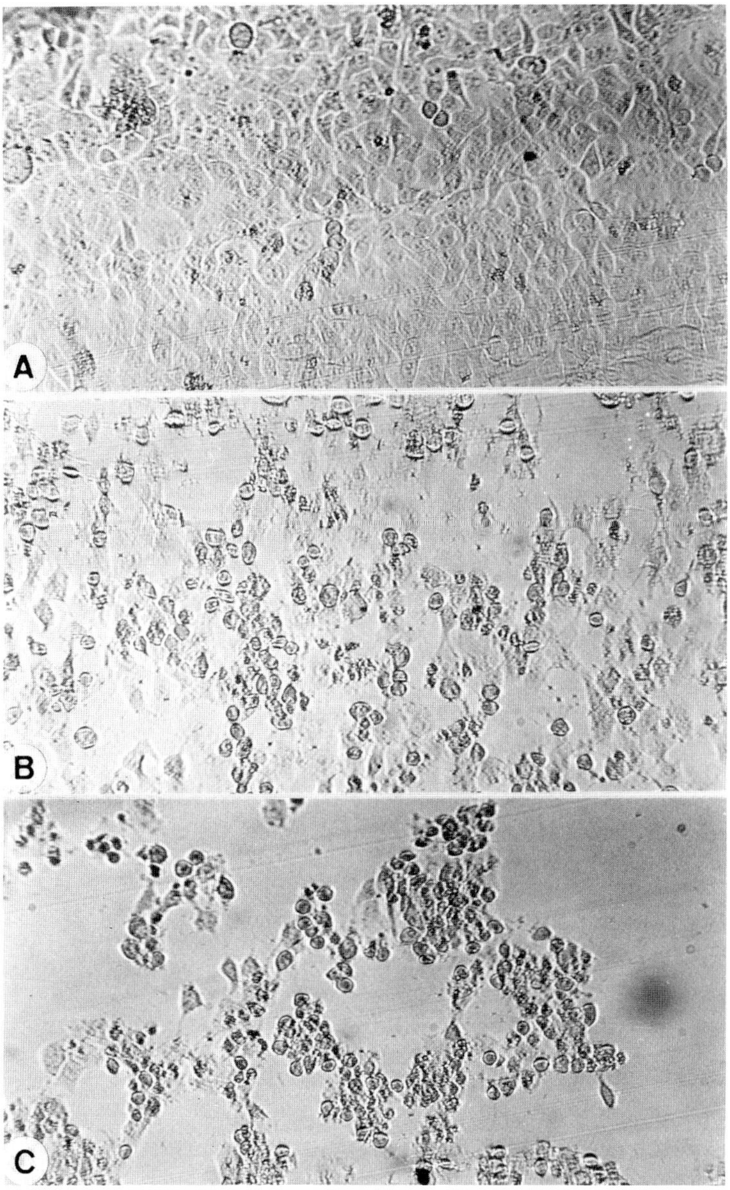

Figure 14.12 Cytopathic effect induced by rhinovirus 14 in HeLa cell culture (100×): *(A)* Uninoculated HeLa culture; *(B)* Early CPE, 24 hours after infection; *(C)* Late CPE, 72 hours after infection

GROWTH IN ORGAN CULTURE

Isolation of certain rhinovirus types can be facilitated by using organ cultures of human embryonic trachea, although the method is not commonly used in a routine virology laboratory. Decreasing ciliary activity and the progressive degenerative changes in an organ culture indicate the presence of a rhinovirus. (For details, see D. A. J. Tyrrell, *Virology Monograph*, 1968.)

Newer Methods for Rapid Rhinovirus Diagnosis

ELISA for antigen detection using a rabbit anti–rhinovirus hyperimmune serum as both the capture and test antibody, or using polymerase chain reaction amplification followed by a filter hybridization test can be used. These are not commonly used in a clinical laboratory.

Identification of Isolates

BIOCHEMICAL PROPERTIES

The similarity of CPE induced by the rhinoviruses and enteroviruses sometimes necessitates determining whether the isolate is an enterovirus. Since the infectivity of a rhinovirus suspension treated at pH 3.0 is often reduced by 2–4 logs as compared to 1 without treatment, this property provides a useful marker for separating the rhinoviruses from the enteroviruses. The latter are stable at low pH. To eliminate ether-sensitive viruses that are also acid pH sensitive, it is desirable to test for both ether-stability and acid-lability. The procedures used are as follows:

Ether-Stability

1. Add 0.5 ml diethyl ether to 0.5 ml of an unknown virus suspension, shake vigorously, let stand at room temperature for 1–2 hours, and shake intermittently, or allow mixture to stand at 4°C for 18–24 hours.
2. Transfer the mixture to an open glass petri dish. Allow the ether to evaporate.
3. Make serial 10-fold dilutions from 10^{-1} to 10^{-6} of both ether-treated and untreated virus suspensions.
4. Inoculate each dilution into a set of 2–4 culture tubes.
5. Compare virus infectivity titers of ether-treated and untreated samples.
6. As controls, known ether-sensitive and ether-resistant viruses should be included in the test.

Acid pH Stability Test

1. To 5 ml 0.2 M tris buffer solution (pH 7.4), add drops of concentrated HCl in order to reach pH 3.0. Dilute the solution to a final volume of 20 ml with distilled water.
2. Add 0.2 ml of unknown virus suspension to 1.8 ml of the diluted buffer, pH 3.0. As a control, add 0.2 ml of the virus suspension to 1.8 ml of stock tris buffer, pH 7.4.
3. Keep the mixtures at room temperature for 3 hours.
4. Make serial 10-fold dilutions of the two mixtures and inoculate each dilution into 2–3 culture tubes.
5. Compare virus infectivity titers of acid-treated and untreated samples.
6. Known acid-sensitive and acid-resistant viruses should be included in the test as additional controls.

MOLECULAR BIOLOGY TEST

Genomic amplification has been used to detect and differentiate rhinoviruses and enteroviruses in clinical specimens (as stated above in the enterovirus section).

NEUTRALIZATION TEST IN CELL CULTURE

Both CPE and plaque methods can be used to identify the rhinoviruses by means of serum pools or type-specific antiserum. The growing number of serotypes, however, greatly increases the labor necessary to perform a serologic identification.

Serodiagnosis

The situation is similar to that described for enteroviruses. Without virus isolation, serologic diagnosis is not practical. The neutralization test using a cell culture and the ELISA test for selected virus types are used for testing acute and convalescent sera for antibody rise in research studies.

AGAR OVERLAY MEDIUM FOR PRIMARY CELL CULTURE IN CAPPED FLASKS

Nutrient Medium

Distilled water, demineralized, sterile	60.0 ml
Earle's BSS 10X	18.0 ml
Calf serum (heat inactivated)	3.6 ml
$NaHCO_3$ (7.5%)	5.4 ml

Neutral red (1:1,000)	3.0 ml
Penicillin (200,000 u/ml)	0.2 ml
Streptomycin (500,000 μg/ml)	0.1 ml
Total	90.0 ml

Using sterile precautions:

1. Mix each solution according to order listed above.
2. Warm to 37°C in a water bath before use.

Agar (3%)

Agar (Difco-Noble)	2.7 g
Distilled water, demineralized	90.0 ml

1. Mix agar with demineralized water and melt in boiling water bath.
2. Autoclave at 15 lb for 15 minutes.
3. Cool to 43°C before use.

Complete Overlay Medium

1. When ready for overlay, mix equal volumes of A (nutrient medium) at 37°C and B (3% agar) at 43°C by pouring nutrient medium into the agar. Use immediately.
2. After flasks are overlaid and capped tightly, avoid exposure to light and incubate in a regular incubator without supply of CO_2.

Preparation of Neutral Red Solution (1:1,000)

Neutral red (dye content 92%)	1 g
Distilled water, demineralized	1,000 ml

1. Weigh neutral red powder and transfer to a 1-liter volumetric flask.
2. Add demineralized water slowly and mix well until dye dissolves.
3. Make up a final volume of 1,000 ml with water.
4. Dispense into smaller convenient volumes, 50 or 100 ml per bottle.
5. Autoclave at 15 lb for 15 minutes.
6. Cap tightly and store at room temperature.

SUPPLEMENTARY READING FOR ENTEROVIRUSES

Auvinen, P.; Ziegler, T.; Skern, T.; Kuechler, E.; Stanway, G.; and Hyypia, T. Identification of rhinoviruses by cDNA probes. *J. Virol. Meth.* 27:61–68 (1990).

Bruce, C. B.; Chadwick, P.; and Al-Nakib, W. Detection of rhinovirus RNA in nasal epithelial cells by in situ hybridization. *J. Virol. Meth.* 30:115–26 (1990).

Chonmaitree T.; Ford, C.; Sanders, C.; and Lucia, H. L. Comparison of cell cultures for rapid isolation of enteroviruses. *J. Clin. Microbiol.* 26:2576–80 (1988).

Couch, R. B. Rhinoviruses. In Fields, B. N., and Knipe, D. M., eds., *Fields Virology*, pp. 607–29. New York: Raven, 1990.

Dagan, R., and Menegus, M. A. A combination of four cell types for rapid detection of enteroviruses in clinical specimens. *J. Med. Virol.* 19:219–28 (1986).

Dearden, C. J., and Al-Nakib, W. Direct detection of rhinoviruses by an enzyme-linked immuno-sorbent assay. *J. Med. Virol.* 23:179–89 (1987).

Enders, J. F.; Weller, T. H.; and Robbins, F. C. Cultivation of the Lansing strain of poliomyelitis virus in cultures of various human embryonic tissues. *Science* 109.85–87 (1949).

Forsyth, M.; Al-Nakib, W.; Chadwick, P.; Stanway, G.; Hughes, P. J.; Leckie, G.; Almond, J. W.; and Tyrrell, D. A. Rhinovirus detection using probes from the 5′ and 3′ end of the genome. *Arch. Virol.* 107:55–63 (1989).

Gama, R. E.; Horsnell, P. R.; Hughes, P. J.; North, C.; Bruce, C. B.; Al-Nakib, W.; and Stanway, C. Amplification of rhinovirus specific nucleic acids from clinical samples using the polymerase chain reaction. *J. Med. Virol.* 28:73–77 (1989).

Hamparian, V. V.; Colonno, R. J.; Cooney, M. K.; Dick, E. C.; Gwaltney, J. M., Jr.; Hughes, H. S.; Jordan, W. S., Jr.; Kapikian, A. Z.; Mogabgab, W. J.; and Monto, A. A collaborative report: Rhinoviruses extension of the numbering system from 89 to 100. *Virology* 159:191–92 (1987).

Hsiung, G. D. Further studies on characterization and grouping of ECHO viruses-comparative virology. *Ann. N.Y. Acad. Sci.* 101:413–22 (1962).

———. The impact of cell culture sensitivity on rapid viral diagnosis: A historical perspective. *Yale J. Biol.* 62:79–88 (1989).

Hsiung, G. D., and Melnick, J. L. Comparative susceptibility of kidney cells from different monkey species to enteric viruses (poliomyelitis, coxsackie, and ECHO groups). *J. Immun.* 78:137–46 (1957).

Johnston, S. L. G., and Siegel, C. S. Presumptive identification of enteroviruses with RD, HEP-2 and RMK cell lines. *J. Clin. Microbiol.* 28:1049–50 (1990).

La Monica, N., and Racaniello, V. R. Differences in replication of attenuated and neurovirulent polioviruses in human neuroblastoma cell line SH-SY5Y. *J. Virol.* 63:2357–60 (1989).

Landry, M. L.; Madore, P.; Fong, C. K. Y.; and Hsiung, G. D. Use of guinea pig embryo cell cultures for isolation and propagation of group A coxsackieviruses. *J. Clin. Microbiol.* 13:588–93 (1981).

Lim, K. A., and Benyesh-Melnick, M. Typing of viruses by combination of antiserum pools. Application to typing of enteroviruses (coxsackie and echo). *J. Immun.* 84:309–17 (1960).

Lipson, S. M.; Walderman, R.; Castello, P.; and Szabo, K. Sensitivity of rhabdomyosarcoma and guinea pig embryo cell cultures to field isolates of difficult-to-cultivate group A coxsackieviruses. *J. Clin. Microbiol.* 26:1298–1303 (1988).

MacNaughton, M. R. The structure and replication of rhinoviruses. *Current Topics in Microbiology and Immunology* 97:1–26 (1982).

Melnick, J. L. Portraits of viruses: The picornaviruses. *Intervirology* 20:61–100 (1983).

Melnick, J. L.; Wenner, H. A.; and Phillips, C. A. Enteroviruses. In Lennette, E. H., and Schmidt, N. J., eds., *Diagnostic Procedures for Viral, Rickettsial and Chlamydial Infections*, 5th ed., pp. 471–534. Washington, D.C.: American Public Health Association, 1979.

Melnick, J. L., and Wimberly, I. L. Lyophilized combination pools of enterovirus equine antisera: New LBM pools prepared from reserves of antisera stored frozen for two decades. *Bull. WHO* 63:543–50 (1985).

Menegus M., and Hollick, G. E. Increased efficiency of Group B coxsackie-virus isolation from clinical specimens by use of BGM cells. *J. Clin. Microbiol.* 15:945–48 (1982).

Olive, D. M.; Al-Mufti, S.; Al-Mulla, W.; Khan, M. A.; Pasca, A.; Stanway, G.; and Al-Nakib, W. Detection and differentiation of picornaviruses in clinical samples following genomic amplification. *J. Gen. Virol.* 71:2141–47 (1990).

Rotbart, H. A. Nucleic acid detection systems for enteroviruses. *Clin. Microbiol. Rev.* 4:156–68 (1991).

Tyrrell, D. A. J. Rhinoviruses, *Virology Monogr.* 2:67–124 (1968).

15

ORTHOMYXOVIRIDAE AND PARAMYXOVIRIDAE

G. D. HSIUNG

Influenza viruses are classified as Orthomyxoviridae, while parainfluenza viruses, mumps virus, and the Newcastle disease virus of poultry are members of Paramyxoviridae (table 15.1). Influenza viruses are usually associated with respiratory disease in all age groups, A strains with pandemics, and A and B strains with epidemics, but the C strains are infrequently associated with disease in humans. Parainfluenza viruses are usually associated with epidemic-endemic respiratory infections in children.

ORTHOMYXOVIRIDAE: THE INFLUENZA VIRUSES

Members of the Orthomyxoviridae are single-stranded RNA viruses; influenza A and B virions consist of an eight-segmented RNA genome and the influenza C virus genome has seven segments. All influenza viruses are enclosed in a lipoprotein envelope. The influenza virus is about 100 nm in diameter (figure 15.1). Budding from the plasma membrane is seen in influenza virus–infected cells. Virus infectivity is readily destroyed by ether treatment, and the virions are relatively heat-labile.

Free virions of influenza A and B exhibit two well-defined spikes: hemagglutinin (H) and neuraminidase (N), on the envelopes. Influenza C has a single glycoprotein possessing the activity of both hemagglutinin and neurominidase. The hemagglutinin is responsible for the attachment of the virus to cellular receptors, and the neuraminidase is a viral enzyme that cleaves sialic acid from the receptors on the surface of infected cells and promotes virus release from infected cells. The influenza virus A strains have

ISBN 0-300-05845-4.

Table 15.1 Members of Orthomyxoviridae and Paramyxoviridae

Family	Genus	Common Species
Orthomyxoviridae	Influenza virus	Influenza virus type A Influenza virus type B Influenza virus type C
Paramyxoviridae	*Paramyxovirus*	Parinfluenza virus types 1–5* Mumps virus Newcastle disease virus of fowl
	Morbillivirus	Measles virus Canine distemper virus Rinderpest virus of cattle
	Pneumovirus	Respiratory syncytial virus (RSV) of humans Bovine RSV Pneumonia virus of mice

*Parainfluenza virus type 5 is not an official designation.

been designated according to their H and N antigens, for example, H1N1, H2N2, and H3N2. Their ability to adsorb onto the surface of RBC, the hemagglutination phenomenon, is a distinct property of this virus group. Erythrocyte agglutination occurs at 4°C or 22°C but not at 37°C (table 15.2). Guinea pig RBC are preferable for testing new influenza virus A and B isolates, whereas day-old chick RBC are more satisfactory for influenza virus C hemagglutination.

Influenza viruses are capable of multiplying in a variety of animals as well as in various cell culture systems. For certain influenza virus strains, embryonated eggs are still the most susceptible host. Since some virus strains in this group do not regularly induce distinct CPE, the hemadsorption technique is used to recognize their presence in infected cell cultures (see below). The presence of only minimum CPE during virus replication may be due to the action of protective inhibitor substances, known as interferon, produced as a result of cell infection. Interferon was originally described by A. Isaacs and J. Lindenman in the influenza virus–infected chick embryo cell system.

Specimen Source and Inoculation

Throat washings and throat or nasopharyngeal swabs collected within the first 24–48 hours of illness are used for the isolation of influenza viruses. Tissues collected at autopsy, especially lungs, can be used for influenza virus isolation. Embryonated chicken eggs 10–11 days old are the most sensitive host system for the recovery of influenza virus isolates (for the egg inoculation procedure, see chapter 5). A canine kidney cell line, MDCK, has been found highly sensitive for the isolation of influenza virus when the medium contains trypsin (see below).

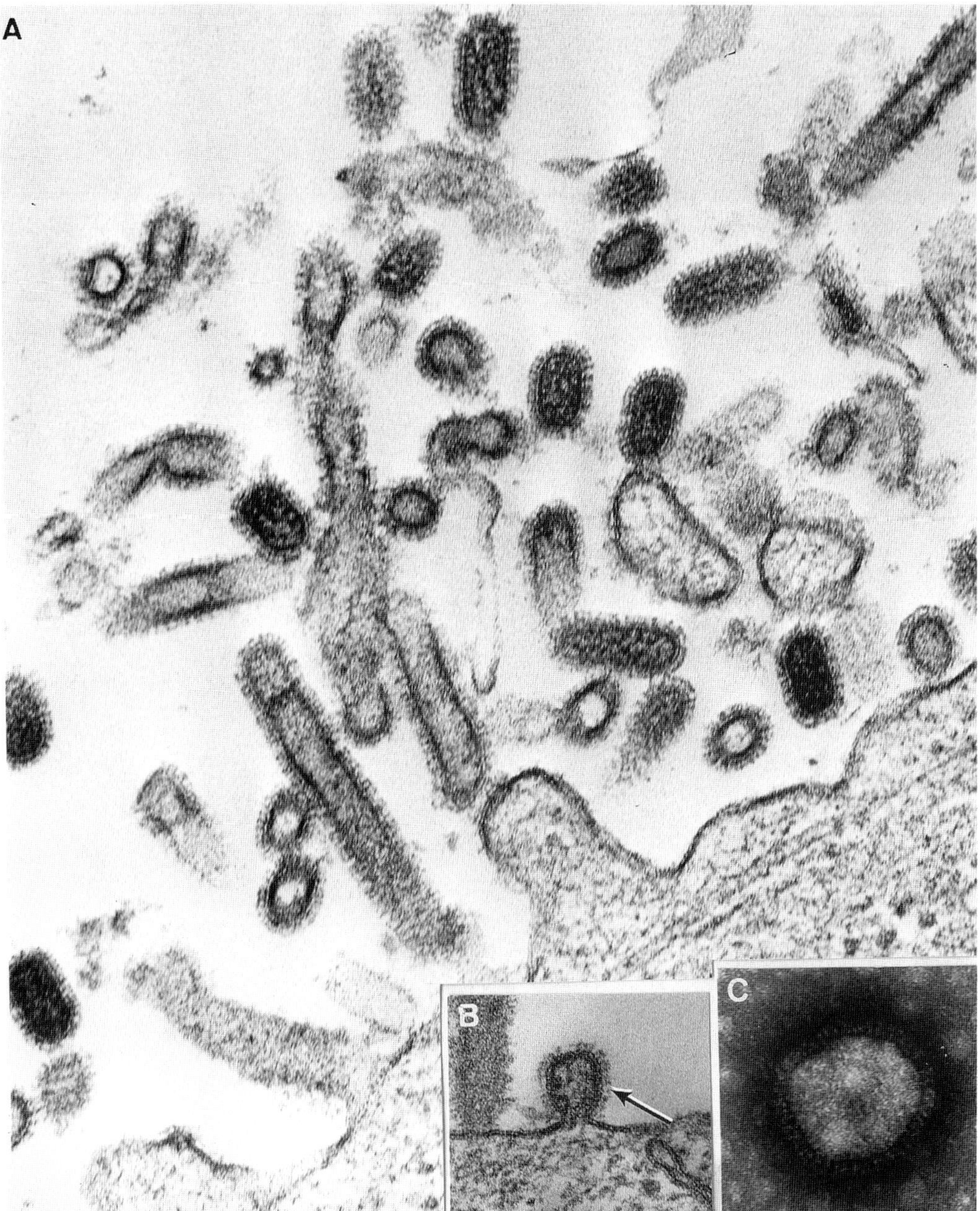

Figure 15.1 *(A)* Electron micrograph of an influenza A virus–infected monkey kidney cell showing virions at the extracellular space (100,800×); *(B)* Virion budding at the cell membrane (100,800×); *(C)* Influenza virion stained with PTA (168,000×)

Table 15.2 Biologic Properties of Influenza and Parainfluenza Viruses

Viruses	Host Systems		Hemagglutination at Various Temperatures Using RBC Obtained from				
			Chick		Guinea Pig		Human (Type O)
	Embryonated Eggs	Tissue Culture	4°C	22°C	4°C	22°C	22°C
Orthomyxoviridae							
Influenza virus type A	+	+/−	+	+	+	+	+
Influenza virus type B	+	+	+	+	+	+	+
Influenza virus type C	+	−	+	−	−	−	−
Paramyxoviridae							
Parainfluenza virus type 1 (HA-2)	+	+	+	+	+	+	+
Parainfluenza virus type 2 (CA)	−	+	+	+	+	+	+
Parainfluenza virus type 3 (HA-1)	−	+	+	+	+	+	+
Parainfluenza virus type 4 (M-25)	−	+	−	−	−	+[a]	+
Parainfluenza virus type 5 (SV_5-SA-DA)[b]	+/−	+	+	+	+	+	+
Mumps	+	+/−	+	+	+	+	+
NDV[c]	+	+	+	+	+	+	+

+ virus-induced change; − no change
a. Hemaggutination also at 35°–37°C
b. Not officially classified
c. Newcastle Disease Virus

Virus Isolation and Identification

HEMADSORPTION IN CELL CULTURE

Primary monkey kidney cell cultures made from rhesus, cynomolgus, or African green monkeys are the most sensitive system for the isolation of influenza viruses. Since calf serum commonly contains antibodies to influenza viruses, all cultures maintained in a medium containing calf serum need to be removed before inoculation. For better results, all cultures should be washed with PBS two to three times to remove residual serum. Because CPE may not be observed in the influenza virus–infected cells, hemadsorption test is used. (Procedures are detailed in the parainfluenza virus section below.) MDCK, a canine kidney passaged cell-line, can be used if the maintenance medium contains trypsin (2 μg/ml).

Some strains of the influenza virus, especially influenza B, may produce some cellular changes in infected cell cultures. The infected cells become progressively granular, swollen, and round; later they become pyknotic and fragmented, and finally the whole cell sheet is destroyed. Again, when guinea pig RBC are added, hemadsorption also occurs.

EMBRYONATED EGG INOCULATION

Although hemadsorption in cell cultures is used for the isolation of influenza viruses, embryonated eggs still constitute the most susceptible host system for some strains of influenza A. For primary isolation, specimens should be inoculated by the amniotic route, and for the propagation of virus stock, the allantoic route should be used. (Procedures for both the amniotic and allantoic routes of inoculation are detailed in chapter 5. The procedure for testing the harvested egg fluids for hemagglutination is given in chapter 9.)

Identification of Specific Virus Types

IMMUNOFLUORESCENT STAINING FOR RAPID IDENTIFICATION

Once hemadsorption is observed in inoculated cell cultures, the culture fluid is removed and the infected cells are detached from the monolayers either by trypsin or scraping. The cell spot, stained with monoclonal antibody for influenza type A or type B by direct immunofluorescent test (for details, see the parainfluenza section) is commonly used for the rapid typing of influenza viruses.

HEMADSORPTION-INHIBITION METHOD

When grown well in cell cultures, some strains of influenza virus can be identified by the hemadsorption-inhibition method (for the procedure, see below).

HEMAGGLUTINATION-INHIBITION TEST

Most of the influenza viruses isolated in embryonated eggs can be identified by HI test for specific virus type or strain identification (chapter 9). In the HI test, both allantoic fluids (ALF) and amniotic fluids (AMF) can be used as HA antigens and can be tested against a known type-specific influenza virus antiserum. Upon initial isolation of the influenza virus in eggs, the AMF often yields higher titers of HA antigen than the ALF. Guinea pig erythrocytes in PBS should be used for the HA or HI test since chick erythrocytes often are not agglutinable by the primary isolates of influenza virus A. Note that nonspecific inhibitors may be present in the animal sera from which antisera are prepared, thus interfering with the identification procedure. (For removal of nonspecific inhibitors in sera, see chapter 9.)

Serodiagnosis

Serologic tests have been used for the detection and diagnosis of influenza virus infections. The diagnosis of influenza may be made when a significant antibody rise is observed to the group specific antigen consisting of membrane protein and nucleoprotein (*S* antigen). The CF test is most frequently used for the determination of serum antibodies to *S* antigens (nucleocapsid), which are common to members of the same influenza type. Antibody rise to a specific virus strain (*V* antigens consisting of the *H* and *N* antigens) is tested by the HI test. In the HI test for the influenza virus, it is necessary to destroy nonspecific inhibitors present in the serum. A receptor-destroying enzyme (RDE) is used to remove inhibitors in human serum (chapter 9). The procedures for HI and CF tests are described in chapters 9 and 10, respectively.

PARAMYXOVIRIDAE: THE PARAINFLUENZA VIRUSES

The human parainfluenza viruses 1, 2, 3, and 4 are important and common causes of lower respiratory tract infections in infants and young children, and may occur in epidemic proportions. Adults, when infected, manifest only mild disease. The viruses were first recognized by the use of the hemadsorption technique in cell cultures.

Although sharing many properties with the influenza viruses (table 15.2), parainfluenza viruses are distinct in certain characteristics. They are larger, 150–300 nm in diameter (figures 15.2, 15.3); each virion has a large, nonsegmented inner helix; and a fusing factor is located on their envelope. Commonly seen in infected cells is a budding from the plasma membrane. Virus infectivity is destroyed by ether treatment.

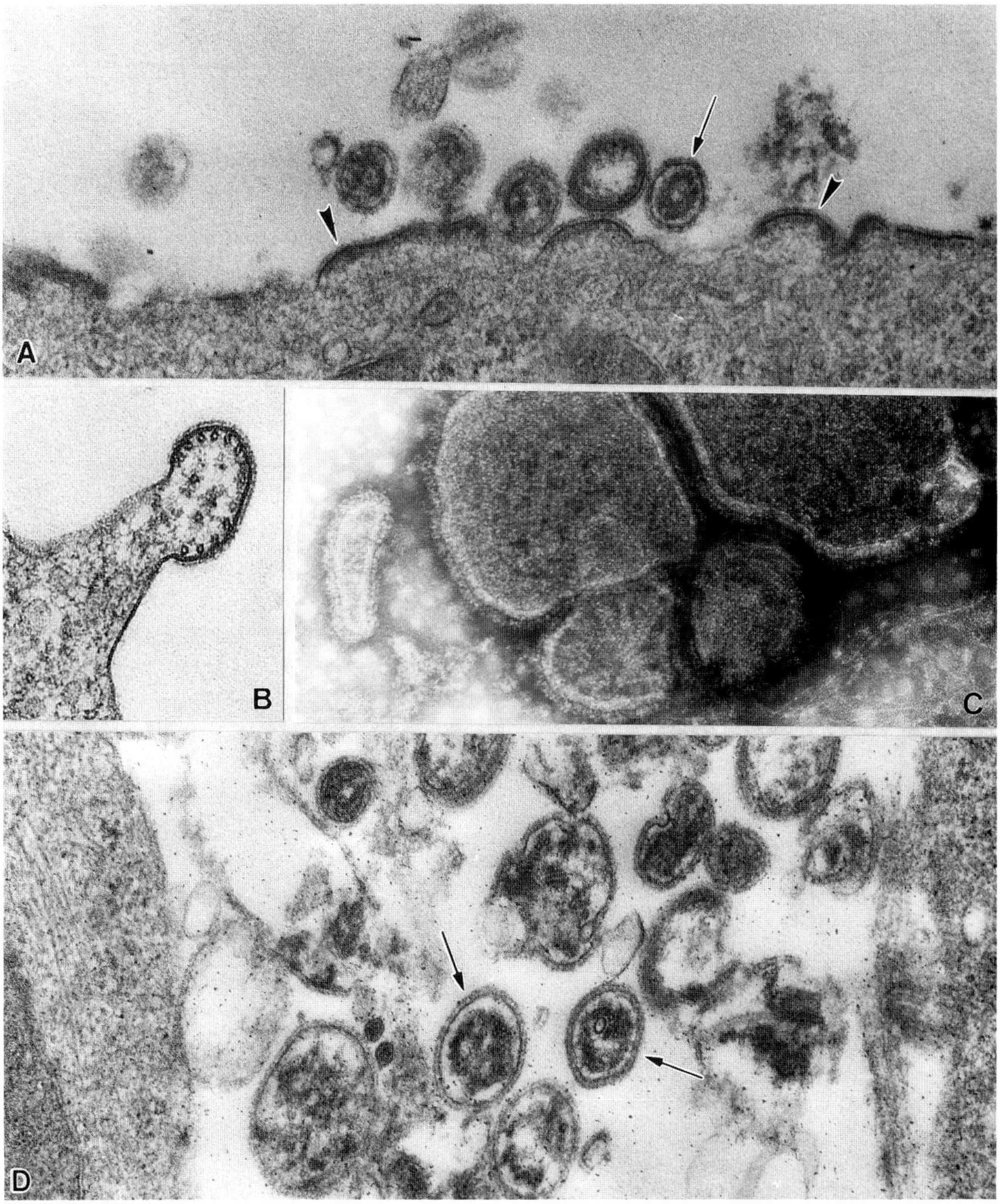

Figure 15.2 Electron micrographs of parainfluenza virus–infected RhMK cell: *(A)* Thin section of a parainfluenza virus type 2 infected cell showing an extracellular virion (arrow) and budding (arrowheads) from plasma membrane (60,000×); *(B)* Parainfluenza virus type 3 virion in the process of budding at the cell membrane (70,400×); *(C)* Parainfluenza virus type 3 virions negatively stained with PTA (100,800×); *(D)* Parainfluenza virus type 2 extracellular spherical virions (arrows) (60,000×)

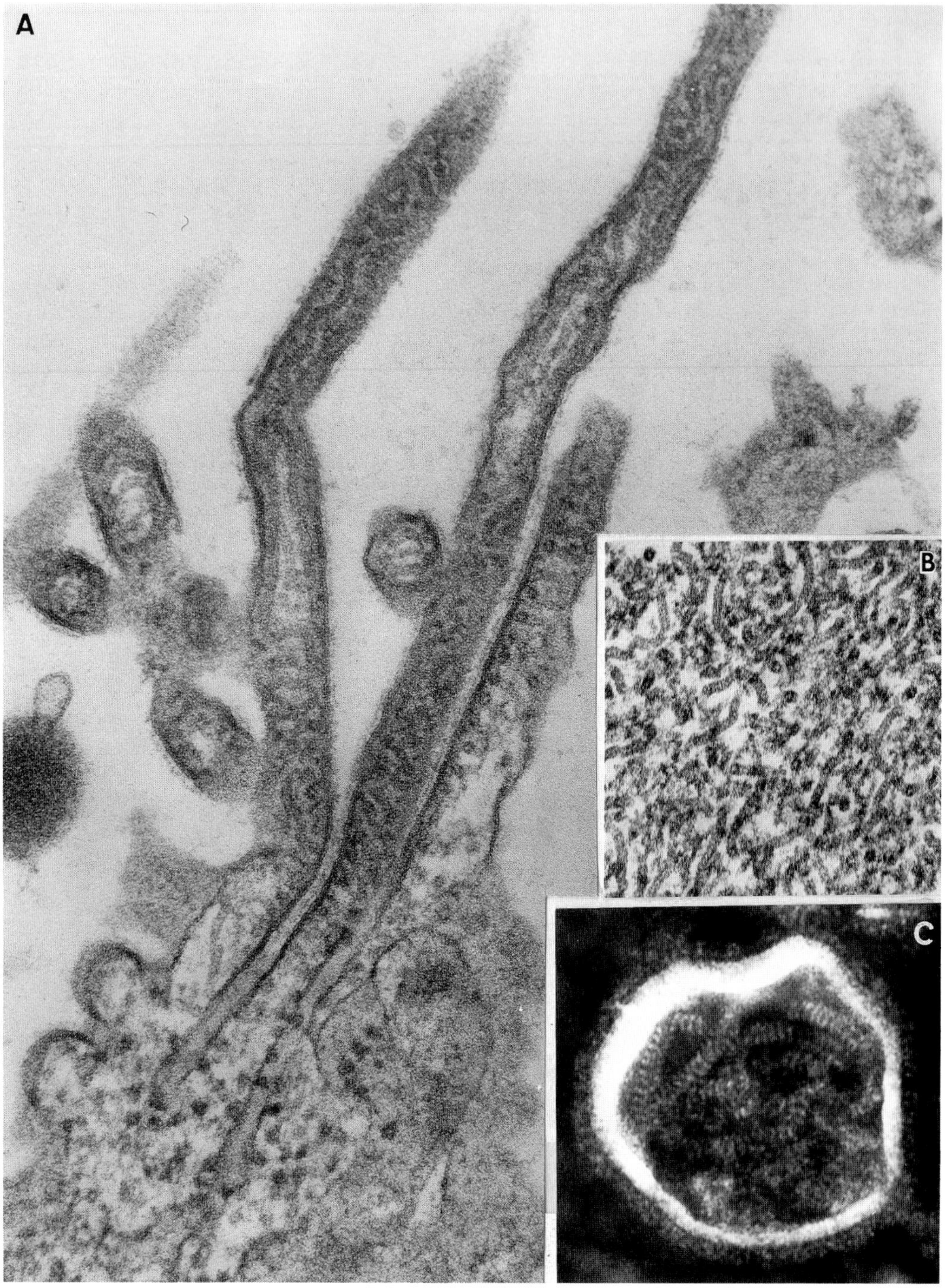

Figure 15.3 Electron micrograph of a parainfluenza virus–(SV_5) infected GPE cell: *(A)* Filamentous-form virions (SV_5) budding at a cell surface (100,800×) (courtesy F. Bia); *(B)* Helical nucleocapsids in the cytoplasm (79,200×); *(C)* Virion (DA), negatively stained with PTA revealing internal nucleocapsid structure (126,150×). (Courtesy J. D. Wright)

Specimen Source and Inoculation

Nasal or throat washings or a pharyngeal swab immersed in 2 ml of a transport medium (0.5% gelatin or bovine albumin in Hanks' BSS) are suitable specimens for isolation. If possible, specimens should be tested fresh, without prior freezing. Primary monkey kidney cell cultures are the most sensitive cell culture system for the isolation of the parainfluenza viruses. The culture medium should not contain calf serum, which may have specific antibodies against parainfluenza viruses. Thus, all maintenance medium should be removed from cultures and washed before inoculation. For example, a culture medium in the absence of calf serum and maintained at pH 6.8 is a favorable condition for the replication of parainfluenza viruses. Incubation at 33°C is more favorable than at 37°C. A rhesus monkey cell line, LLC-MK_2, treated with a trypsin-EDTA solution may be used for the isolation of parainfluenza viruses when the primary kidney cell cultures are not available. Passaged cell lines are not as sensitive as primary cell cultures for virus isolation but can be used for serial passages of virus stocks.

Virus Isolation and Identification

HEMADSORPTION IN CELL CULTURES

Ordinarily, parainfluenza virus–infected cell cultures do not exhibit any distinctive cellular changes. When a freshly obtained guinea pig RBC suspension is added onto the infected cultures, however, the RBC adsorb onto the cells, resulting in the hemadsorption phenomenon, as shown in figure 15.4. Caution should be taken when aged guinea pig RBC are used since nonspecific hemadsorption often occurs in an uninfected cell culture (figure 15.5) and should be distinguished from that resulting from a specific viral infection. Furthermore, hemadsorption usually takes place at 4°C or 22°C since the RBC will elute when incubated at 37°C, except parainfluenza virus 4 in which hemadsorption takes place at 37°C. Once a culture shows a positive hemadsorption, the culture fluid is subcultured into a fresh culture to confirm virus isolation and to permit further identification. The procedure for the hemadsorption technique is:

1. Remove fluids from the inoculated cultures and collect into a sterile tube, then add 0.2 ml of the 0.5% guinea pig RBC suspension to each inoculated culture.
2. Place tubes in a refrigerator at 4°C in a horizontal position (for parainfluenza 4, 35°–37°C is preferable), cell monolayer down, for 20 minutes. Make sure that the RBC suspension is in contact with the infected cells.
3. Remove unadsorbed RBC by washing the culture with cold PBS.
4. Read the tubes microscopically for hemadsorption and record the results as −, +, ++, +++, ++++, according to the concentration of

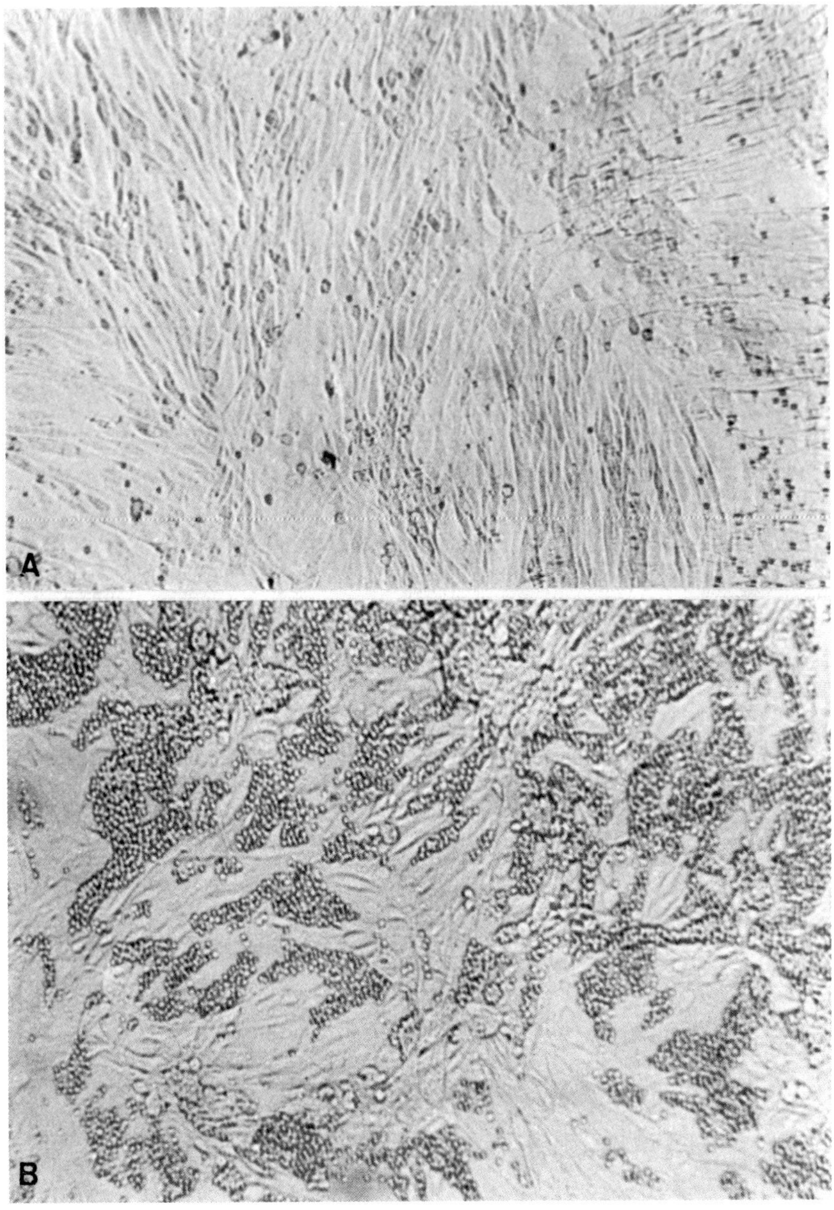

Figure 15.4 Hemadsorption of a parainfluenza virus type 1–infected RhMK culture (100×): *(A)* Uninfected RhMK culture; *(B)* RhMK culture infected with parainfluenza virus 1 showing hemadsorption of guinea pig RBC

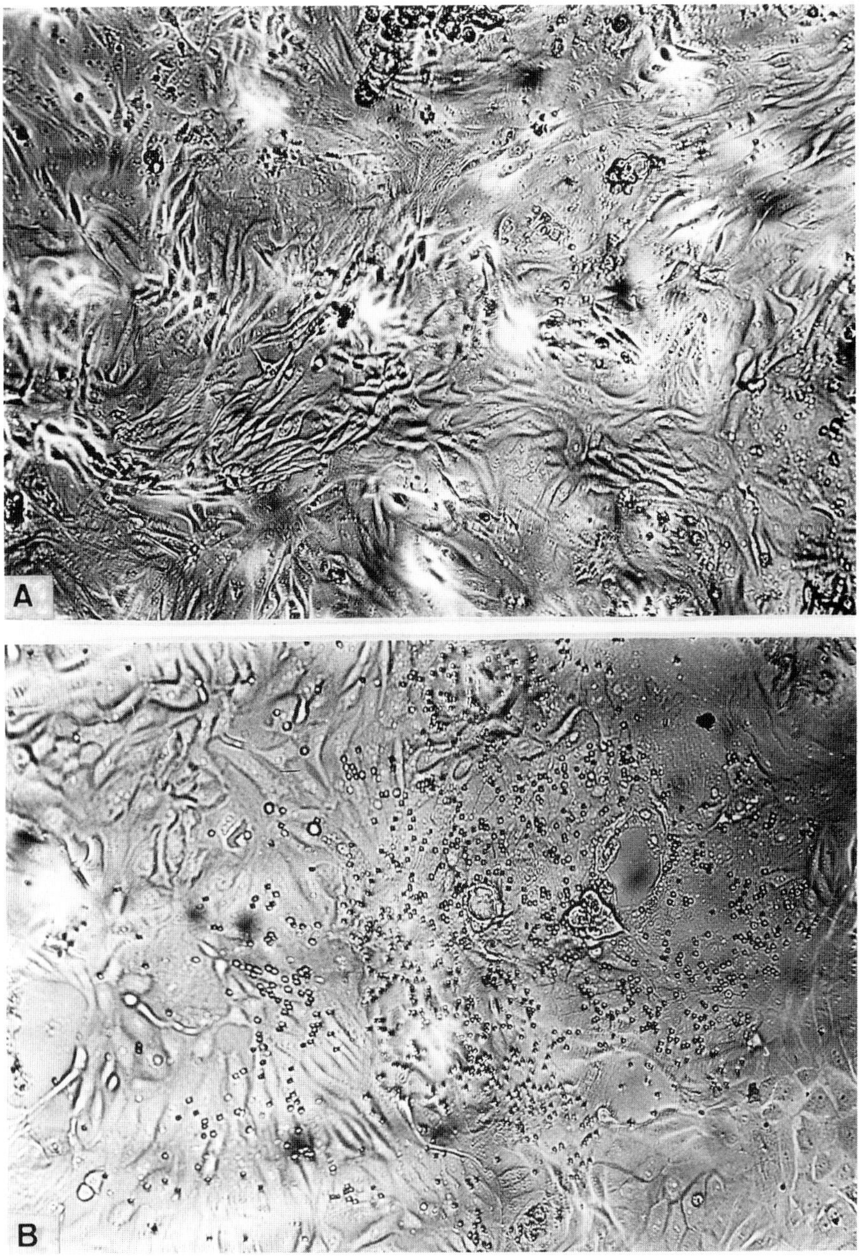

Figure 15.5 Nonspecific hemadsorption in an uninfected RhMK culture (100×): *(A)* Uninfected RhMK culture; *(B)* Nonspecific hemadsorption resulting when aged guinea pig RBC were used. Compare with figure 15.4.

adsorbed RBC. An example of a positive hemadsorption is shown in figure 15.4.

5. Subculture the fluid from the culture tube showing hemadsorption into fresh cultures for further identification (see below).

CYTOPATHIC EFFECT: SYNCYTIAL CELL FORMATION

Certain strains of parainfluenza virus, particularly parainfluenza virus 3, produce multinucleated syncytial cells in infected Hep-2 or HeLa cells, as shown in figure 15.6. These cultures will also exhibit hemadsorption when a guinea pig RBC suspension is added.

IMMUNOFLUORESCENT STAINING FOR RAPID IDENTIFICATION

1. Once hemadsorption occurs in an infected cell culture, remove the supernatant fluid; the infected cells are detached by trypsin.
2. Remove trypsin by centrifugation.
3. Suspend cell pellet with 0.2 ml PBS.
4. Make cell spots by placing a drop of cell suspension onto well slides.
5. Perform IF, as explained in chapter 7. Monoclonal antibodies for parainfluenza viruses 1, 2, 3, and 4 are available for rapid virus identification (see color plate 10).

HEMADSORPTION-INHIBITION METHOD

The parainfluenza viruses are routinely identified by direct immunofluorescent staining using conjugated monoclonal antibody. They can also be identified by the hemadsorption-inhibition method, which may be performed in two ways. The first type of hemadsorption-inhibition test is performed by adding antiserum onto an infected culture before the addition of an RBC suspension. The inhibition of hemadsorption identifies the virus type. The second type of hemadsorption-inhibition test is a neutralization test in cell culture systems that is used in select cases and is described below.

1. Inoculate serial 10-fold dilutions of the virus suspension into tissue culture tubes. After 3–5 days, test for hemadsorption with guinea pig RBC. Calculate the $TCID_{50}$ endpoint.
2. Prepare 2-, 4-, or 10-fold dilutions of antiserum (all sera should be heat-inactivated at 56°C for 30 minutes).
3. Add an equal volume of a constant dilution of the unknown virus containing 30–50 $TCID_{50}$ (50% hemadsorption tissue culture dose, in this instance) to each of the antiserum dilutions. Mix well.
4. Allow the virus-serum mixtures and the control virus suspension to remain at room temperature for 1 hour.
5. Inoculate 0.2 ml amounts of the virus-serum mixtures into culture tubes, using three or four tubes for each mixture.

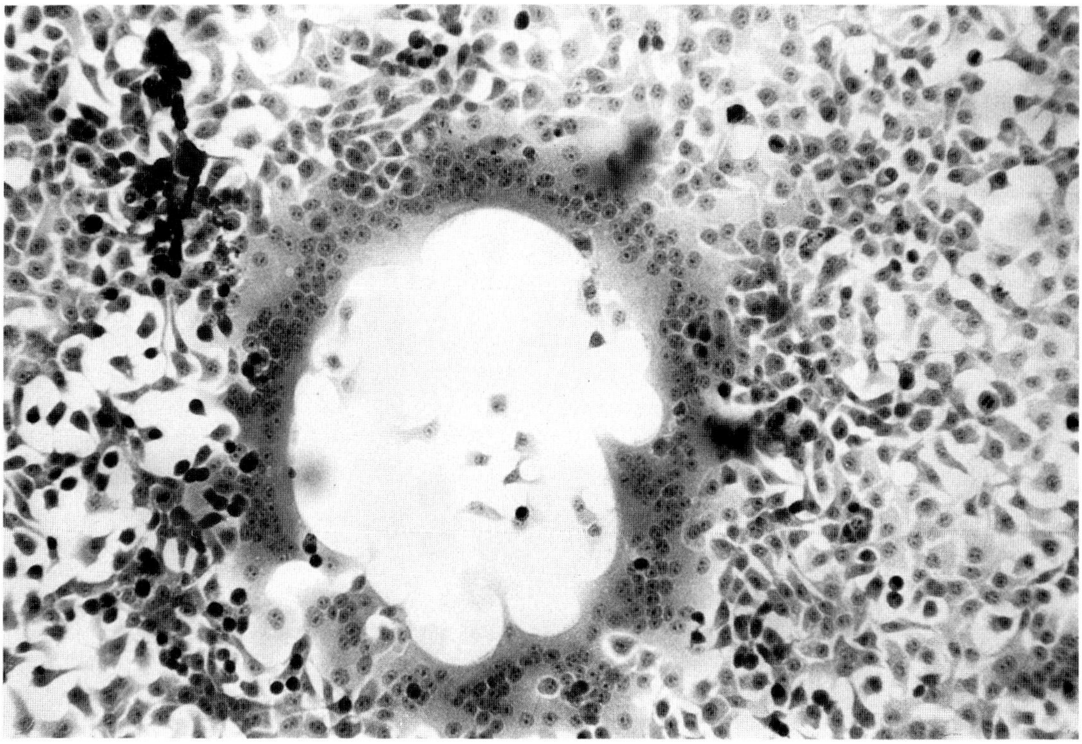

Figure 15.6 Syncytial cell formation induced by parainfluenza virus type 3 in Hep-2 cell culture three days after infection (H&E, 100×)

6. Inoculate 0.1 ml of the serial 10-fold dilutions of the virus suspension into each tube, using three or four tubes per dilution for the virus control.
7. Test all cultures for hemadsorption 3–5 days after inoculation by adding 0.2 ml of a 0.5% freshly obtained guinea pig RBC suspension to each tube. Refrigerate for 20 minutes and read. (It is advisable to test the virus control tubes inoculated with 30–50 $TCID_{50}$ first; if hemadsorption is observed in these challenge virus tubes, all other tubes are ready for testing.)
8. Complete inhibition of hemadsorption by a known antiserum is considered a positive serum neutralization test and indicates the identity of the virus. The serum titer is the highest dilution inhibiting multiplication of the challenge virus so that no hemadsorption is detected.

Serodiagnosis

Infection due to a parainfluenza virus causes heterotypic antibody responses to other members of the group, including mumps virus. Thus, it is difficult to rely on serodiagnosis. Virus isolation is still the best means of etiologic diagnosis. Paired serum samples, however, can be assayed together in the same test, either by HI or NT. A 4-fold or greater rise in antibody titer is considered significant. It should be interpreted with caution owing to the occurrence of heterotypic rises.

MUMPS VIRUS

Mumps is a systemic infection characterized by fever, swelling of the parotid glands, and frequent involvement of the central nervous system. Specimens for virus isolation include saliva, urine, and cerebrospinal fluid.

Virus Isolation and Assay

DIRECT INOCULATION

1. Remove fluids from primary monkey kidney cell culture tubes and inoculate 0.3–0.5 ml of undiluted, freshly obtained urine or other specimens directly into each culture tube. (Seven- to 8-day-old embryonated eggs may also be used.)
2. For inoculated cultures, adsorb for 2 hours at 37°C.
3. Replace inoculum with fresh medium.
4. For egg inoculation procedure, see chapter 5.

URINE CONCENTRATION METHOD

1. Centrifuge 15 ml urine at 500 × *g* at 4°C for 10 minutes to remove debris.
2. Collect 10 ml of the supernatant and recentrifuge at 100,000 × *g* for 90 minutes in an ultracentrifuge.
3. Resuspend the pellet in 1.0 ml of Hanks' BSS and inoculate into culture tubes or embryonated eggs, 0.1–0.2 ml per tube/embryonated egg.

TEST FOR VIRUS ACTIVITY

With inoculated cultures, check CPE and test for hemadsorption 3–7 days after inoculation (figure 15.7). Once hemadsorption is observed in inoculated cultured cells, rapid identification can be performed with cell spots made from infected cells and monoclonal antibody against mumps virus by the IF technique. For inoculated embryonated eggs, harvest AMF and ALF 5–7 days after inoculation and test for hemagglutinin in the egg fluids as described above for the influenza virus.

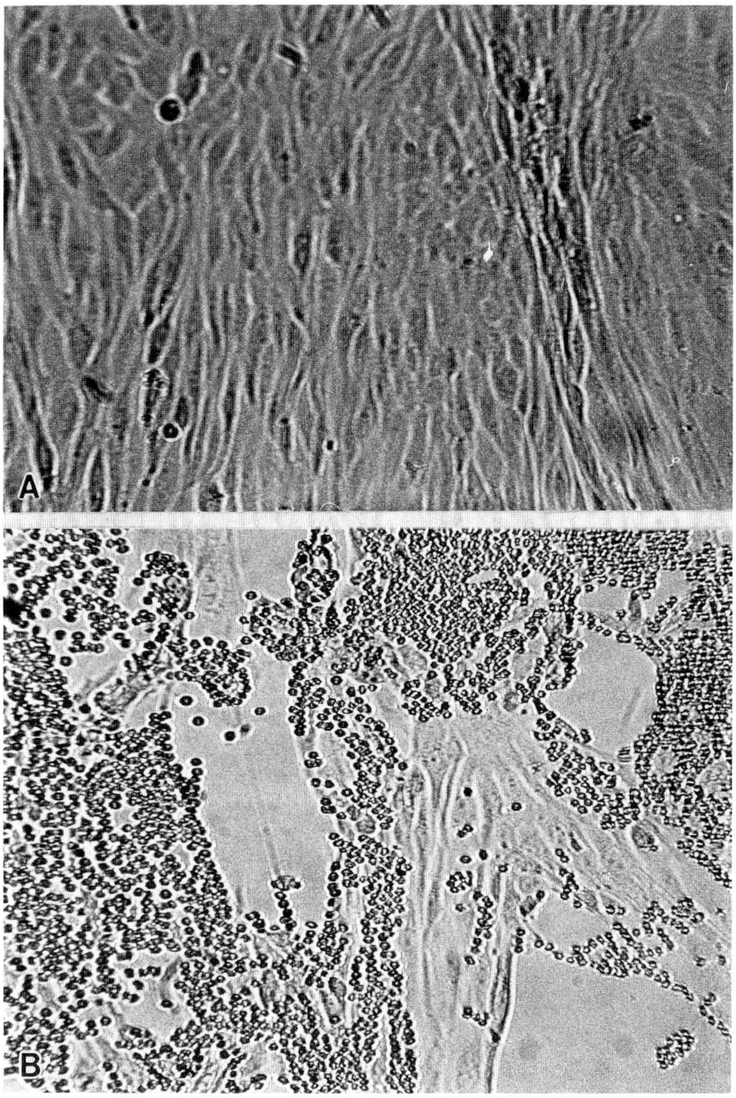

Figure 15.7 Hemadsorption of a mumps virus–infected RhMK culture (100×): *(A)* Uninfected RhMK culture; *(B)* RhMK culture infected with mumps virus showing hemadsorption of guinea pig RBC

Serodiagnosis

Mumps virus isolation procedures are not routinely done because of the slow growth rate of the virus. Serodiagnosis has therefore been commonly performed. A significant antibody rise in paired sera by the CF or HI method is an indication of mumps virus infection, although heterologous antibody rise often occurs with infection of other parainfluenza virus types.

SUPPLEMENTARY READING

Bartoloma, N. Y., and Forbes, A. B. Successful use of shell vial centrifugation and sixteen to eighteen hour immunofluorescence staining for detection of influenza A and B in clinical specimens. *Am. J. Clin. Path.* 92:487–90 (1989).

Castells, E.; George, V. G.; and Hierholzer, J. C. NCI-H292 as an alternative cell line for the isolation and propagation of the human paramyxoviruses. *Arch. Virol.* 115:277–88 (1990).

Chomel, J. J.; Thouvenot, D.; Onno, M.; Kaiser, C.; Gourreau, J. M.; and Aymard, M. Rapid diagnosis of influenza infection of NP antigen using an immunocapture ELISA test. *J. Virol. Meth.* 25:81–92 (1989).

Dowdle, W. R., and Robinson, R. Q. Non-specific hemadsorption by rhesus monkey kidney cells. *Proc. Soc. Exp. Biol. Med.* 121:193–98 (1966).

Espy, M. J.; Smith, T. F.; Harmon, M. W.; and Kendal, A. P. Rapid detection of influenza virus by shell vial assay with monoclonal antibodies. *J. Clin. Microbiol.* 24:677–79 (1986).

Frank, A. L.; Couch, R. B.; Griffis, C. A.; and Baxter, B. D. Comparison of different tissue cultures for isolation and quantitation of influenza and parainfluenza viruses. *J. Clin. Microbiol.* 10:32–36 (1979).

Galinski, M. S. Paramyxoviridae: Transcription and replication. *Adv. Virus Res.* 39:129–62 (1991).

Grandien, M.; Petter, C. A.; Gardner, P. S.; Linde, A.; and Stanton, A. Rapid viral diagnosis of acute respiratory infections: Comparison of enzyme-linked immunosorbent assay and the immunofluorescence technique for detection of viral antigens in nasopharyngeal secretions. *J. Clin. Microbiol.* 22:751–60 (1985).

Hirst, G. K. The agglutination of red cells by allantoic fluid of chick embryos infected with influenza virus. *Science* 94:22–23 (1941).

Hsiung, G .D. Parainfluenza 5 virus: Infection in man and animals. *Progr. Med. Virol.* 14:241–74 (1972).

Kilbourne, E. D., ed. *Influenza*. New York: Plenum, 1987.

McQuillin, J.; Madeley, C. R.; and Kendal, A. P. Monoclonal antibodies for the rapid diagnosis of influenza A and B virus infections by immunofluorescence. *Lancet* 2:911–14 (1985).

Minnich, L. L., and Ray, G. C. Early testing of cell cultures for detection of hemadsorbing viruses. *J. Clin. Microbiol.* 25:421–22 (1987).

Monto, A. S.; Moassab, H. F.; and Bryan, E. R. Relative efficacy of embryonated eggs and cell culture for isolation of contemporary influenza viruses. *J. Clin. Microbiol.* 13:233–35 (1981).

Olsen, M. A.; Shuck, K. M.; Sambol, A. R.; Flor, S. M.; O'Brien, J.; and Cabrera, B. J. Isolation of seven respiratory viruses in shell vials: A practical and highly sensitive method. *J. Clin. Microbiol.* 31:422–524 (1993).

Palese, P., and Schulman, J. L. Mapping of the influenza virus genome: Identification of the hemagglutination and the neuraminidase genes. *Proc. Natl. Acad. Sci. U.S.A.* 73:2142–46 (1976).

Palese, P., and Young, J. F. Variation of influenza A, B, and C viruses. *Science* 215:1468–74 (1982).

Pothier, P.; Denoyel, G. A.; Ghim, S.; Prudhomme de Saint Maur, G.; and Freymuth, F. Use of monoclonal antibodies for rapid detection of influenza A virus in nasopharyngeal secretions. *Europ. J. Clin. Microbiol.* 5:336–39 (1986).

Ray, C. G., and Minnich, L. L. Efficiency of immunofluorescence for rapid detection of common respiratory viruses. *J. Clin. Microbiol.* 25:355–57 (1987).

Swenson, P. D., and Kaplan, M. H. Rapid detection of influenza virus in cell culture by indirect immunoperoxidase staining with type-specific monoclonal antibodies. *Diagnos. Microbiol. Infect. Dis.* 7:265–68 (1987).

Vainionpaa, R.; Marusyk, R.; and Salmi, A. The Paramyxoviridae: Aspects of molecular structure, pathogenesis and immunity. *Adv. Virus Res.* 37:211–42 (1989).

Walls, H. H.; Johansson, K. H.; Harmon, M. W.; Halonen, P. E.; and Kendal, A. P. Time-resolved fluoroimmunoassay with monoclonal antibodies for rapid diagnosis of influenza infections. *J. Clin. Microbiol.* 24:907–12 (1986).

Warner, J. L.; Whitehurst, N. J.; Downs, T.; and Graves, D. G. Production of monoclonal antibodies against parainfluenza 3 virus and their use in diagnosis by immunofluorescence. *J. Clin. Microbiol.* 22:535–38 (1985).

16

OTHER MEMBERS OF PARAMYXOVIRIDAE

G. D. HSIUNG

Measles and respiratory syncytial (RSV) viruses are classified in the Paramyxoviridae family as morbillivirus (measles virus) and pneumovirus (RSV) (table 16.1). They resemble the parainfluenza viruses morphologically, but biologically and antigenically, they are distinctly different. These two viruses are similar in size and shape to the parainfluenza viruses. Although measles virus hemagglutinates monkey erythrocytes, no hemagglutinin has been described for RSV. Furthermore, no neuraminidase can be detected in the measles virus.

MEASLES VIRUS

Measles is an exanthematous disease that has a high morbidity rate in young children and a high mortality rate among infants as a result of a secondary bacterial infection particularly due to gram positive cocci.

Like the parainfluenza viruses, the measles virus is a single-stranded RNA virus consisting of a nucleoprotein helix enclosed by a lipoprotein envelope (figure 16.1). The size of the measles virion ranges from 150 to 300 nm. Virus infectivity is inactivated by ether. The infectious virus is relatively unstable with respect to temperature and to pH below 5 or above 10.

The measles virus can be cultivated in a variety of primate and nonprimate cells. Characteristic multinucleated giant cells are observed in infected cultures both in MK and serial cell lines. Although measles virus–infected culture fluids contain hemagglutinin and react with erythrocytes from simian primates, including patas and green monkeys, they do not agglutinate

ISBN 0-300-05845-4.

Table 16.1 Biologic Properties of Measles and Respiratory Syncytial Viruses of Paramyxoviridae

Genus (Virus)	Cell Culture System	CPE	Eosinophilic Inclusion	Hemagglutinin in Tissue Culture Fluid	Identification Method of Choice
Morbillivirus (measles)	MK and HEK	Multinucleated syncytial cells	Intracytoplasmic and intranuclear	+*	IF, HI, NT
	Hep-2, A549	Syncytial cells	–	+	
Pneumovirus (respiratory syncytial)	Hep-2, MK	Multinucleated syncytial cells	Occasionally intracytoplasmic	–	IF
	A549	Syncytial cells	+	–	

\+ virus-induced change; – no change
*Monkey erythrocytes only; no enzymatic reaction can be detected

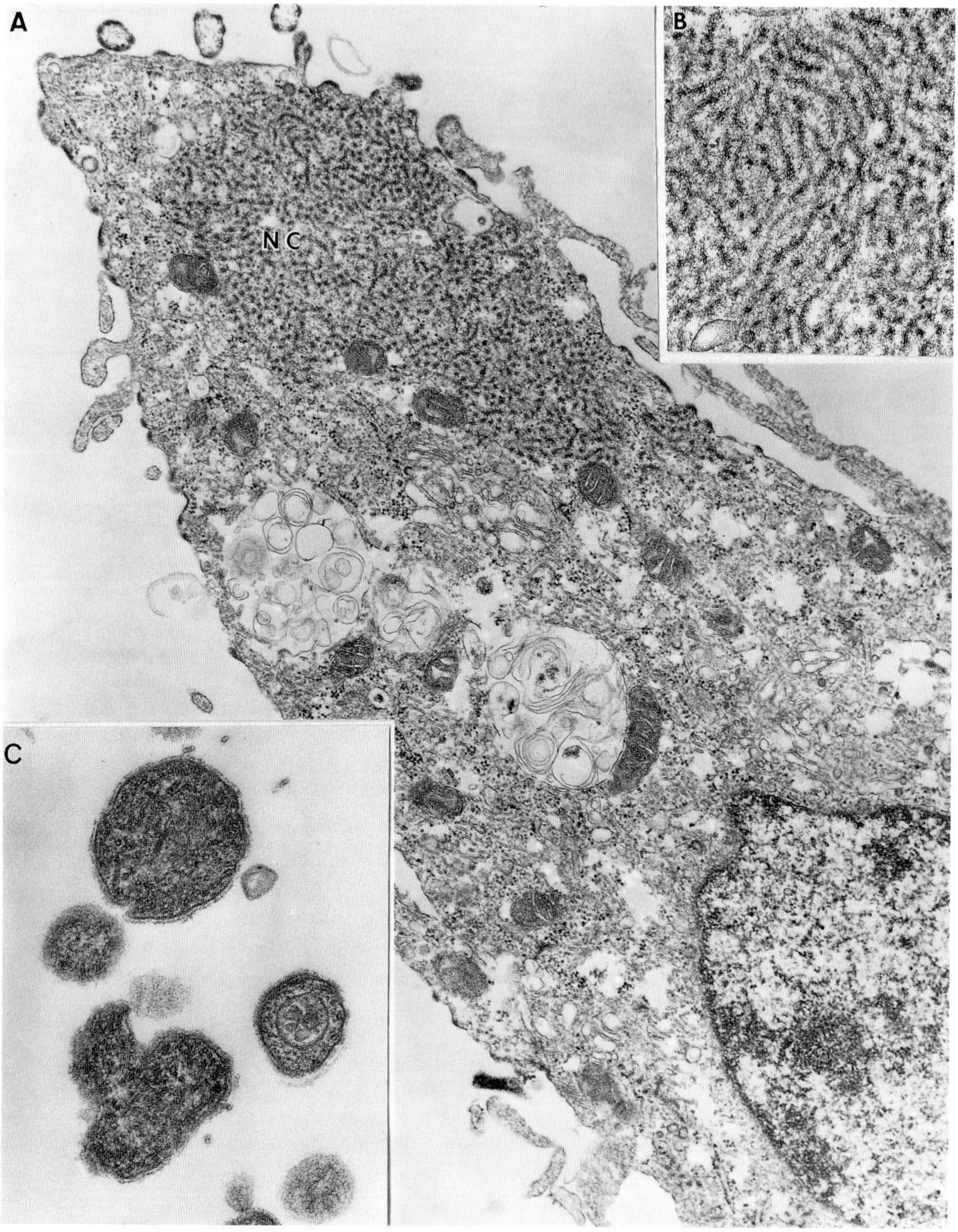

Figure 16.1 Electron micrograph of a measles virus–infected Vero cell: *(A)* Measles virus inclusion containing helical nucleocapsids (NC) in the cytoplasm (19,680×); *(B)* High magnification of nucleocapsids (30,000×); *(C)* Measles virions in the extracellular space (48,000×)

nonsimian RBC, including RBC from humans. This characteristic is useful for separating the measles virus from other paramyxoviruses. In hematoxylin-eosin-stained preparations of primary MK or HEK cultures infected with measles virus, eosinophilic intranuclear and intracytoplasmic inclusions are easily seen. The measles virus induces characteristic multinucleated giant cells in Hep-2 cell cultures, including cells in mitosis. The sensitivity of passaged cell lines, however, may vary from time to time.

Specimen Source and Inoculation

Nasopharyngeal secretions (NPS), throat washings, urine, and whole blood or buffy coat of patients, if specimens are collected not later than 24 hours after the appearance of a rash, have been used for virus isolation. Whole blood specimens should be processed within six hours after collection or stored at room temperature. Separate leukocytes from heparinized blood for inoculation. Primary African green monkey or rhesus monkey kidney cells, human embryonic kidney cells (if available), and A549 cell cultures are susceptible host systems for primary virus isolation—although laboratory strains can grow in the Hep-2 and Vero cell lines. Both tube and shell vial cultures are each inoculated with specimens. Tube cultures are inoculated after removing the medium, allowing one hour adsorption, then refeeding with a maintenance medium containing 2% fetal bovine serum. Shell vial cultures are inoculated after removing the medium, then centrifuged at 700 × *g* for one hour. Refeed with a similar maintenance medium. Recent reports indicate that measles virus can be isolated more efficiently by the cocultivation of the patient's peripheral blood mononuclear cells with phytohemagglutinin (PHA)-stimulated peripheral blood mononuclear cells or B95–8 cells (EB virus–transformed marmoset lymphocytes). in shell vials.

Virus Isolation, Propagation, and Assay

CYTOPATHIC EFFECT IN CELL CULTURES WITH FLUID MEDIUM

The measles virus induces typical foamylike cellular changes in primary human kidney or MK cell cultures (figure 16.2*B*). Human embryonic kidney cultures are apparently the most suitable host system for primary virus isolation. Cultures should be maintained as long as possible, for four to six weeks, with frequent changes of medium. It is also advisable to transfer infected culture fluids, diluted 1:10, to fresh culture tubes at weekly intervals, carrying them through several blind passages before a negative result is reported. After laboratory adaptation, the measles virus also replicates in Hep-2 cultures, producing multinucleated syncytial cells (figure 16.3).

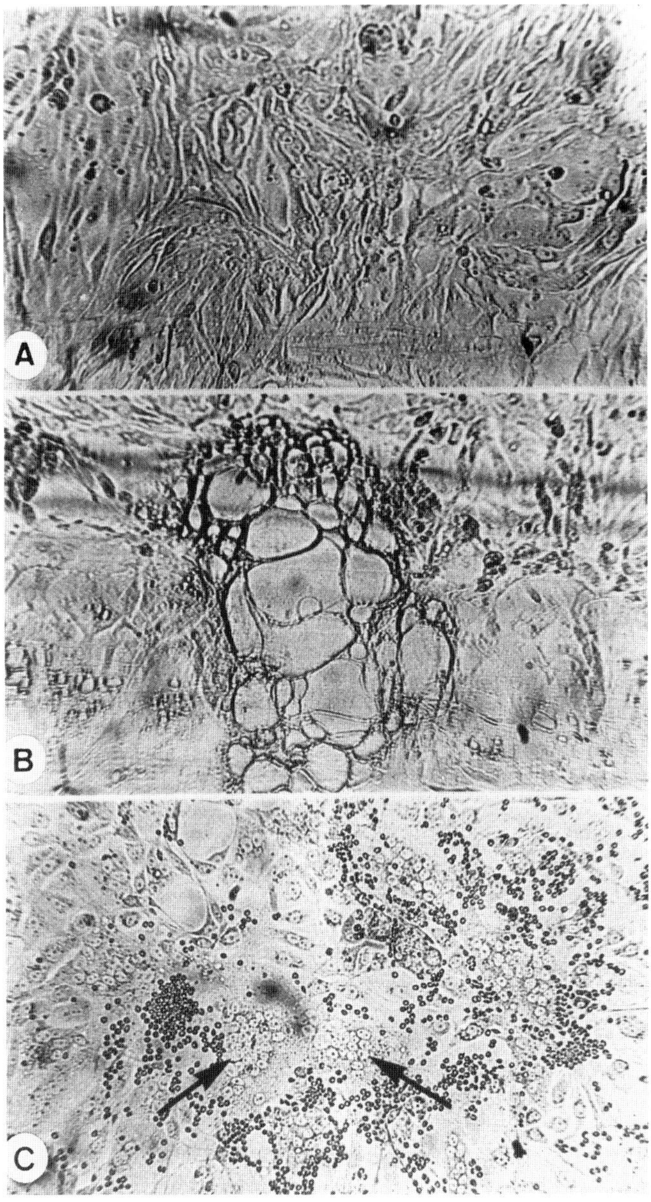

Figure 16.2 Syncytial cell formation and hemadsorption of monkey RBC by measles virus–infected RhMK cell cultures (100×): *(A)* Uninfected RhMK culture; *(B)* Foamy appearance of CPE caused by measles virus; *(C)* Multinucleated cells (arrows) and hemadsorption of monkey RBC by measles virus–infected cells

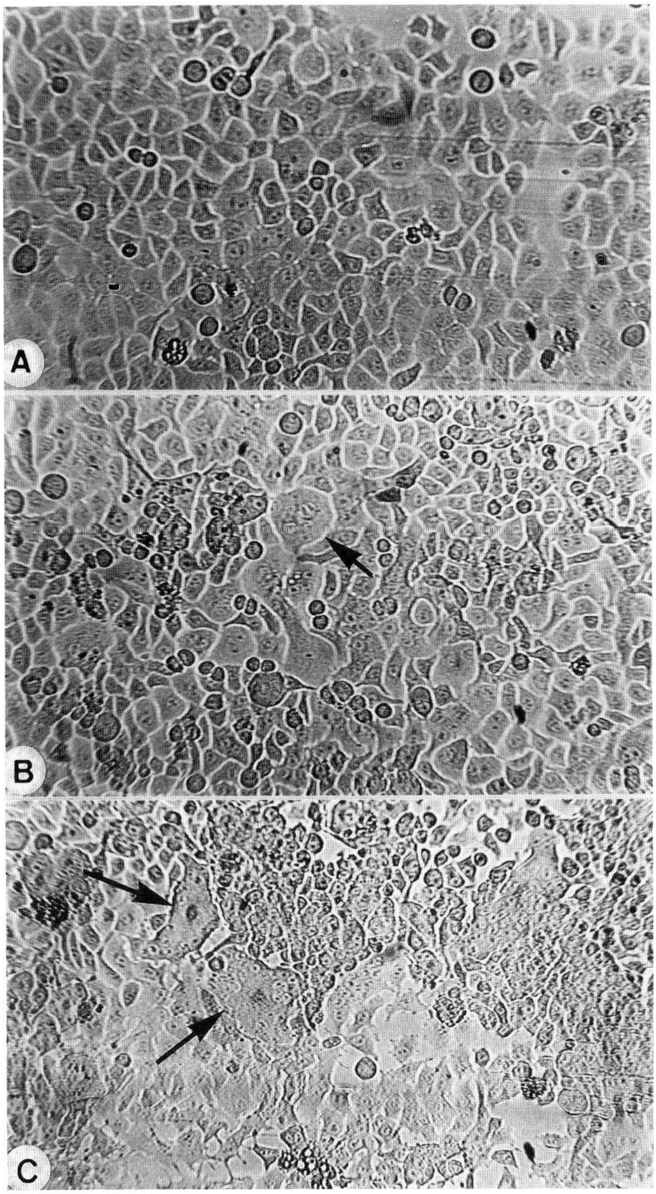

Figure 16.3 Syncytial cell formation induced by measles virus in Hep-2 cell cultures (100×): *(A)* Uninfected Hep-2 cell culture: *(B)* Early appearance of multinucleated cells (arrow); *(C)* Advanced appearance of multinucleated cells (arrows)

RAPID DETECTION OF MEASLES VIRUS IN SHELL VIAL CULTURES BY IMMUNOFLUORESCENT STAINING

After 24–36 hours of incubation, aspirate the medium from the shell vial cultures before CPE appears. Fix the cells on the coverslip with cold acetone for 5–10 minutes and stain with monoclonal antibody to the measles virus by indirect FA method (see chapter 7). Positive and negative samples are tested in parallel as controls.

HEMADSORPTION IN CELL CULTURES

Measles virus–infected cells show hemadsorption upon the addition of 0.5% monkey RBC (figure 16.2*C*). Since guinea pig and human RBC are not similarly hemadsorbed, this reaction provides a useful indicator that the cultures are infected with the measles virus and not another paramyxovirus.

CYTOPATHOLOGY IN MEASLES VIRUS–INFECTED CELLS

The measles virus induces characteristic intranuclear and intracytoplasmic eosinophilic inclusions in infected primary monkey kidney cell cultures (figure 16.4). These can be demonstrated when infected cells are fixed and stained with hematoxylin-eosin.

PLAQUE FORMATION UNDER AGAR OVERLAY

The measles virus produces distinct pinpoint plaques in primary monkey kidney cell cultures. Although this method is not used for routine isolation purposes, it can be used for a quantitative assay of the measles virus.

Identification of Measles Virus Isolates

The immunofluorescent staining technique is the most rapid method for the identification of virus isolates in cell cultures even without evidence of CPE. The availability of monoclonal antibody to the measles virus facilitates greatly the identification of measles-infected cells and has been used successfully for the rapid detection of measles in shell vial cultures with A549 cells, as stated above (see color plate 9).

NEUTRALIZATION TEST IN CELL CULTURE

The test can be set up in primary MK, A549, or in Hep-2 tube cultures with 100 $TCID_{50}$ as the challenge dose against a known antiserum. The measles virus induces CPE slowly, and there is a direct relation between dilution of virus and the degeneration time of cells in culture. Thus, the observation period should range from two to three weeks, especially when doses of less than 100 $TCID_{50}$ are used. Since CPE usually develops very slowly, the more sensitive hemadsorption-inhibition method described in the chapter on

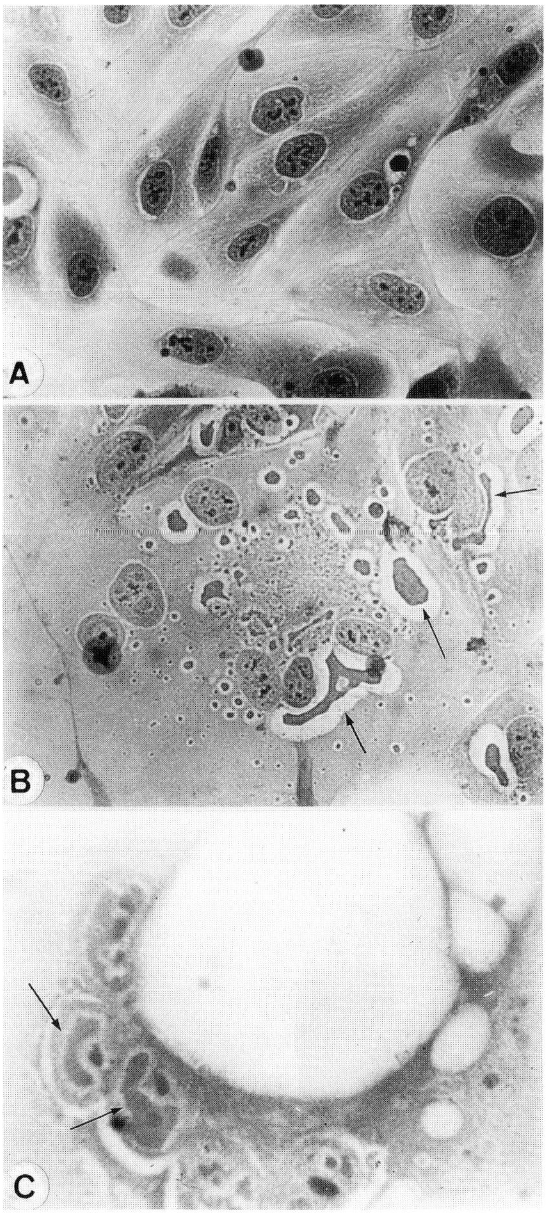

Figure 16.4 Eosinophilic intracytoplasmic and intranuclear inclusions induced by measles virus in HEK cell cultures (H&E): *(A)* Uninfected HEK cell culture (400×); *(B)* Intracytoplasmic inclusions (arrows) induced by measles virus (400×); *(C)* Intranuclear inclusions (arrows) induced by measles virus (970×)

Paramyxoviridae (chapter 15), and the plaque-reduction neutralization method (chapter 6) can be used to identify measles virus isolates, if necessary.

HEMAGGLUTINATION, HEMAGGLUTINATION-INHIBITION, AND HEMOLYSIN PRODUCTION BY MEASLES VIRUS

The hemagglutinin (HA) titer in measles virus–infected cultures is often low. Several methods have therefore been used to concentrate the antigen in order to yield a sufficiently high titer of HA. Monkey erythrocytes, preferably from African green monkeys, resuspended to 0.5% in saline are used for the measles virus HA titrations. The cells are allowed to settle at 37°C for one hour. If hemagglutination is observed, the HI test can be used for the final identification of the virus. Provided the HA antigen is sufficiently potent, further incubation at 37°C for four hours may demonstrate specific hemolysin (for HA and HI procedures, see chapter 9).

ELECTRON MICROSCOPY

The association of the measles virus with subacute sclerosing panencephalitis (SSPE) was recognized through the visualization of nucleocapsids with characteristic measles virus morphology in brain biopsy materials following thin sectioning. Electron microscopic examination of measles virus–infected cells reveals intranuclear and intracytoplasmic nucleocapsids (see figure 16.1) that correspond to the viral inclusions seen by light microscopy. Extracellular virions can be seen occasionally (figure 16.1*C*).

Viral Antigen and Genome Detection

Since virus isolation is time-consuming and difficult, recent reports indicate that the direct detection of viral antigen in nasal secretions by immunofluorescence can be made for rapid diagnosis. Monoclonal antibody against the measles virus is commercially available (for details on immunofluorescent staining see chapter 7).

Detection of measles virus genomic sequences in SSPE brain tissues by PCR has been reported.

Serodiagnosis

Acute and convalescent sera are used for serodiagnostic purposes. ELISA, HI, CF, IF, and neutralization tests can all be used in determining the measles virus antibody. The HI test remains the most widely used method for measuring specific antibodies.

HEMAGGLUTINATION-INHIBITION TEST

Since human serum contains natural agglutinins for monkey RBC, the human sera being tested must be adsorbed with 50% monkey RBC in saline at 4°C overnight before use. The sera should also undergo kaolin treatment. Virus-serum mixtures are incubated at room temperature for one hour before the addition of monkey RBC, 0.5% in saline. For more sensitive results, the mixtures are incubated at 37°C for one hour after the addition of RBC. (For the HI procedure, see chapter 9).

RESPIRATORY SYNCYTIAL VIRUS

RSV has been associated with upper and lower respiratory tract illnesses, especially in infants and young children; it occasionally affects adults. Infections are generally severe in the former and very mild in the latter. RSV is a significant cause of hospitalization among infected infants and young children.

RSV is a pleomorphic virus, spherical, and with a diameter of 150–300 nm, similar to the measles virus (figure 16.5). It does not multiply in embryonated eggs but grows in cell cultures of a variety of animal species as well as in several cell lines, including monkey kidney, Hep-2, A549, and HL cell cultures. The virus envelope surrounds the helical nucleocapsid. In cell cultures, the virus produces characteristic syncytial cells (figure 16.6), from which its name is derived. Neither hemagglutinin nor neuraminidase activity has yet been discovered. Stained preparations of infected cells reveal eosinophilic intracytoplasmic inclusions.

Specimen Source and Inoculation

RSV is best recovered from nasopharyngeal aspirates or nasal washings collected as early as possible after the onset of illness. RSV is extremely labile and poorly withstands freezing and thawing. Specimens for virus isolation should therefore be inoculated directly into tube or shell vial cultures when received by the laboratory. Before inoculation, remove the culture medium, allow adsorption for tube cultures or perform centrifugation for shell vial cultures, then refeed with a maintenance medium.

Virus Isolation and Identification

For RSV isolation, Hep-2 cell cultures at the subconfluent stage and primary RhMK cell cultures are very sensitive. The use of shell vial centrifugation cultures for virus isolation is advisable, especially when specimens tested are negative by direct FA or ELISA methods. Methods for isolation and identification

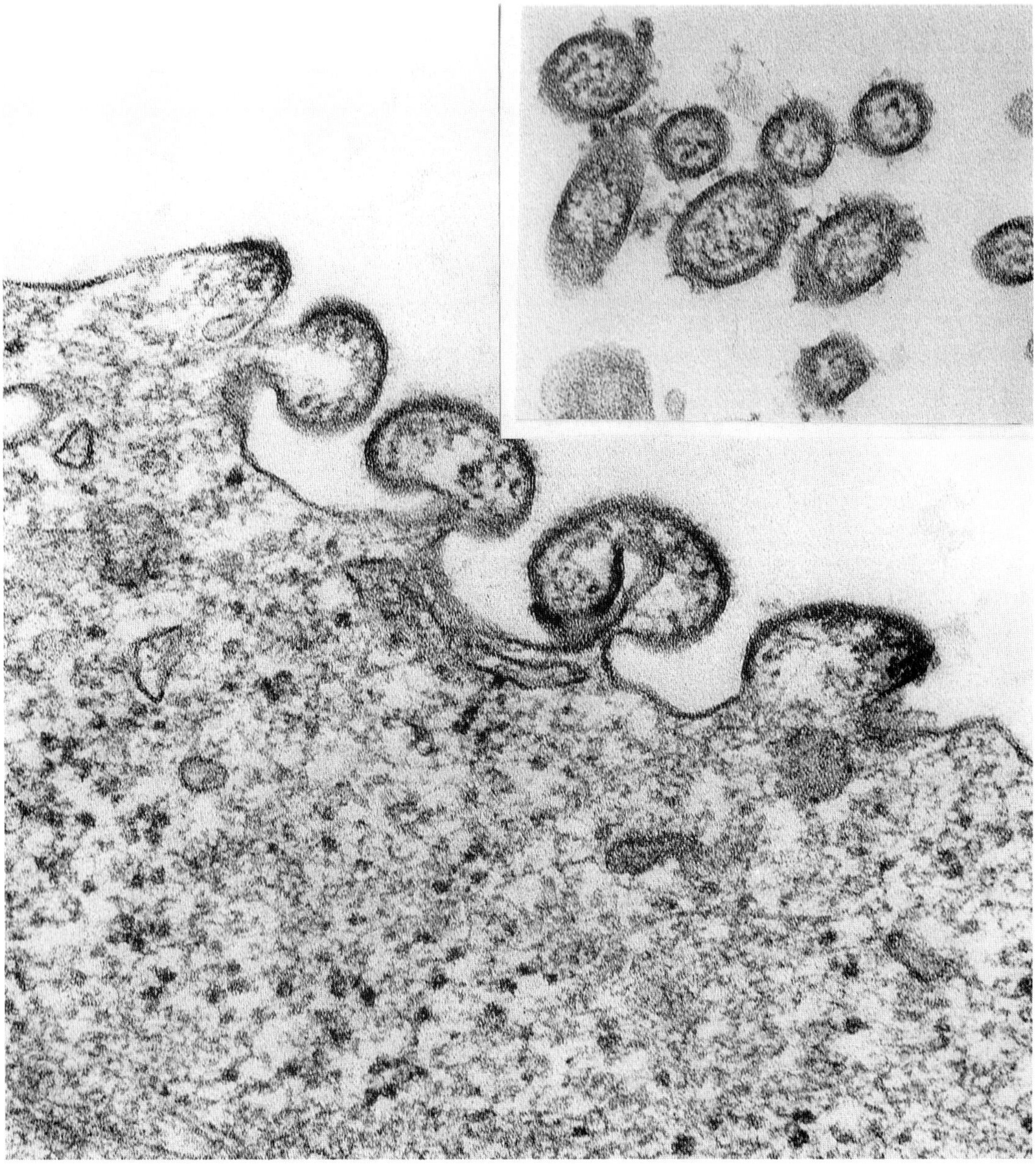

Figure 16.5 Electron micrograph of respiratory syncytial virus (RSV) in an A549 cell showing several virions in the process of budding at the cell membrane (100,800×). Inset: Cross section of extracellular virions (100,800×)

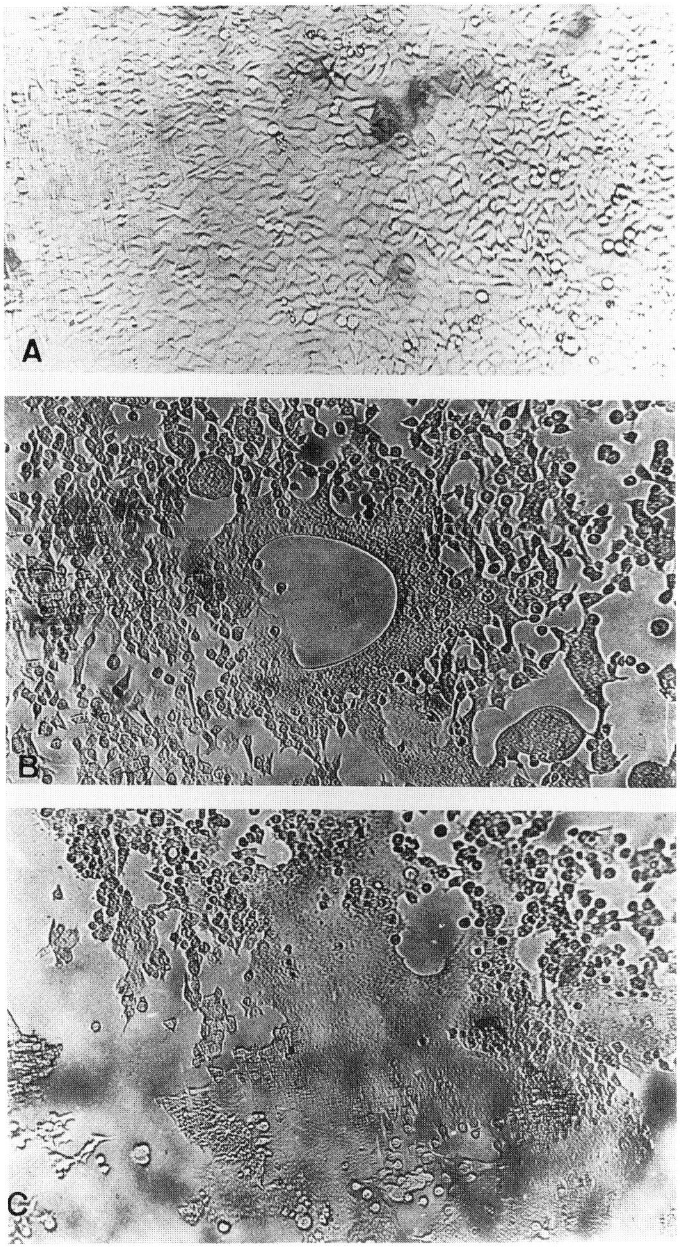

Figure 16.6 Syncytial cell formation induced by RSV in Hep-2 cell cultures (100×): *(A)* Uninoculated cell control; *(B)* RSV-infected cells at three days; *(C)* RSV-infected cells at five days

are similar to those described for the measles virus, except that Hep-2 monolayer cell cultures are the most sensitive host systems for RSV isolation. The characteristic syncytial formation is most easily detected in Hep-2 cells, provided glutamine is supplemented in the medium.

RSV can be rapidly identified by the immunofluorescent staining technique using monoclonal antibody to the virus (for the procedure, see chapter 7; see color plate 11). Virus identification can be made by the neutralization test in the cell culture, but it is tedious.

Rapid Detection of RSV Antigen in Nasopharyngeal Secretions

The rapid detection of RSV antigen directly in clinical specimens of young children is important for clinical management and the appropriate use of antiviral drugs. Currently, the demonstration of RSV antigens in specimens of nasopharyngeal secretions using a direct immunofluorescent test is the most successful method for the rapid diagnosis of RSV infections using monoclonal antibody. The specimen quality, however, is of the utmost importance for adequate IF test, which requires a sufficient number of appropriate cells from the site of infection. In addition, detection of RSV by ELISA tests (Testpack RSV and Ortho RSV) are also rapid and specific, although the sensitivity may not be comparable to that of IF test. Both IF and ELISA tests for antigen detection in clinical specimens do not require infectious virus, and the results are available on the same day. Nevertheless, there is a continuing need for RSV isolation using sensitive cell cultures to ensure maximum detection of the virus. As such, it continues to be a valuable adjunct in the clinical laboratory.

Serodiagnosis

An ELISA has been developed for the detection of serum antibodies against RSV. Antibody titers detected by ELISA are 100 times higher than those tested by CF, and two to four times higher than the plaque reduction neutralization test.

SUPPLEMENTARY READING FOR MEASLES VIRUS

Black, F. L. Measles. In Evans, A. S., ed., *Viral Infections of Humans, Epidemiology and Control,* pp. 451–69. New York: Plenum, 1989.

Forthal, D. N.; Blanding, J.; Aarnaes, S.; Peterson, E. M.; dela Maza, L.; and Tilles, J. G. Comparison of different methods and cell lines for isolating measles virus. *J. Clin. Microbiol.* 31:695–97 (1993).

Godec, M. S.; Asher, D. M.; Swoveland, P. T.; Eldadah, Z. A.; Feinstone, S. M.; Goldfarb, L. G.; Gibbs, C. J.; and Gajdusek, D. C. Detection of measles virus genomic sequences in SSPE brain tissue by the polymerase chain reaction. *J. Med. Virol.* 30:237–44 (1990).

Kao, C. L.; Lin, J. H.; Lee, C. N.; and Lee, C. Y. Rapid diagnosis of measles virus infection with immunofluorescence staining technique. *Formosan Med. Assoc.* 86:222–26 (1987).

Kleiman, M. B.; Blackburn, C. K. L.; Zimmerman, S. E.; and French, M. L. V. Comparison of enzyme-linked immunosorbent assay for acute measles with hemagglutination inhibition, complement-fixation, and fluorescent antibody methods. *J. Clin. Microbiol.* 14:147 (1981).

Minnich, L. L.; Goodenough, F.; and Ray, G. G. Use of immunofluorescence to identify measles virus infections. *J. Clin. Microbiol.* 29:1148–50 (1991).

Olding-Stenkvist, E., and Bjorvatn, E. Rapid detection of measles virus in skin rashes by immunofluorescence. *J. Infect. Dis.* 134:463–69 (1976).

Smaron, M. F.; Saxon, E.; Wood, L.; McCarthy, C.; and Morello, J. A. Diagnosis of measles by fluorescent antibody and culture of nasopharyngeal secretions. *J. Virol. Meth.* 33:223–29 (1991).

SUPPLEMENTARY READING FOR RESPIRATORY SYNCYTIAL VIRUS

Anestad, G.; Breivik, N.; and Thorasen, T. Rapid diagnosis of respiratory syncytial virus and influenza A virus infection by immunofluorescence: Experience with a simplified procedure for the preparation of cell smears from nasopharyngeal secretion. *Acta Path. Microbiol. Immun. Scand. (B)* 91:267–71 (1983).

Bromberg, K.; Tannis, G.; Daidone, B.; Clarke, L.; and Sierra, M. Comparison of Hep-2 cell culture and Abbott respiratory syncytial virus enzyme immunoassay. *J. Clin. Microbiol.* 25:434–36 (1987).

Chao, R. K.; Fishaut, M.; Schwartzman, J. D.; and McIntosh, K. Detection of respiratory syncytial virus in nasal secretions from infants by enzyme-linked immunosorbent assay. *J. Infect. Dis.* 139:483–86 (1979).

Freymuth, F.; Quibriac, M.; Petitjean, J.; Amiel, M. L.; Pothier, P.; Denis, A.; and Duhamel, J. F. Comparison of two new tests for rapid diagnosis of respiratory syncytial virus infections by enzyme-linked immunosorbent assay and immunofluorescence technique. *J. Clin. Microbiol.* 24:1013–16 (1986).

Halstead, D. C.; Todd, S.; and Fritch, G. Evaluation of five methods for respiratory syncytial virus detection. *J. Clin. Microbiol.* 28:1021–25 (1990).

Kao, C. L.; McIntosh, K.; Fernie, B.; Talis, A.; Pierik, L.; and Anderson, L. Monoclonal antibodies for the rapid diagnosis of respiratory syncytial virus infection by immunofluorescence. *Diagn. Microbiol. Infect. Dis.* 2:199–206 (1984).

Kaul, T. N.; Welliver, R. C.; and Ogra, P. L. Comparison of fluorescent-antibody, neutralizing-antibody, and complement-enhanced neutralizing antibody assays for detection of serum antibody to respiratory syncytial virus. *J. Clin. Microbiol.* 13:957–62 (1981).

Marquez, A., and Hsiung, G. D. Influence of glutamine on multiplication and cytopathic effect of respiratory syncytial virus. *Proc. Soc. Exp. Biol. Med.* 124:95–99 (1967).

Masters, H. B.; Weber, K. O.; Groothuis, J. R.; Wren, C. G.; and Lauer, B. A. Comparison of nasopharyngeal washings and swab specimens for diagnosis of respiratory syncytial virus by EIA, FAT, and cell culture. *Diagn. Microbiol. Infect. Dis.* 8:101–5 (1987).

Ostler, E.; Egeres, R. M.; and Wiedbrauk, D. L. Improved isolation of respiratory syncytial virus using HL cells. Eighth Annual Clin. Virol. Sym. 1992.

Ray, C. G., and Minnich, L. Efficiency of immunofluorescence for rapid detection of common respiratory viruses. *J. Clin. Microbiol.* 25:355–57 (1987).

Smith, M. C.; Creutz, C.; and Huang, Y. T. Detection of respiratory syncytial virus in nasopharyngeal secretions by shell vial techniques. *J. Clin. Microbiol.* 29:463–65 (1991).

Subbarao, E. K.; Whitehurst, N. J.; and Waner, J. L. Comparison of two enzyme-linked immunosorbent assay (EIA) kits with immunofluorescence and isolation in cell culture for detection of respiratory syncytial virus (RSV). *Diagn. Microbiol. Infect. Dis.* 8:229–34 (1987).

Takimoto, S.; Grandien, M.; Ishida, M. A.; Pereira, M. S.; Paiva, T. M.; Ishimaru, T.; Makita, E. M.; and Martinez, C. H. Comparison of enzyme-linked immunosorbent assay, indirect immunofluorescence assay and virus isolation for detection of respiratory virus in nasopharyngeal secretions. *J. Clin. Microbiol.* 29:470–74 (1991).

Thomas, E. E., and Book, L. E. Comparison of two rapid methods for detection of respiratory syncytial virus (RSV) (Testpack RSV and Ortho RSV ELISA) with direct immunofluorescence and virus isolation for the diagnosis of pediatric RSV infection. *J. Clin. Microbiol.* 29:632–35 (1991).

Welliver, R. C. Detection, pathogenesis and therapy of respiratory syncytial virus infections. *Clin. Microbiol. Rev.* 1:27–39 (1988).

17

RUBELLA VIRUS

G. D. HSIUNG

Rubella virus is classified as a nonarthropod-borne virus in the Togaviridae family, being the only member of the genus *Rubivirus*. It is the etiologic agent of German measles, a common and benign disease of children. Clinically, the patient starts with mild upper-respiratory symptoms followed by enlarged lymph nodes of the neck and finally a generalized rash that lasts two to three days. This disease, however, is of great significance when it occurs in pregnant women, especially during the first trimester, since it frequently gives rise to serious defects in the infant, known as congenital rubella syndrome (CRS).

The rubella virion is a spherical RNA virus and is approximately 60 nm in diameter (figure 17.1). Negative staining preparations reveal an envelope, but the nucleocapsid symmetry is obscure. Small spikes, 5–6 nm in length, project from the outer envelope. These contain hemagglutinin. Like other viruses with lipid envelopes, the rubella virus is rapidly inactivated by ether, chloroform, or sodium deoxycholate. The virus is relatively labile but can be stored at −70°C for prolonged periods.

The rubella virus replicates in a variety of primary cell cultures and passaged diploid cell strains, but in general cellular degeneration is not readily observed in the infected cultures owing to the induction of interferon. Thus, in infected primary African green monkey kidney cell cultures, evidence of virus growth can be determined by an interference phenomenon, that is, interference with the growth of a challenge virus. Cytopathic effect, however, has been observed in several other cell systems infected with the rubella virus. These include a baby hamster kidney cell line (BHK-21), a rabbit kidney cell line (RK-13), and an African green monkey kidney cell line

ISBN 0-300-05845-4.

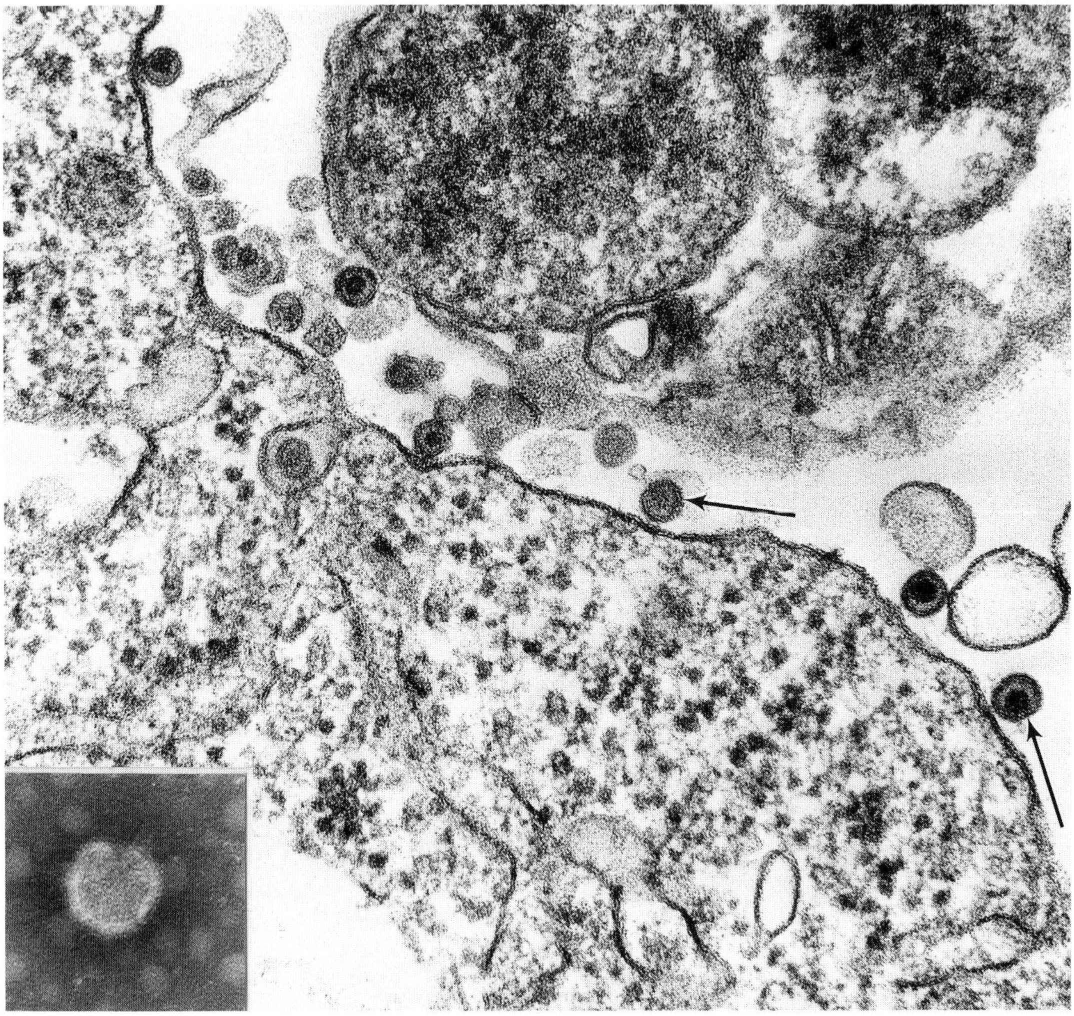

Figure 17.1 Electron micrograph of a rubella virus–infected Vero cell showing mature virions at the extracellular space (100,800×). Inset: Virions stained with PTA (168,000×)

(Vero cells) because these cells do not produce interferon efficiently. After two or three passages in BHK-21 cell cultures, a hemagglutinin titer can be detected in the infected culture fluid. The ability of cell lines to demonstrate rubella virus–induced CPE varies greatly from cell line to cell line and from laboratory to laboratory.

Specimen Source and Inoculation

Throat swabs or washings, acute phase blood, cerebrospinal fluid, urine, amniotic fluid, and placental and fetal tissues are all useful in isolating the rubella virus. Primary African green monkey kidney cell cultures, RK-13, and BHK-21 cells are commonly used to inoculate specimens, although other cell lines, including Vero and BSC-1 green monkey kidney cell lines, can also be used.

Virus Isolation, Propagation, and Assay

INTERFERENCE IN AFRICAN GREEN MONKEY KIDNEY CELL CULTURES

Primary AGMK cultures infected with rubella virus show no evidence of CPE. A rubella virus–infected culture, however, will not support the growth of added echovirus type 11 as a challenge virus. This interference phenomenon was one of the methods originally used for the detection of rubella virus, as the inhibition of echovirus 11–induced CPE indicates the presence of an interfering agent. Other virus types, including poliovirus 1, coxsackie virus A-9, vesicular stomatitis virus (VSV), and Newcastle disease virus (NDV), have also been used as challenge viruses. Despite the difficulties encountered with the interference method, which requires an additional step for detection, the interference phenomenon in primary AGMK cultures is still the most reliable and sensitive procedure for detecting the presence of the rubella virus.

CYTOPATHIC EFFECT IN RABBIT KIDNEY CELLS

Primary rabbit kidney cells or passaged cell lines have been used in the isolation of the rubella virus: CPE has been observed in RK-13 cell cultures (figure 17.2). The microplaques produced are more easily seen under a phase-contrast microscope. The appearance of distinct CPE often requires seven days or more of cultivation, and occasionally two or three serial blind passages of culture fluid may be necessary.

PLAQUE FORMATION IN PASSAGED CELL LINES

The rubella virus induces clear sharp plaques in Vero cells and RK-13 cell cultures. Plaques appear seven to eight days after infection; they are round with ragged edges and measure 2–3 mm in diameter, increasing in size after prolonged incubation. This method is not used for primary virus isolation but is used frequently for virus assay and antibody detection.

ISOLATION OF VIRUS FROM FETAL TISSUES

Isolation of the rubella virus from fetal specimens is most successfully accomplished by the cultivation or cocultivation of intact infected cells. The fetal tissues are trypsinized to obtain a cell suspension, and the intact cells are either seeded in a growth medium for propagation into cell sheets or are cocultivated with susceptible cell monolayers. The cultured cells and fluids thus obtained provide good sources for virus stocks.

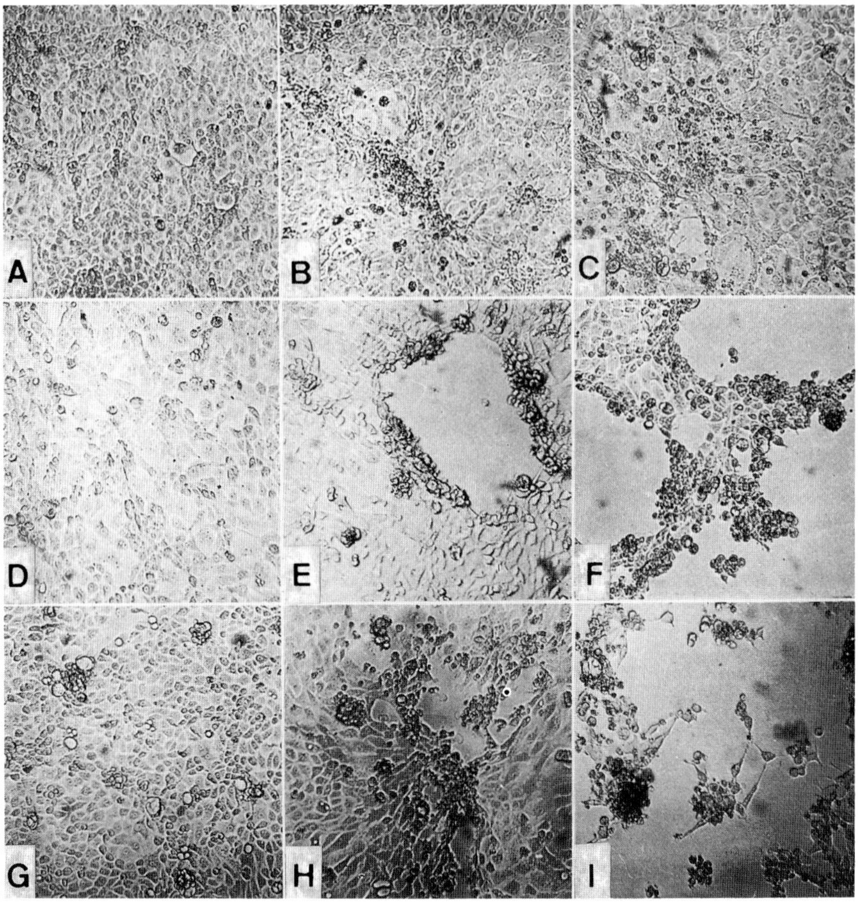

Figure 17.2 Cytopathic effect induced by rubella virus in rabbit kidney cell line cultures (100×). (Reproduced with permission, J. Furesz et al., *Can. J. Microbiol.* 15:71 [1969])

A–C. RK-13 microcultures: *(A)* Uninoculated cultures; *(B)* five days; *(C)* eight days after inoculation with HP-78 rubella virus

D–F. RK1-H microcultures: *(D)* Uninoculated cultures; *(E)* five days; *(F)* eight days after inoculation with the same virus

G–I. RK1-FC microcultures. *(G)* Uninoculated cultures; *(H)* five days; *(I)* eight days after inoculation with the same virus

Identification of Isolates

IMMUNOFLUORESCENCE OR IMMUNOPEROXIDASE STAINING METHOD

These methods provide more rapid alternatives to the neutralization test.

1. Subculture rubella virus isolates into shell vials containing coverslips with the same cell type in which they were isolated.

2. Five to 7 days after inoculation, remove culture fluids and air dry the infected cells on coverslips at room temperature.
3. Fix the infected cells on coverslips with cold acetone. Air dry at room temperature. (These fixed cells can be stored at −60°C until used.)
4. Before the test, inactivate known immune and preimmune rabbit sera at 56°C for 30 minutes and adsorb with 20% normal mouse brain to reduce nonspecific staining.
5. For the identification of rubella virus, the indirect IP or IF methods (chapter 7) can be used. The fluorescent or enzyme-conjugated anti–rabbit globulin prepared in goat can be obtained commercially.

If clinical material is being examined specifically for the rubella virus and the isolate is propagated in Vero or RK-13 cells, identification can be made by immunofluorescence using specific antiserum to rubella virus. Confirmation should be made by subculture of the infected material into primary AGMK cell cultures. The interference neutralization test in the same cell line is performed using echovirus 11 as the challenge virus.

NEUTRALIZATION TEST IN TISSUE CULTURE

This test is the most specific method for the identification of a suspected rubella virus isolate.

1. The isolate is diluted in serial 10-fold dilutions up to 10^{-6}, and each dilution is inoculated into three tubes, 0.1 ml per tube, for the determination of infectivity titers.
2. Add an equal volume of a constant amount of known rubella antiserum to each dilution of virus and allow the mixture to stand at 37°C for 1 hour.
3. Inoculate 0.2 ml of each mixture into a culture tube, three to four tubes per dilution.
4. If the cell system is one in which rubella virus causes CPE, such as RK-13 or Vero cell, of which interferon is not induced, the inoculated cultures are read at intervals until CPE appears in cultures that are infected with rubella virus only (without antiserum).
5. If an interferon-producing cell system is used, such as the AGMK cell cultures, the challenge virus is added routinely at 7 days after the inoculation of virus-antiserum mixtures. A 100–1,000 $TCID_{50}$ of echovirus 11 is commonly used as a challenge dose and is added to each culture tube, with or without rubella antiserum. A positive neutralization test is indicated when culture tubes inoculated with virus-serum mixture show no interference, that is, when CPE of echovirus 11 is evident.

HEMAGGLUTINATION INHIBITION TEST

In the preparation of rubella HA antigen, it is necessary to remove inhibitors from the serum that is used in the maintenance medium for rubella virus–infected cell cultures. The maintenance medium for BHK-21 cell cultures consists of 2% fetal bovine serum, which can be treated with kaolin to remove nonspecific inhibitors in the serum. Therefore, rubella HA antigen obtained from serum containing BHK-21 cell culture fluid can be extracted directly with Tween 80 and ether treatment. The antigen from rubella-infected BHK-21 cultures maintained with 0.4% bovine albumin can be used without further treatment. Rubella HA antigen is stable indefinitely when stored at −70°C. Procedures for the HA and HI tests are described in chapter 9, but to test for the presence of rubella virus hemagglutinin, 0.5% day-old chick, adult goose, or pigeon RBC in PBS are used (goose RBC give the highest HA titer), and the reaction takes place at 4°C or 22°C, but not at 37°C. Calcium is essential for the agglutination of RBC by the rubella virus. The inhibition of chick RBC agglutination by the virus in the presence of immune serum identifies rubella virus.

ELECTRON MICROSCOPY

Electron microscopic examination of rubella virus–infected cells reveals that virus replication occurs in the cytoplasm and that virion envelopment takes place by budding at the plasma membrane or into cytoplasmic vacuoles (see figure 15.1). Examination of negatively stained virions by electron microscope is not a routine procedure, since the structure of the rubella virus particle in a nonpurified preparation is not easily distinguishable from cellular material.

Serodiagnosis

Since the procedures used for rubella virus isolation can be time-consuming and tedious, serologic tests remain the mainstay in the diagnosis of acquired rubella or congenital rubella infections. Several methods have been established to detect rubella virus antibody, including ELISA, latex agglutination (LA), immunofluorescence, and HI. Rubella antibodies can be demonstrated in a few days after the appearance of the rash, at which time antibodies rise rapidly. Most of the serodiagnostic tests are commercially available in kit form from various suppliers.

HEMAGGLUTINATION-INHIBITION TEST

Until recently, HI in microtiter plates was the method most widely used in the laboratory for the diagnosis of rubella virus infection. The rubella HI antibody titer that accurately correlates with clinical protection has been established, but not for the other "more sensitive" assays, such as ELISA and LA. A rubella HI antibody titer of 1:8–1:10 is evidence of immunity. Although the HI test described in chapter 9 can be applied, the impact of variables on HI test results is often encountered. Thus the type of RBC, the concentration of antigen, the composition of the diluent, the method used for removing nonspecific inhibitors, and the temperature of incubation can all affect the outcome. The following procedures describe the variables that have been identified as critical when adapting the HI test to rubella virus serodiagnosis.

1. Treatment of Serum: Serum samples maybe inactivated at 56°C for 30 minutes, although noninactivated serum can be used. Serum must be treated with kaolin, preferably with heparin-$MnCl_2$ (for formula, see below), to remove the nonspecific β-lipoprotein inhibitors. Adsorption with chick RBC to remove natural cell agglutinin is also recommended.
2. Choice of Erythrocytes: Although RBC from a variety of species can be used, goose RBC are preferred for testing rubella virus hemagglutinin.
3. Composition of Diluent: Hepes (N-2-hydroxyethyl-piperazine-N′ ethane sulphonic acid)-saline-albumin-gelatin (HSAG diluent) at pH 6.2 is used for rubella HA test (formula below). The reaction is enhanced by the presence of 0.001 M $CaCl_2$.

DETERMINATION OF IGM ANTIBODY

Demonstration of specific IgM antibody to rubella virus is a valuable aid in the serologic diagnosis of rubella infection. Several techniques are available to separate IgM. These include adsorption with staphylococcal protein A to remove IgG or sucrose gradient centrifugation to separate IgM from IgG. Additional techniques using IgM antiglobulin are available in the indirect IF or ELISA test.

REAGENTS

$MnCl_2$ 1 M Solution

$MnCl_2 \cdot 4H_2O$	39.6 g
Distilled water, demineralized	200.0 ml

Sterilize by Millipore membrane filtration and store in the dark at 4°C. Discard if brown precipitate appears. To prepare a working solution of heparin-$MnCl_2$ solution, mix equal parts of heparin (5,000 units per ml) and 1 M $MnCl_2$. Store at 4°C and use in two weeks.

Hepes Saline 5X Stock Solution

Hepes	29.80 g
NaCl	40.95 g

$CaCl_2 \cdot 2H_2O$	0.74 g
Distilled water, demineralized	1,000 ml

Sterilize by filtration through Millipore membrane and store at 4°C.

Bovine Serum Albumin 2X Stock Solution

Bovine albumin powder (Fraction v)	20.0 g
Distilled water, demineralized	1,000 ml

Sterilize by filtration through Millipore membrane and store at 4°C.

Gelatin 10X Stock Solution

Gelatin	25.0 g
Distilled water, demineralized	1,000 ml

Dissolve the gelatin and sterilize the solution at 15 lb for 15 minutes. Store at 4°C.

HSAG Diluent Working Solution

Hepes saline 5X stock solution	200 ml
Bovine serum albumin 2X stock solution	500 ml
Gelatin 10X stock solution	100 ml
Distilled water, demineralized	200 ml

Adjust the pH to 6.2. This solution can be stored at 4°C for 2 months.

SUPPLEMENTARY READING

Ankerst, I.; Christensen, P.; Kjellen, L.; and Kronvall, G. A routine diagnostic test for IgA and IgM antibodies to rubella virus: Adsorption of IgG with staphylococcus aureus. *J. Infect. Dis.* 130:268 (1974).

Banatvala, J. E., and Best, J. M. Rubella. In Brown, F., and Wilson, Sir G., eds., *Topley and Wilson's Principles of Bacteriology, Virology and Immunology,* 7th ed., 4:271–302. London: Edward Arnold (1984).

Champsur, H.; Fattal-German, M.; and Arranhado, R. Sensitivity and specificity of viral immunoglobulin M determination by indirect enzyme-linked immunosorbent assay. *J. Clin. Microbiol.* 26:328–32 (1988).

Cooper, L. Z. The history and medical consequences of rubella. *Rev. Infect. Dis.* 7:S2–10 (1985).

Enders, G.; Knotek, F.; and Pacher, U. Comparison of various serological methods and diagnostic kits for the detection of acute, recent and previous rubella infection, vaccination, and congenital infections. *J. Med. Virol.* 16:219–32 (1985).

Grangeot-Keros, L.; Pillot, J.; Daffos, F.; and Forestier, F. Prenatal and postnatal production of IgM and IgA antibodies to rubella virus studied by antibody capture immunoassay. *J. Infect. Dis.* 158:138–43 (1988).

Gravell, M.; Procett, P. H.; Gutenson, O.; and Ley, A. C. Detection of antibody to rubella virus by enzyme-linked immunosorbent assay. *J. Infect. Dis.* 136:S300–303 (1977).

Herrmann, K. L. Available rubella serologic tests. *Rev. Infect. Dis.* 7:S108–12 (1985).

Meegan, J. M.; Evans, B. K.; and Horstmann, D. M. Comparison of the latex agglutination test with the hemagglutination test, enzyme-linked immunosorbent assay, and neutralization test for detection of antibodies to rubella virus. *J. Clin. Microbiol.* 16:644–49 (1982).

Morbidity Mortality Weekly Report (MMWR). 39(RR-15): 1–18 (1990).

Morgan-Capner, P. Diagnosing rubella may be difficult in patients without symptoms. *Br. Med. J.* 299:338–39 (1989).

Schmidt, N. J.; Ho, H. H.; and Chin, J. Application of immunoperoxidase staining to more rapid detection and identification of rubella virus isolates. *J. Clin. Microbiol.* 13:627–30 (1981).

Skvorc-Ranko, R.; Lavoie, H.; St. Denis, P.; Valleneuve, R.; Gagnou, M.; Chicoine, R.; Boucher, M.; Guimond, J.; and Dontigny, Y. Intrauterine diagnosis of cytomegalovirus and rubella infections by amniocentesis. *Can. Med. Ass. J.* 145:649–54 (1991).

18

REOVIRIDAE

G. D. HSIUNG

Reoviridae includes the genera of reoviruses and the rotaviruses. Reoviruses occur widely in humans and a variety of animal species. The first human reovirus was isolated from the stool of a healthy child and was identified as echovirus type 10. They have been isolated occasionally from individuals with a variety of respiratory and enteric illnesses; thus, the name reovirus was proposed. At present, the role of reoviruses in human disease is not clear. Rotaviruses are the predominant etiologic agent of epidemic and endemic gastroenteritis in children and can also cause disease in adults. These viruses were first recognized in 1973–74. Because of clinical importance, rotaviruses will be discussed before reoviruses.

ROTAVIRUSES

The name of the virus is derived from the Latin word *rota,* meaning wheel, reflecting its shape as seen under EM. There are now three groups of rotaviruses: groups A, B, and C. Most rotaviruses that infect humans belong to group A. They cause acute diarrhea in infants and young children and occasionally in the elderly. Group B rotaviruses have been associated with adult human diarrhea originally recognized in China. Group C rotaviruses appears to have a worldwide distribution as an agent associated with diarrhea in children and adults. Rotavirus transmission occurs by the fecal-oral route. A brief incubation period (one to two days) is followed by the sudden onset of gastroenteritis characterized by fever, vomiting, and diarrhea.

Morphologically, the rotavirus exhibits a wheellike structure, 70–75 nm

ISBN 0-300-05845-4.

in diameter, with two capsid layers (figure 18.1). The rotavirus possesses 11-segmented double-stranded RNA genomes. The electrophoretic profiles of the genome segments of the three groups of rotaviruses are shown in figure 18.2: group A, 4:2:3:2; group B, 4:2:2:3; and group C, 4:3:2:2.

Although the rotavirus in stool is stable at room temperature for several months, propagation of the virus to high concentrations in cell cultures is difficult. Most physicochemical characterizations have been accomplished using virions present in stool specimens in which sufficient numbers can be obtained by ultracentrifugation and concentration.

Rotaviruses isolated from several animal species have shown a serological relation with human rotaviruses (HRV). Antigenic crossings are demonstrated between HRV and rotaviruses isolated from pigs, calves, mice, and monkeys.

Specimen Source for Virus Detection

Stool samples collected during the acute phase of illness are the specimen of choice for EM visualization of the virus. Duodenal mucosa at biopsy and the intestinal contents, when available, are also suitable choices for EM examination. Cell culture isolation methods are mainly done in research laboratories.

Virus Recognition and Assay

DIRECT DETECTION BY ELECTRON MICROSCOPY AND IMMUNOELECTRON MICROSCOPY

A 20% stool suspension, prepared in Hanks' BSS with a 0.5% bovine serum albumin, free of antibody to human rotavirus (HRV), or a liquid stool specimen (which need not be diluted) is centrifuged at 1,000 × *g* for 30 minutes at 4°C to remove large debris and bacteria. Supernatant fluids can be stained with PTA (chapter 13) and examined directly by EM. In certain instances, supernatant fluids are sedimented at high-speed centrifugation (100,000 × *g*) for one hour. The pellet is resuspended in distilled water and after negative staining examined by EM. The lowest limit of detection by this procedure is approximately 10^5 virions per gram of stool. If the number of virions in a stool specimen is very high, a water suspension of the stool can be examined directly without concentration procedures, in which case the specimen can be dried on a grid, washed with distilled water, and negatively stained.

The immunoelectron microscopy techniques that have been developed are essential in identifying the rotavirus. The partially purified, concentrated stool extract is mixed with appropriate dilutions of antiserum against rotavirus. After incubation (one hour at room temperature), the mixture is applied on EM grids, negatively stained with PTA solution, and examined with an electron microscope (see chapter 13 for the detailed procedure). Figure 18.1*D* illustrates an aggregated rotavirus–antibody complex.

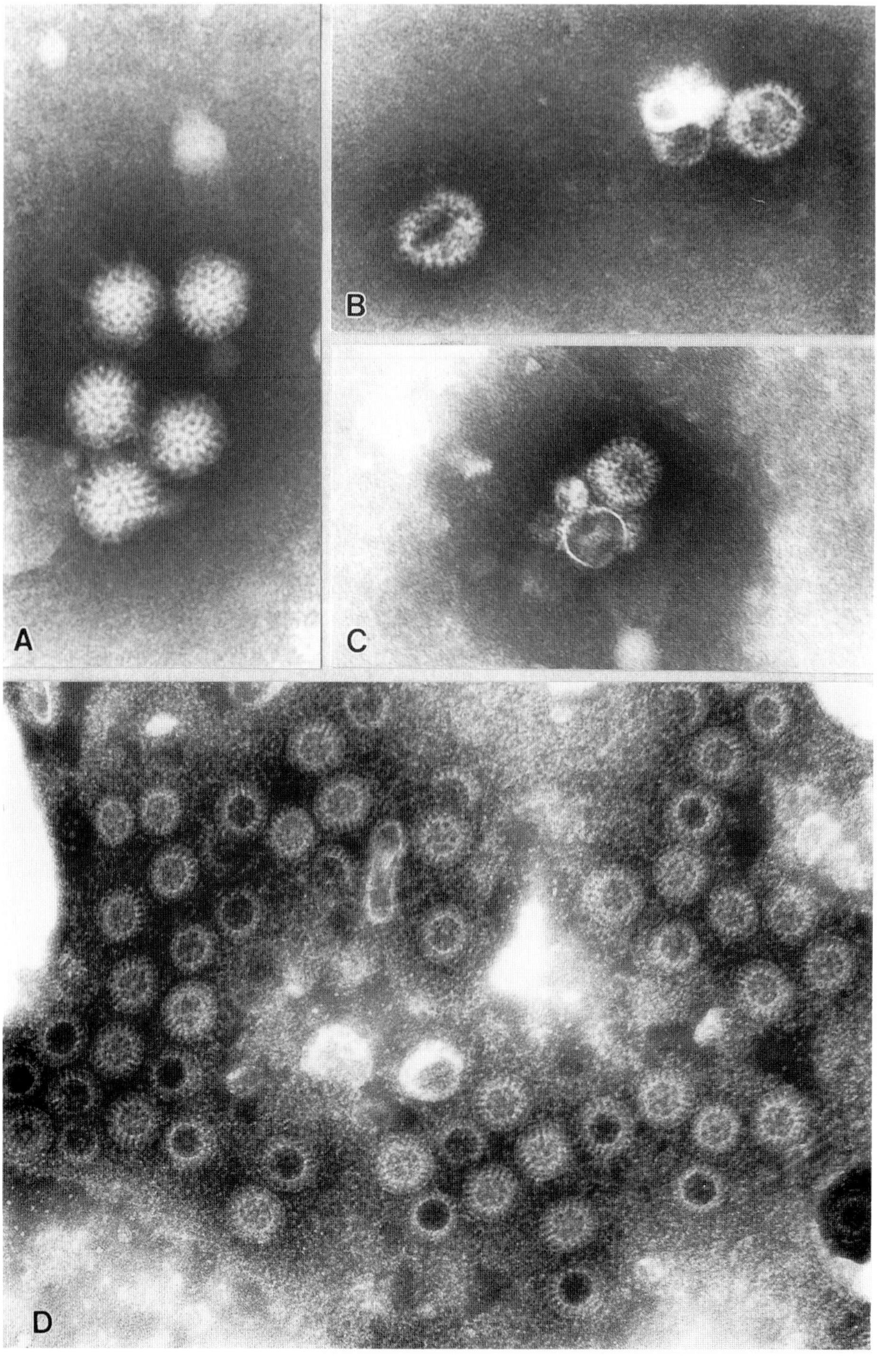

Figure 18.1 *(A)* Electron micrograph of rotavirus virions stained with PTA. Rotavirus virions in a stool sample of a case of infant diarrhea, without concentration procedure, negatively stained with PTA (168,000×); *(B* and *C)* Rotavirus virions in stool samples of adult diarrhea, negatively stained with PTA (168,000×); *(D)* Infant diarrhea rotavirus virions in the presence of homologous antibody (IEM), negatively stained with PTA (100,800×)

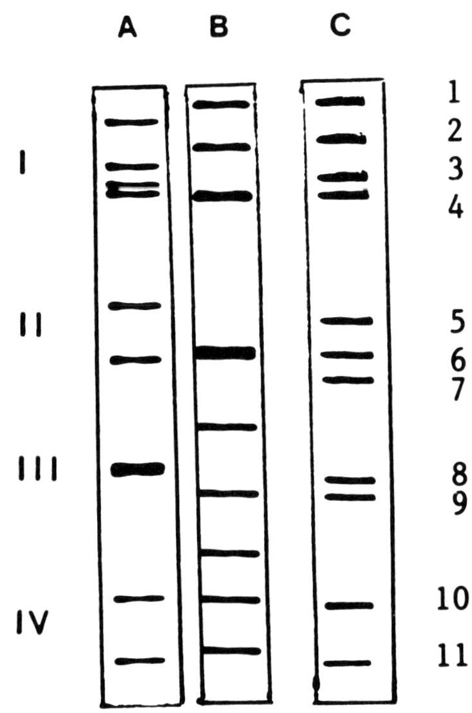

Figure 18.2 Electrophoretic profiles of RNA segments of rotavirus groups *A*, *B*, and *C* after electrophoretic separation in polyacrylamide gels. *A* is SA-11 strain, *B* is J-1 ADRV, and *C* is Cowden strain L-1049 of human adult diarrhea rotavirus. The roman numerals on the left indicate the distribution of the different molecular weight groups. The arabic numerals on the right indicate the different RNA segments. (Modified from S. Nakata et al., *J. Inf. Dis.* 154:452 [1986]; and M. E. Penaranda et al., *J. Inf. Dis.* 160:394 [1989])

Scanning EM techniques have also been applied to cases of rotavirus infection. Examination of villi obtained from the small intestine during acute infection (approximately 48 hours after the onset of the disease) reveals changes in appearance that are characteristic of a rotavirus infection.

DETECTION OF ROTAVIRUS IN CELL CULTURE

Isolation of the human rotavirus from clinical specimens is not as yet done in diagnostic laboratories. Primary African green monkey or cynomolgus monkey kidney cells or cell lines, including MA104 and LLC-MK_2, support the growth of the rotavirus if trypsin is supplemented in the serum-free medium. Pretreatment of the stool specimen with trypsin and rolling cell cultures also help in the cultivation of the rotavirus. Human embryonic kidney cell cultures showed limited growth for certain virus strains. Centrifugation of cell cultures inoculated with stool filtrates appears to increase sensitivity. Twenty-four hours after incubation at 37°C, infected cells can be detected by the IF technique using a specific antibody. Freshly trypsinized cells appear to be more sensitive to infection than established confluent monolayer cultures. Actively multiplying cells are apparently more susceptible to rotavirus infection.

DETECTION OF ROTAVIRUS BY ELISA

Both direct and indirect ELISA methods have been adapted for the detection of rotavirus virions in stool samples. In the direct test, polyvinyl microtiter plates are precoated with a rotavirus-specific antibody (capture antibody). After attachment of rotavirus virions to this specific antibody, a secondary rotavirus antibody, which has been conjugated with an enzyme, is added. Then, the substrate is added, which is in turn converted to a visible color by the enzyme. The intensity of color observed is directly proportional to the amount of rotavirus antigen present in the stool sample. (The specific methodology is outlined and illustrated in detail in chapter 8.) Commercial ELISA kits for testing group A rotaviruses are available, but groups B and C rotaviruses will not be detected by these kits.

DETECTION OF ROTAVIRUS BY LATEX AGGLUTINATION (LA) TEST

The LA kit (such as Rota-Stat, Rotalex, Rotavirus Latex, and Virogen Rotatest) is used according to manufacturer's instructions. Briefly, the procedure is as follows:

1. Dilute the stool specimen 1:10 with a special buffer. Let it stand for 30 minutes at room temperature, mix, then centrifuge at 2,000 rpm for 20 minutes.
2. Add 1 drop of the supernatant to 1 drop of latex particles on a glass slide. Latex particles are either precoated with rabbit anti-NCDV (Nebraska Calf Diarrhea Virus, a bovine rotavirus) or nonimmune globulin.
3. Mix the drops by tilting the glass slide to the front and back for 2 minutes. The test is positive when distinct agglutination is observed while the control suspension remains milky.

Identification of Isolates

IMMUNOELECTRON MICROSCOPY (IEM)

Immunoelectron microscopy was the method by which rotavirus infection in humans was first recognized. An 0.8 ml suspension of stool filtrate rich in rotavirus virions is mixed with 1:5 dilutions of acute or convalescent serum. The virus-serum mixture is incubated at room temperature for one hour, then overnight at 4°C to obtain the maximum amount of antibody-virus aggregates (figure 18.1*D*) (for details on IEM, see chapter 13).

IMMUNOFLUORESCENT TEST

As mentioned above, immune serum can be added to rotavirus-infected cell cultures, and the infected cells can be detected by the indirect IF staining technique.

Serodiagnosis

The rotavirus antibody can be detected and measured in microtiter plates by the ELISA. In chapter 8, rotavirus antigen was used as an example for ELISA methods. In general, serological tests for antibody studies are not used in the routine diagnosis of rotavirus infections.

REOVIRUSES

The reoviruses, like the enterovirus group, are common inhabitants of the human intestinal tract and can be recovered from fecal specimens, but infrequently from throat swabs or nasal secretions of persons with mild upper respiratory illness.

Reovirus virions are approximately 70 nm in diameter (figure 18.3), icosahedral in shape, with a double-layer capsid. Each consists of 92 capsomeres and a double-stranded RNA core with ten segments. Since this is the first group of animal viruses shown to possess a double-stranded RNA genome, reoviruses have been studied more intensively by the molecular virologists than by those concerned with human disease. Viruses of this group are ether-resistant and relatively stable but vary in their pathogenicity for suckling mice.

Unlike enteroviruses, reoviruses are capable of multiplying and inducing CPE in a variety of nonprimate cells derived from pig, calf, dog, cat, guinea pig, and hamster. Serologically, they may be divided into three antigenic groups based upon neutralization and hemagglutination-inhibition reactions. All reoviruses share a common group complement-fixing antigen. Reovirus types 1 and 2 agglutinate human type-O erythrocytes at high titers, whereas agglutination by type 3 is best with bovine erythrocytes at 4°C. Thus, differentiation of type 3 from types 1 and 2 may be made on this basis (table 18.1).

Specimen Sources and Inoculation

Stools, rectal swabs, and occasionally throat swabs or nasal secretions are used for virus isolation. Primary kidney cell cultures of both primates and nonprimates are highly susceptible to infection with reoviruses.

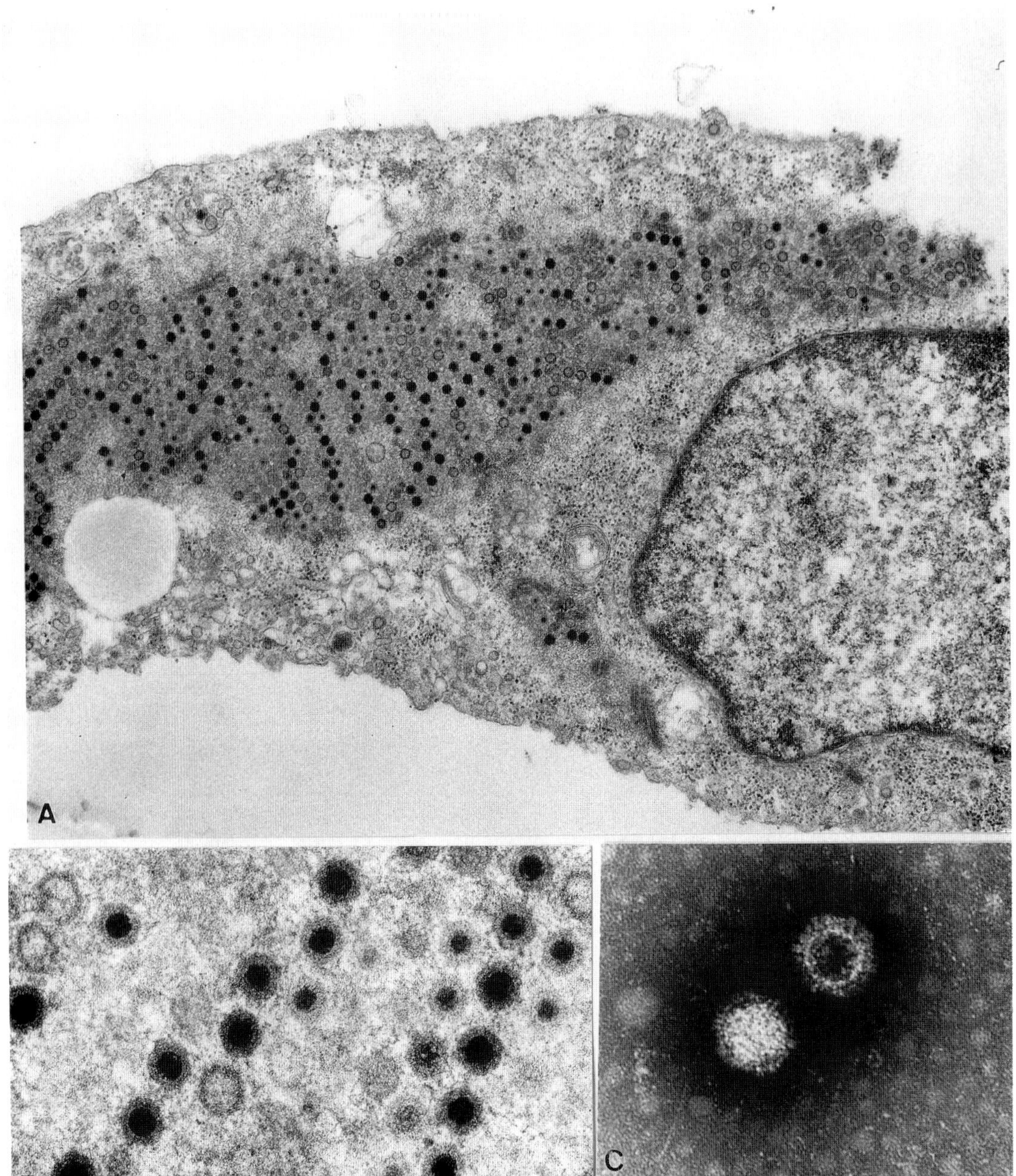

Figure 18.3 Electron micrographs of reovirus-infected RhMK cell: *(A)* Thin section of reovirus type 1–infected RhMK cell showing a large viral aggregate containing many virions in the cytoplasm (19,680×) (C. K. Y. Fong and G. D. Hsiung, *Virology and Rickettsiology,* 1978, vol. 1, pt. 1, 62); *(B)* Higher magnification of a portion in *A* showing individual virions (88,000×); *(C)* Reovirus virions stained with PTA (168,000×)

Virus Isolation, Propagation, and Assay

CYTOPATHIC EFFECT IN FLUID CULTURES

The cytopathic effects produced by reoviruses show cellular granulation but not the characteristic rounding up of cells associated with enterovirus cytopathology. Recognition of their presence requires some experience since typical reovirus CPE may be confused with the nonspecific cellular degeneration of aging cultures. Infected cells do not dislodge readily from the monolayer. Confirmation of the specificity of the cellular changes may be obtained by reproducing them after passage into other culture types, especially those derived from certain nonprimate species. Figure 18.4 illustrates the CPE induced by reovirus type 1 in cat and guinea pig kidney cell cultures, which

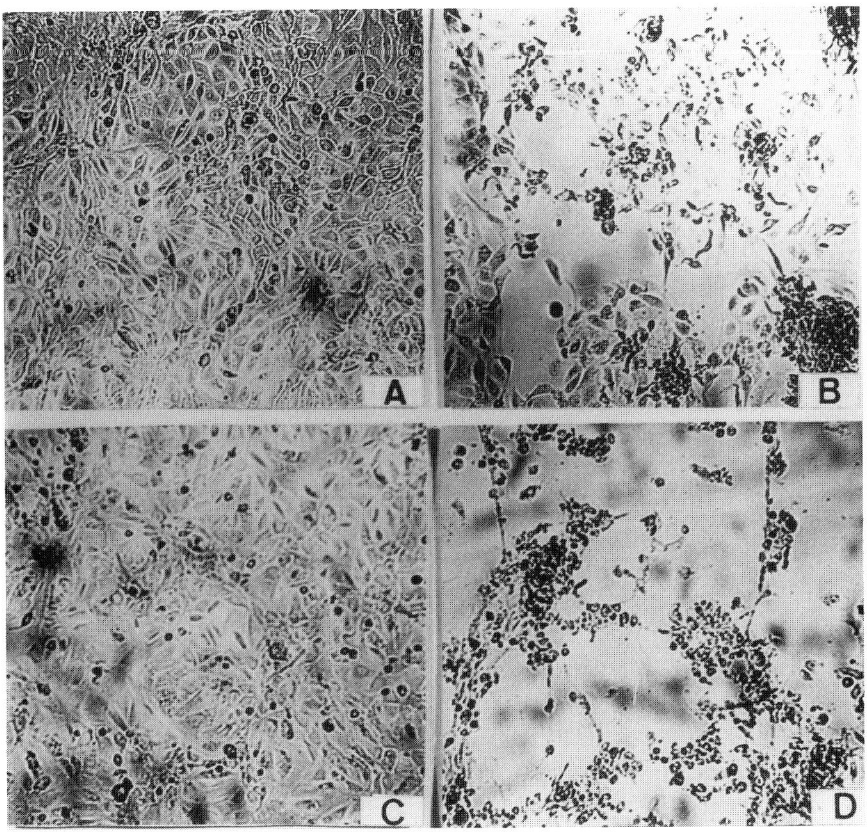

Figure 18.4 Cytopathic effect induced by reovirus in nonprimate kidney cell cultures (100×) (G. D. Hsiung, *Proc. Soc. Exp. Biol. Med.* 99:387 [1958]). *(A)* Uninfected cat kidney cell culture, seventeen days old; *(B)* Cat kidney cell culture infected with reovirus type 1, thirteen days after inoculation; *(C)* Uninfected guinea pig kidney cell culture, twelve days old; *(D)* Guinea pig kidney cell culture infected with reovirus type 1, five days after inoculation

Table 18.1 Hemagglutination Properties of Reoviruses

Serotype	Source of Virus Isolation	Hemagglutination with RBC: Human (type-O)	Bovine	Guinea pig
1	Man, calf, monkey	+	−	−
2	Man, chimpanzee, calf, monkey	+	−	−
3	Man, calf, mouse	P*	+	−

\+ virus-induced agglutination of RBC; − no agglutination of RBC indicated
*Partial hemagglutination or low hemagglutinin titer

are not commonly sensitive to human viruses. CPE is usually more evident in the second passage but may not be readily apparent until the fourth or fifth passage. Fetal bovine serum should be used for maintaining cultures in order to avoid interference by reovirus antibodies that may be present in calf serum.

HEMAGGLUTINATION

Reovirus-infected cells do not show hemadsorption upon the addition of human type-O erythrocytes. Yet fluids from the same infected culture show significant hemagglutinin titers with human type-O RBC but not with guinea pig erythrocytes (table 18.1). Such agglutination occurs at 4°C, 22°C, and 37°C. For practical purposes, a 0.75% suspension of human type-O RBC in 0.85% NaCl is satisfactory for agglutination of reovirus types 1 and 2, but bovine RBC are more suitable for reovirus type 3.

CYTOPATHOLOGY IN ACRIDINE ORANGE AND H&E-STAINED PREPARATIONS

Owing to the unusual nonspecific appearance of CPE induced by the reoviruses, fixed and stained preparations are helpful for confirmation of the isolates. Characteristic reovirus intracytoplasmic eosinophilic inclusions in hematoxylin-eosin-stained preparations (figure 18.5, color plate 19) are distinguishable from those produced by other groups of viruses. The presence of these perinuclear inclusions permits a presumptive diagnosis of a reovirus infection. These inclusions also can be easily recognized after infected cells are stained with acridine orange (color plate 20). In the latter preparation, cytoplasmic yellowish-green inclusions are characteristic of the double-stranded RNA reovirus.

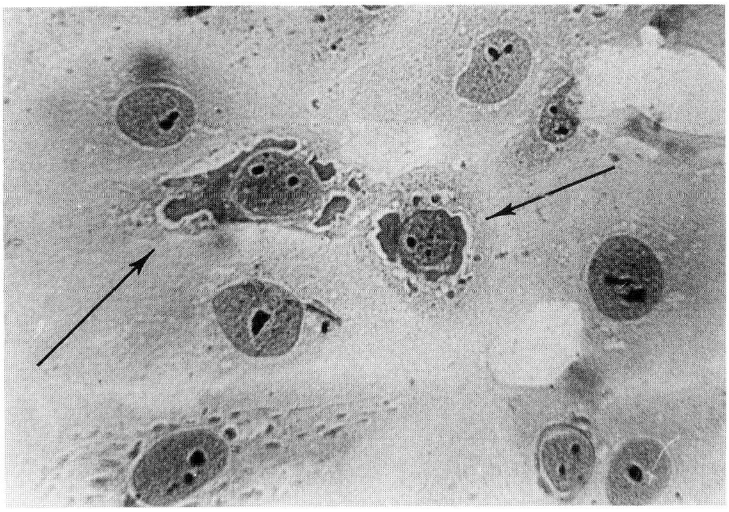

Figure 18.5 Intracytoplasmic eosinophilic inclusions (arrows) induced by reovirus type 1 in RhMK cells (H&E, 400×)

Identification of Isolates

COMPLEMENT-FIXATION TEST

The three reovirus serotypes share a group antigen that can be detected by the CF test, but this method cannot identify the individual serotypes.

HEMAGGLUTINATION-INHIBITION TEST

The HI test is used for identification of specific reovirus serotypes. The typing sera should be treated with 25% kaolin in 0.85% NaCl. After kaolin treatment, the sera should be adsorbed with washed, packed, human type-O erythrocytes to remove any natural agglutinins to the human RBC that are present in the sera. (Although procedures for the HI test are the same as described in chapter 9, note that 0.85% NaCl is used instead of PBS for the human type-O erythrocyte suspension and that the RBC are allowed to settle at 22°C for best results.)

NEUTRALIZATION TEST

Inhibition of CPE or the plaque-reduction neutralization test can be performed in rhesus monkey kidney cell cultures or guinea pig embryo cells using 100 $TCID_{50}$ as the challenge dose against type-specific antisera.

ELECTRON MICROSCOPY

Reovirus virions have a distinctive morphology. It is fairly easy to recognize their presence using negatively stained virus preparations (figure 18.3*C*). In infected cells, the reovirus develops characteristic intracytoplasmic inclusions. Electron microscopic examination of fixed and embedded cells reveals virus aggregates in the cytoplasm of infected cells (figure 18.3) corresponding to the inclusions seen by light microscopy (figure 18.5).

Serodiagnosis

The diagnosis of most reovirus infections has been determined through the HI test because of its simplicity and sensitivity as well as its specificity. All human sera should be heated at 56°C for 30 minutes and treated with kaolin before use, as described above, using the test procedure described in chapter 9. Since antibodies to reoviruses are commonly present in human sera, for diagnostic purposes, a 4-fold rise in paired sera is considered positive.

An enzyme-linked immunosorbent assay has been developed for the measurement of reovirus immunoglobulin IgG, IgA, and IgM levels in human sera.

SUPPLEMENTARY READING FOR ROTAVIRUS

Bishop, R. F.; Davidson, G. P.; Holms, I. H.; and Ruck, B. J. Virus particles in epithelial cells of duodenal mucosa from children with acute non-bacterial gastroenteritis. *Lancet* 2:1281–83 (1973).

Brandt, C. D.; Kim, H. W.; Rodriguez, W. J.; Thomas, L.; Yolken, R. H.; Arrobio, J. O.; Kapikian, A. Z.; Parrott, R. H.; and Chanock, R. M. Comparison of direct electron microscopy, immune electron microscopy and rotavirus enzyme-linked immunosorbent assay for detection of gastroenteritis viruses in children. *J. Clin. Microbiol.* 13:976–81 (1981).

Dennehy, P. H.; Gauntlett, D. R.; and Spangenberger, S. E. Choice of reference assay for the detection of rotavirus in fecal specimens: Electron microscopy versus enzyme immunoassay. *J. Clin. Microbiol.* 28:1280–83 (1990).

Estes, M. K.; Graham, D. Y.; and Dimitrov, D. H. The molecular epidemiology of rotavirus gastroenteritis. *Prog. Med. Virol.* 29:1–22 (1984).

Flewett, T. H.; Bryden, A. S.; and Davies, H. Virus particles in gastroenteritis. *Lancet* 2:1497 (1973).

Fujii, R.; Kuzuya, M.; Hamano, M.; Yamada, M.; and Yamazaki, S. Detection of human group C rotavirus by an enzyme-linked immunosorbent assay using monoclonal antibodies. *J. Clin. Microbiol.* 30:1307–11 (1992).

Gerna, G.; Sarasini, A.; DiMatteo, A.; Parea, M.; Torsellini, M.; and Battaglia, M. Rapid detection of human rotavirus strains in stools by single-sandwich enzyme-linked immunosorbent assay systems using monoclonal antibodies. *J. Virol. Meth.* 24:43–56 (1989).

Grahnquist, L.; Stintzing, G.; Alam, A. N.; and Svensson, L. Rapid detection of rotavirus in faeces by a slide latex agglutination test as compared with an enzyme-linked immunosorbent assay. *J. Diarrh. Dis. Res.* 5:178–80 (1987).

Hasegawa, A.; Matsuno, S.; Inouye, S.; Kono, R.; Tsurukobo, Y.; Mukoyama, A.; and Saaito, Y. Isolation of human rotaviruses in primary cultures of monkey kidney cells. *J. Clin. Microbiol.* 16:387–90 (1982).

Hermann, J. E., and Blacklow, N. R. Gastroenteritis viruses. In Schmidt, N. J., and

Emmons, R. W., eds., *Diagnostic Procedures for Viral, Rickettsial and Chlamydial Infections,* 6th ed., pp. 925–56. Washington, D.C.: American Public Health Association, 1989.

Hung, T.; Chen, G.; Wang, C.; Chow, Z.; Chao, T.; Ye, W.; Yao, H.; and Meng, K. Rotavirus-like agent in adult non-bacterial diarrhea in China. *Lancet* 2:1078–79 (1983).

Jensen, A. Evaluation of two commercial kits for rapid detection of human rotavirus in feces: Rotalex, a latex agglutination test and rotavirus ELISA kit. *Acta Path. Microbiol. Immun. Scand, (B), Microbiol.* 93:159–60 (1985).

Kapikian, A. Z.; Kim, H. W.; Wyatt, R. G.; Rodriguez, W. G.; Ross, S.; Cline, W. L.; Parrott, R. H.; and Chanock, R. M. Reovirus-like agent in stools' association with infantile diarrhea and development of serological tests. *Science* 185:1049–53 (1974).

Kinney, J. S.; Viscidi, R. P.; Vonderfecht, S. L.; Eiden, J. J.; and Yolken, R. H. Monoclonal antibody assay for detection of double-stranded RNA and application for detection of group A and non-group A rotaviruses. *J. Clin. Microbiol.* 27:6–12 (1989).

Lipson, S. M.; Leonardi, G. P.; Salo, R. J.; Schutzbank, T. E.; and Kaplan, M. H. Occurrence of nonspecific reactions among stool specimens tested by the Abbott TestPack rotavirus enzyme immunoassay. *J. Clin. Microbiol.* 28:1132–34 (1990).

Lipson, S. M., and Zelinsky-Papez, K. A. Comparison of four latex agglutination (LA) and three enzyme-linked immunosorbent assays (ELISA) for the detection of rotavirus in fecal specimens. *Am. J. Clin. Path.* 92:637–43 (1989).

Nakata, S.; Estes, M. K.; Graham, D. Y.; Loosle, R.; Hung, T.; Wang, S.; Saif, L. J.; and Melnick, J. L. Antigenic characterization and ELISA detection of adult diarrhea rotaviruses. *J. Inf. Dis.* 154:448–55 (1986).

Pai, C. H.; Shahrabadi, M. S.; and Ince, B. Rapid diagnosis of rotavirus gastroenteritis by a commercial latex agglutination test. *J. Clin. Microbiol.* 22:846–50 (1985).

Pedley, S.; Bridger, J. C.; Brown, J. F.; and McCrea, M. A. Molecular characterization of rotaviruses with distinct group antigens. *J. Gen. Virol.* 64:2093–2101 (1983).

Pedley, S.; Bridger, J. C.; Chasey, D.; and McCrea, M. A. Definition of two new groups of atypical rotaviruses. *J. Gen. Virol.* 67:131–37 (1986).

Penaranda, M. E.; Cubitt, W. D.; Sinarachatanaut, P.; Taylor, P. N.; Linanonsakul, S.; Saif, L.; and Glass, R. I. Group C rotavirus infections in patients with diarrhea. *J. Inf. Dis.* 160:392–97 (1989).

Wyatt, R. G.; James, H. D.; Pittman, A. L.; Hoshino, Y.; Greenberg, H. B.; Kalica, A. R.; Flores, J.; and Kapikian, A. Z. Direct isolation in cell culture of human rotaviruses and their characterization into four serotypes. *J. Clin. Microbiol.* 18:310–17 (1983).

Yolken, R. H.; Kim, H. W.; Clem, T.; Wyatt, R. G.; Kalica, A. R.; Chanock, R. M.; and Kapikian, A. Z. Enzyme linked immunosorbent assay (ELISA) for detection of human reovirus-like agent of infantile gastroenteritis. *Lancet* 2:263–67 (1977).

SUPPLEMENTARY READING FOR REOVIRUS

Gomatos, P. J.; Tamm, I.; Dales, S.; and Franklin, R. M. Reovirus type 3: Physical characteristics and interactions with L cells. *Virology* 17:441–54 (1962).

Hsiung, G. D. Some distinctive biological characteristics of ECHO-10 virus. *Proc. Soc. Exp. Biol. Med.* 99:387–90 (1958).

Joklik, W. Structure and function of the reovirus genome. *Microbiol. Rev.* 45:483–501 (1981).

Richardson, S. C.; Bishop, R. F.; and Smith, A. L. Enzyme-linked immunosorbent assays for measurement of reovirus immuno-globulin G, A, and M levels in serum. *J. Clin. Microbiol.* 26:1871–73 (1988).

Sabin, A. B. Reovirus: A new group of respiratory and enteric viruses formerly classified as ECHO type 10 is described. *Science* 130:1387–89 (1959).

Spencker, F. B.; Weiss, J.; Handrick, W.; Rieske, K.; and Springer, W. Evaluation of a latex agglutination test for rapid detection of rotavirus in faecal specimens. *J. Hyg. Epidemiol. Microbiol. Immun.* 31:65–68 (1987).

Tyler, K. L., and Fields, B. N. Reoviruses. In Fields, B. N., and Knipe, D. M., eds., *Fields Virology,* pp. 1307–28. New York: Raven, 1990.

19

RHABDOVIRIDAE

G. D. HSIUNG

The Rhabdoviridae includes the highly pathogenic rabies virus that is responsible for the central nervous system disease of humans and animals as well as the relatively nonpathogenic vesicular stomatitis virus (vsv). The latter, morphologically similar to the rabies virus, grows well in cell cultures. Since vsv is known to be relatively nonpathogenic to humans, it has been used commonly in the laboratory as a model for elucidating the pathogenesis and morphogenesis of rhabdoviruses.

RABIES VIRUS

The clinical symptoms associated with rabies virus include salivation, excitation, anxiety, difficulty in swallowing, fever, convulsions, and death. Tissues suspected of harboring the rabies virus are sent to specifically designated laboratories for virus isolation and diagnosis. Animals suspected of having the rabies virus infection (for example, skunks, raccoons, and trapped bats), should be sent to state laboratories for diagnosis in order to maintain an epidemiologic surveillance of wildlife.

The rabies virion exhibits a bullet-shaped morphology with cylindrical particles, 60 × 180 nm. Each virion contains a single-stranded RNA nucleocapsid, which is surrounded by a lipid bilayer membrane with surface projections 6–8 nm in length. Virions are observed in association with an intracytoplasmic filamentous matrix (figure 19.1). The virus is relatively stable in the presence of a stabilizing protein, but is sensitive to ether treatment. In fixed tissue sections, the aggregates of nucleocapsids are recognized under

ISBN 0-300-05845-4.

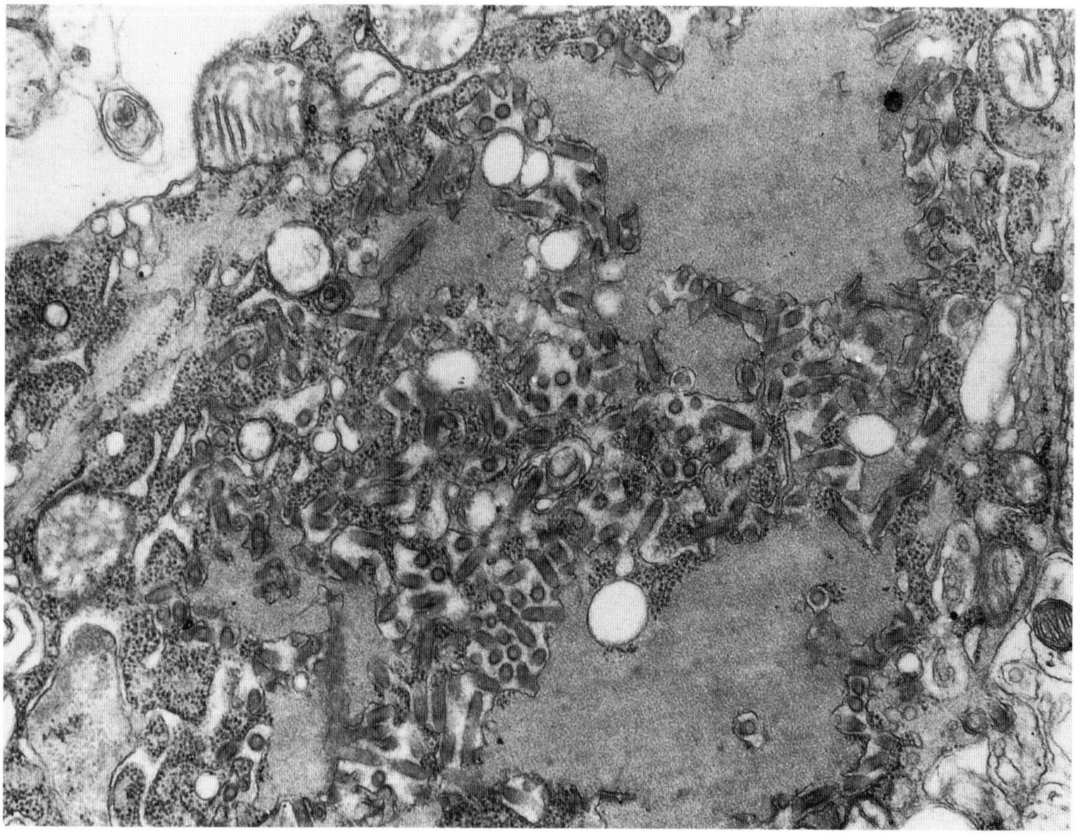

Figure 19.1 Electron micrograph of rabies virus virions in an infected mouse brain cell (25,000×). (F. A. Murphy, Morphology and morphogenesis, chap. 1 in G. M. Baer, ed., *The Natural History of Rabies*, New York: Academic Press, 1975, 33)

light microscopy as intracytoplasmic inclusion bodies or Negri bodies (color plate 18). The detection of Negri bodies in the Ammon's horn of the hippocampus and of rabies virus nucleocapsid proteins in brain tissue by immunofluorescence are important diagnostic tools in the rapid identification of rabies in animals suspected of being infected (color plate 12).

Specimen Sources for Direct Antigen Detection

It is critically important to obtain a rapid diagnosis of the rabies virus infection. Direct antigen detection by immunofluorescent antibody staining is therefore commonly used because of its great sensitivity and specificity and the speed and ease of its performance. There may be a history of exposure to a bite from a dog, cat, wild animal, or bat, or a clinical presentation suggestive of rabies. Corneal impression smears and a skin, nuchal, or brain biopsy can be collected for an antemortem diagnosis in humans. CNS tissues, including medulla (brain stem), cerebellum, and the hippocampus are collected postmortem in humans and infected animals and are examined for evidence of the rabies antigen by direct FA (see below).

Virus Isolation and Propagation

Virus isolation from saliva and infected tissues from humans and animals to confirm a rabies diagnosis requires the inoculation of samples into one- to two-day-old mice. The inoculum contains 20% tissue suspension of brain and/or salivary glands in PBS at pH 7.2 with a protein stabilizer such as 0.75% bovine albumin and antibiotics. The incubation period is from 5–12 days to as long as three weeks. Infected animals show paralysis and convulsions before death. The brain of moribund mice are examined by direct FA (see below) for evidence of rabies infection. Cell culture, including murine neuroblastoma, chicken embryo, and baby hamster cells, is used in the propagation of the rabies virus but are not as sensitive as mouse inoculation for the isolation of street virus strains.

Recognition and Identification of Rabies Virus

DIRECT IMMUNOFLUORESCENCE TEST

For the more rapid identification of the rabies virus antigen, the direct immunofluorescent staining method is preferable. Identification of rabies virus isolated in cell cultures or mouse brain tissues can be made within hours using a type-specific labeled rabies antibody (for the procedure, see below).

The neutralization test for virus identification depends on the use of a known antiserum in order to reduce the virus titer 100-fold or more as compared to a known negative serum. The test is laborious and time-consuming and is only used in selected cases.

DETECTION OF NEGRI BODIES BY STAINING PREPARATIONS

The rabies virus can be detected by direct microscopic examination of H&E-stained infected brain tissue section for Negri bodies (color plate 18), seen as intracytoplasmic eosinophilic structure 2–10 μm, usually round but sometimes irregular in shape.

ELECTRON MICROSCOPY

Examination of autopsy tissues by EM is another valuable method for the direct visualization of the characteristic rabies virions. Figure 19.1 illustrates a section of a rabies virus–infected mouse brain in which the bullet-shaped virions can be correlated with the intracellular inclusion observed in a neuron when examined by light microscopy. Thin sections of VSV in cell culture

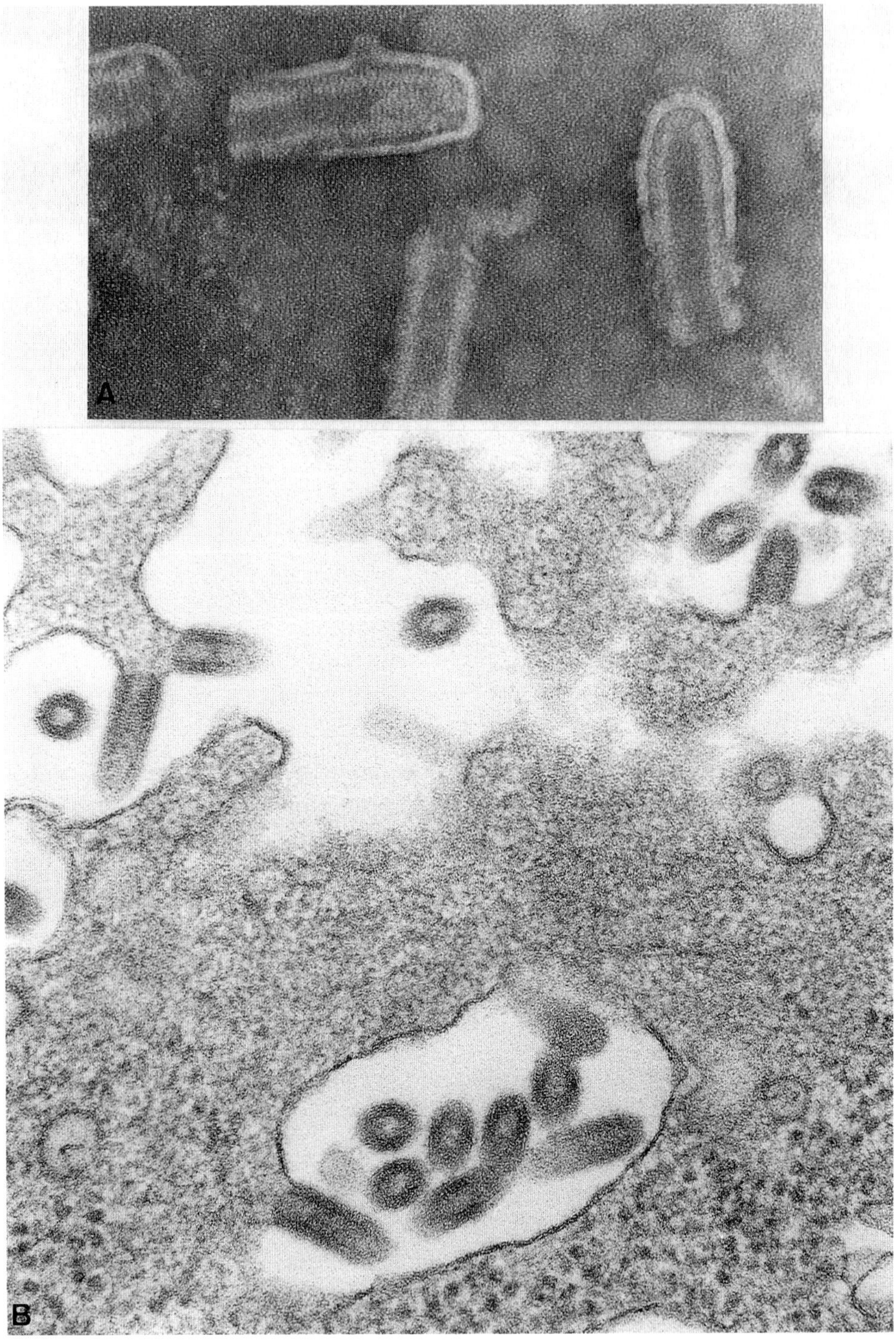

Figure 19.2 Electron micrographs of vesicular stomatitis virus (vsv): *(A)* Virions are bullet-shaped consisting of inner nucleocapsids and outer envelopes with projections, negatively stained with PTA (168,000×); *(B)* Thin section of a vsv-infected chick embryo cell showing virions budding into cytoplasmic vacuoles (100,800×)

and negatively stained virions (figure 19.2) illustrate a morphology similar to that seen in the rabies virus–infected brain cells.

Direct Immunofluorescence Test for Rabies Nucleocapsid Antigen Detection

For detailed procedures of IF (direct method) see chapter 7. Since the use of IF for the detection of the rabies viral antigen was introduced by R. A. Goldwasser and R. E. Kissling in 1958, it has become the preferred routine method for the rapid diagnosis of the rabies virus infection.

1. Fix all specimens on slides (impression smears, corneal epithelial cells, or CNS tissue sections) in cold acetone for 1–4 hours.
2. Add FITC conjugated-monoclonal antibodies to nucleocapsid proteins of rabies virus for 30 minutes at room temperature.
3. Remove excess antibody by washing with PBS
4. Examine under a microscope with UV light source.
5. Rabies virus antigen in infected tissues (cells) appears as bright apple-green (color plate 12). In the brain-impression-stained preparations, bright fluorescent rabies antigens ranging in size and morphology from "dust particles" to bigger aggregates of cytoplasmic inclusions provide a specific diagnosis.

Serodiagnosis

Serological tests can aid in the antemortem diagnosis of rabies, but antibodies may not appear until the sixth to fifteenth day of illness. A neutralization test using newborn mice inoculated intracerebrally is commonly employed, although immunofluorescence tests are replacing other methods. A rapid fluorescent focus inhibition test using rabies virus–infected cell cultures is used to detect antibodies in human sera and CSF after infection and in the sera of vaccinees.

SUPPLEMENTARY READING

Baer, G. M., ed. *Natural History of Rabies*, vol 1. New York: Academic Press, 1975.

Bourhy, H.; Rollin, P.; Vincent, J.; and Sureau, P. Comparative field evaluation of the fluorescent-antibody test, virus isolation from tissue culture, and enzyme immunodiagnosis for rapid laboratory diagnosis of rabies. *J. Clin. Microbiol.* 27:519–23 (1989).

Budzko, D.; Charamella, L.; Jelinek, D.; and Anderson, G. Rapid detection of rabies antibodies in human serum. *J. Clin. Microbiol.* 17:481–84 (1983).

Fekadu, M., and Smith, J. S. Laboratory diagnosis of rabies. In Fishbein, D. B.; Sawyer, L. A.; and Winkler, W. G., eds. *Rabies Concepts for Medical Professionals*, 2d ed., pp. 37–42. Merieux Institute, 1986.

Fishbein, D., and Baer, G. Animal rabies: Implications for diagnosis and human treatment. *Ann. Int. Med.* 109:935–37 (1988).

Goldwasser, R. A., and Kissling, R. E. Fluorescent antibody staining of street and fixed rabies virus antigen. *Proc. Soc. Expl. Biol. Med.* 98:219–23 (1958).

Heberling, R.; Kalter, S.; Smith, J.; and Hildebrand, D. Serodiagnosis of rabies by dot immunobinding assay. *J. Clin. Microbiol.* 25:1262–64 (1987).

Howatson, A. F. Vesicular stomatitis and related viruses. *Adv. Virus Res.* 16:196–256 (1970).

Johnson, K. P.; Swoveland, P. T.; and Emmons, R. W. Diagnosis of rabies by immunofluorescence in trypsin-treated histologic sections. *J. Am. Med. Assoc.* 244:41–43 (1980).

Kaplan, M. M., and Koprowski, H., eds. *Laboratory Techniques in Rabies.* 3d ed. Geneva: WHO Publications, 1973.

Murphy, F. A. Morphology and morphogenesis. In Baer, G. M., ed., *The Natural History of Rabies,* chap. 1, pp. 33–61. New York: Academic Press, 1975.

Perrin, P.; Morgeaux, S.; and Sureau, P. In vitro rabies vaccine potency by ELISA: Advantages of the immunocapture method with a neutralizing anti-glycoprotein monoclonal antibody. *Biologicals* 18:321–30 (1990).

Perrin, P., and Sureau, P. A collaborative study of an experimental kit for rapid rabies enzyme immunodiagnosis (RREID). *Bull. WHO* 65:489–93 (1987).

Reid, F.; Hall, N. H.; Smith, J. S.; and Baer, J. M. Increased immunofluorescent staining of rabies-infected formalin-fixed brain tissue after pepsin and trypsin digestion. *J. Clin. Microbiol.* 18:968–71 (1983).

Rudd, R., and Trimarchi, C. Development and evaluation of an in vitro virus isolation procedure as a replacement for the mouse inoculation test in rabies diagnosis. *J. Clin. Microbiol.* 27:2522–28 (1989).

Sellers, T. F. A new method for staining Negri bodies of rabies. *J. Pub. Hlth.* 17:1080–81 (1927).

Smith, A. L.; Tignor, G. H.; Emmons, R. W.; and Woodie, J. D. Isolation of field rabies virus strains in CER and murine neuroblastoma cell cultures. *Intervirology* 9:359–61 (1978).

Smith, J.; Reid-Sanden, F.; Roumillat, L.; Trimarchi, C.; Clark, K.; Baer, G.; and Winkler, W. Demonstration of antigenic variation among rabies virus isolates by using monoclonal antibodies to nucleocapsid proteins. *J. Clin. Microbiol.* 24:573–80 (1986).

Warrell, M.; Looareesuwan, S.; Manatsathit, S.; White, N. J.; Phuapradit, P.; Vejjajiva, A.; Hoke, C.; Burke, D.; and Warrell, D. Rapid diagnosis of rabies and post-vaccinal encephalitis. *Clin. Expl. Immun.* 71:229–34 (1988).

Winkler, W. G.; Shaddock, J. H.; and Bowman, C. Rabies virus in salivary glands of raccoons *(Procyon lotor)*. *J. Wildlife Dis.* 21:297–98 (1985).

Wunner, W. H.; Larson, J. K.; Dietzchold, B.; and Smith, C. L. The molecular biology of rabies virus. *Rev. Infect. Dis.* 10:S771–84 (1988).

20

RETROVIRIDAE

G. D. HSIUNG

HUMAN RETROVIRUSES

The family of Retroviridae consists of RNA-containing viruses that replicate through a DNA intermediate by a viral encoded enzyme, the reverse transcriptase (RT). It is divided into three subfamilies—Oncovirinae, Lentivirinae, and Spumavirinae (table 20.1). The Oncovirinae includes the oncogenic retroviruses; Lentivirinae includes the "slow" viruses such as visna-maedi virus, equine infectious anemia virus (EIAV), and human immunodeficiency virus (the causative agent of acquired immunodeficiency syndrome [AIDS]); and Spumavirinae consists of the "foamy" viruses that induce persistent infections without classical disease but cause the vacuolization of cultured cells.

The recognition of the retrovirus as etiologic agent of cancer dates back to 1911, when Peyton Rous demonstrated that a chicken sarcoma can be transmitted by the injection of filtered tumor extracts to normal chickens. Later it was discovered that these extracts contained a retrovirus now called Rous sarcoma virus. Forty years later, the first mammalian retrovirus, murine leukemia virus, was isolated by L. Gross in 1951 from an inbred mouse strain. This discovery provided the first reproducible, experimental model for the study of virus-induced leukemia. In the years that followed, retroviruses were shown to cause various types of cancers in animal species ranging from chickens and mice to nonhuman primates.

With the role of retroviruses in carcinogenesis firmly established in animals, great efforts were directed toward the search for human retroviruses in similar diseases. Various approaches were taken to detect retroviruses in

ISBN 0-300-05845-4.

Table 20.1 Classification of Human Retroviridae

Virus Subfamily	Growth Characteristic in Culture	Virus Strain	Investigator (year reported)
Oncovirinae (RNA tumor virus group)	Noncytopathic, transforming	HTLV-I[a]	Poiesz et al. 1980
		HTLV-II	Kalyanaraman et al. 1982
		HTLV-IV	Kanki et al. 1986
		HTLV-V	Manzari et al. 1987
Lentivirinae (slow virus group)	Cytopathic, syncytial cell formation	HIV-1[b] (LAV)	Barre-Sinoussi et al. 1983
		HIV-1[b] (HTLV-III)	Gallo et al. 1984
		HIV-2	Clavel et al. 1986
Spumavirinae (foamy virus group)	Cytoplasmic vacuoles	Human foamy virus	Achong et al. 1971

a. HTLV Human T-cell leukemia or human T-cell lymphotropic virus
b. HIV Human immunodeficiency virus

normal or malignant human tissues. Yet none of these studies yielded successful results. These attempts were hindered in part by the difficulties encountered in growing human lymphocytes in culture.

The discovery by R. C. Gallo and colleagues that the T-cell growth factor (TCGF)—now called interleukin 2 (IL-2)—allows the growth of lymphocytes in cultures was crucial to the subsequent isolation of human retroviruses. The first human retrovirus, the human T-cell leukemia virus, later called human T-cell lymphotropic virus (HTLV-I), was isolated from a patient with cutaneous T-cell lymphoma (table 20.1). The second human retrovirus, also a human T-cell lymphotropic virus (HTLV-II), was similarly isolated from leukocyte cultures derived from a patient with hairy cell leukemia. Following these original successful isolations of human retroviruses, the causative agent of AIDS, human immunodeficiency virus-1 (HIV-1), and then HIV-2, were isolated.

Morphologically, all members of human retroviruses are similar but are not identical. Members of the Retroviridae are single-stranded RNA viruses. Each virion consists of a central electron-dense nucleoid or core that is surrounded by a nucleocapsid of icosahedral symmetry (figure 20.1). Each nucleocapsid is enclosed by an envelope derived from the host cell plasma membrane. Attached to the envelope are numerous glycoprotein knobs that are essential for virus attachment to the host cell. The mature virion is approximately 100 nm in diameter. The virus is often seen budding from the plasma membrane of infected cells (figures 20.2, 20.3).

The retrovirus genome consists of a positive single-stranded RNA molecule. Associated with the viral genome is the enzyme reverse transcriptase

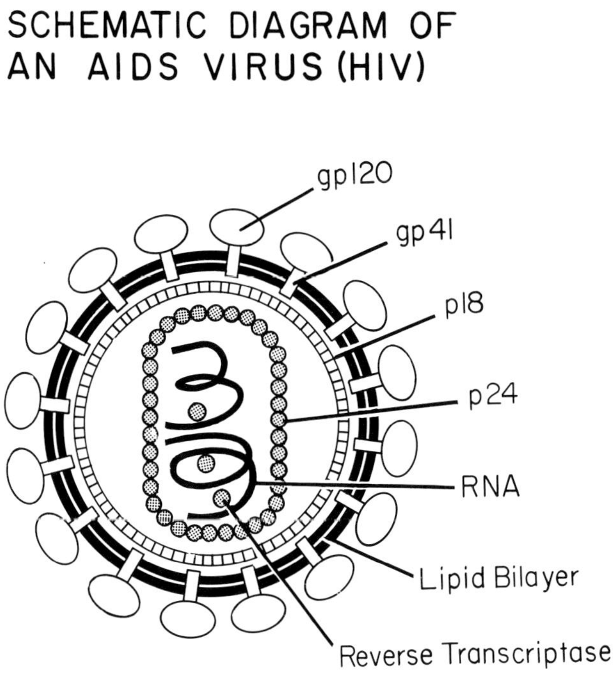

Figure 20.1 Diagram of a human immunodeficiency virus (HIV). (Modified from G. D. Hsiung, *Yale J. Biol. Med.* 60:508 [1987].)

(RT), which enables the virus to make a double-stranded DNA copy of the viral genome. The DNA copy then becomes integrated into the host cell chromosomes (this is called a provirus) and remains latent until it is activated to make new virions.

The genome of all retroviruses contains three major genes: the group-specific core gene (gag), the polymerase gene (pol), and the envelope gene (env) (table 20.2). The gag gene encodes the structural polypeptides (p18 and p24), and the env gene encodes the surface glycoproteins of the virus (gp120 and gp41), whereas the pol gene gives rise to viral RT and other enzymes required for replication. In addition to these genes, the HIV genome contains at least seven additional genes, namely: Vif, Vpu, Vpr, Tev, tat, rev, and nef, which encode proteins that help to control the expression of the virus.

The human retroviruses are fastidious viruses. Because they are highly pathogenic for humans, the virus isolation procedure is undertaken in a facility with strict precautions. The isolation of human retroviruses remains an elaborate, labor-intensive, and lengthy procedure. A flowchart illustrates the various steps used in the isolation of both noncytopathic and cytopathic human retroviruses (figure 20.4).

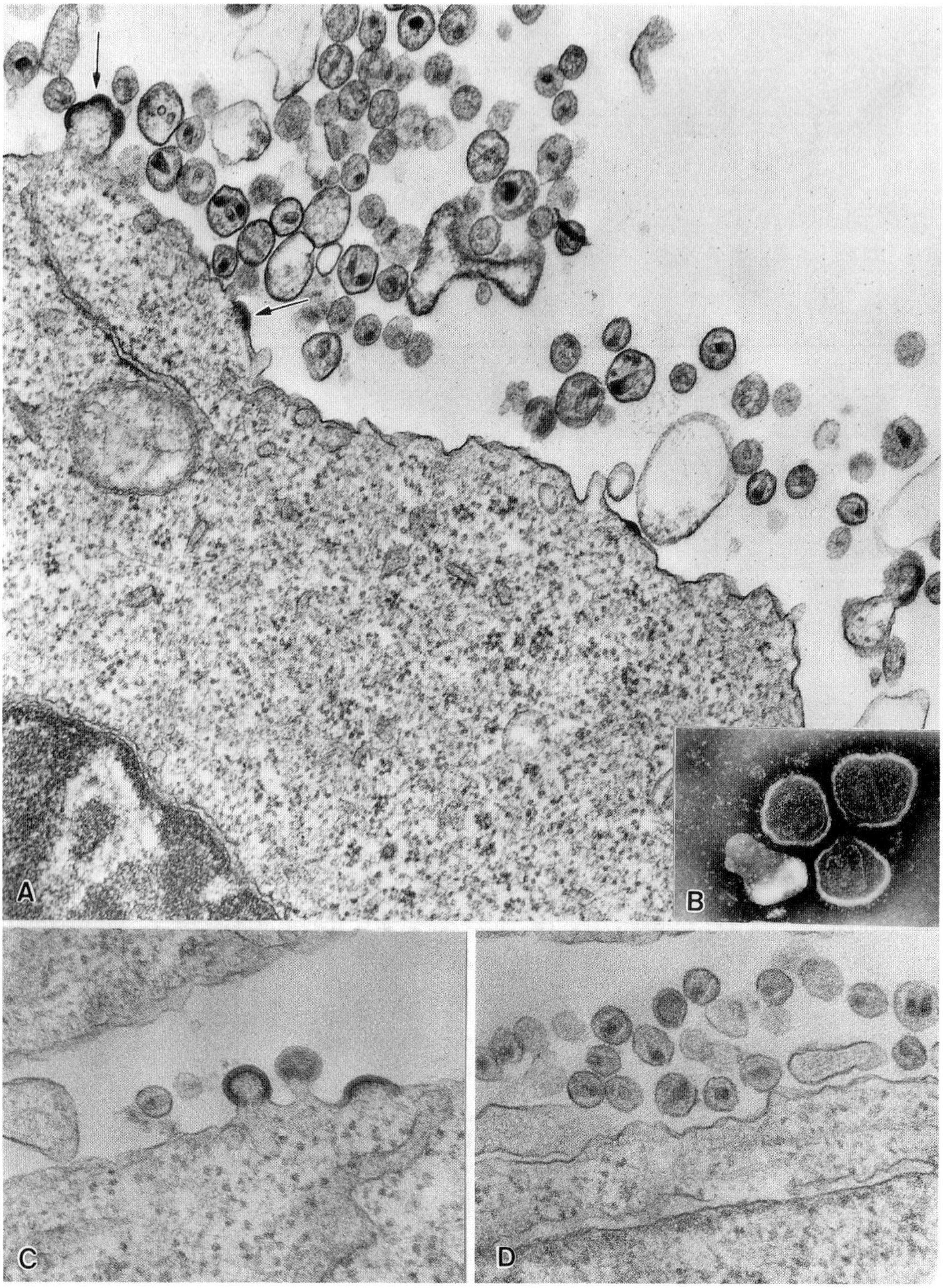

Figure 20.2 Electron micrographs showing HIV-1 in infected H9 and CD_4 cells: *(A)* HIV-1 budding from the plasma membrane of H9-infected cell (arrows) (60,000×) (courtesy D. Medina); *(B)* Inset: Negatively stained HIV-1 virions (courtesy I. L. Chrystie and J. D. Almeida, *AIDS* 2:462 [1988]); *(C)* HIV-1 budding from CD_4 HeLa cell (60,000×); *(D)* HIV-1 released from CD_4 HeLa cell (60,000×)

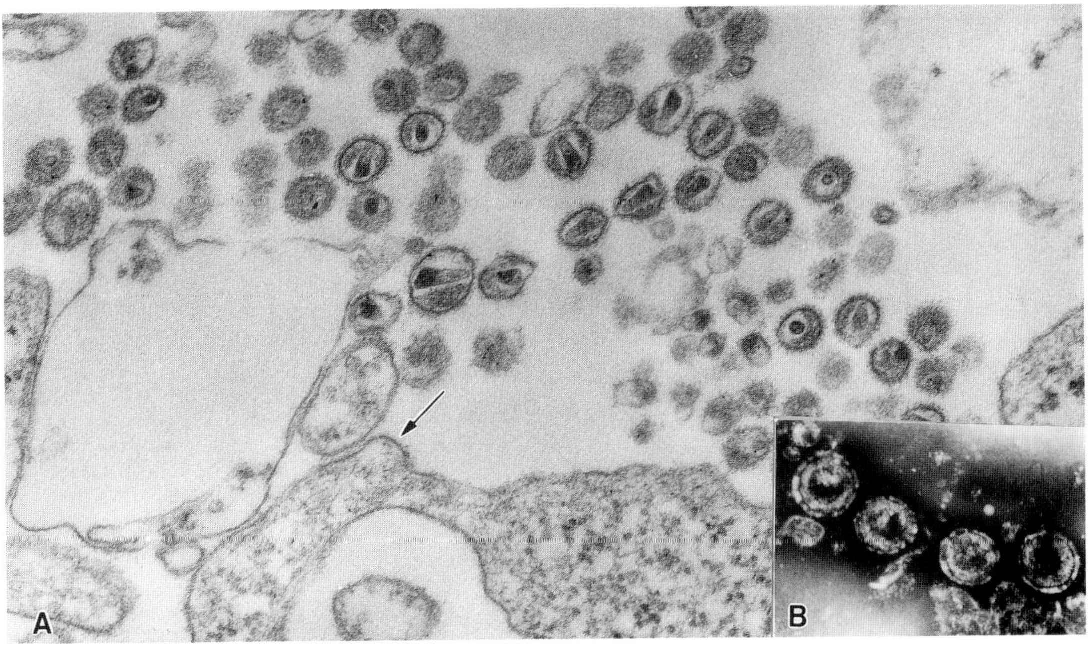

Figure 20.3 Electron micrographs showing HIV-2 in infected H9 cells: *(A)* HIV-2 in H9-infected cells. An arrow points at a budding portion of the plasma membrane (60,000×); *(B)* Inset: Negatively stained HIV-2 virions. (Courtesy I. L. Chrystie and J. D. Almeida, *AIDS* 2:462 [1988])

Table 20.2 Immunoassay for Antibodies to Major Viral Antigens of HTLV-I and HIV-1 as Determined by Western Blot

	Gene Products of Virus Type	
Major Viral Genes	HTLV-I	HIV-1
Group-specific core (gag)	p53	p55
	p24	p24
	p19	p18
	p15	
Polymerase (pol)	p62	p66
	p32	p51
		p31
Envelope (env)	gp61/68	gp160
	gp46	gp120
	gp21	gp41

p53 × 10^3 = molecular weight of each protein; gp glycoprotein

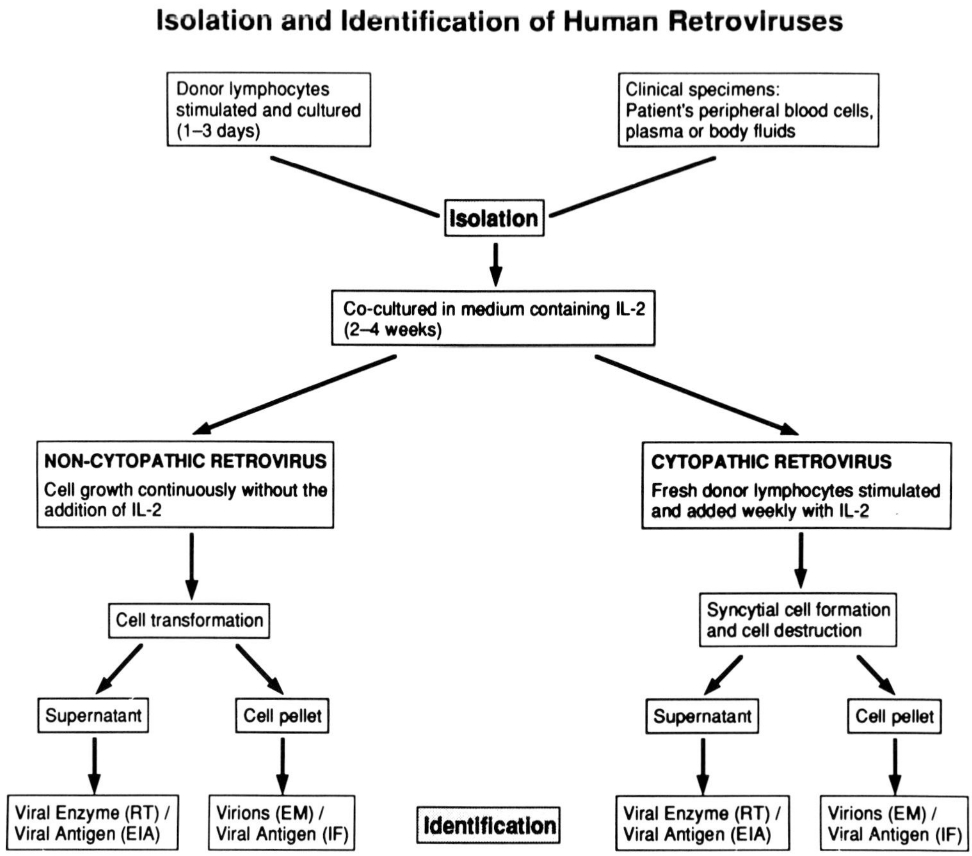

Figure 20.4 Flowchart for isolation and identification of human retroviruses (both noncytopathic retroviruses and cytopathic retroviruses)

HUMAN T-CELL LEUKEMIA VIRUSES

Noncytopathic Retroviruses

HTLV-I and HTLV-II do not induce CPE in infected cells, instead they induce cell proliferation. Thus, a stable cell line is commonly established, for example, the MT-2 cell line was produced after coculturing human cord blood with lymphocytes from a patient infected with HTLV-I. These cells show no evidence of cytopathology but clumping together (figure 20.5*C*). In addition, these transformed cells carry retrovirus virions when examined under EM (figure 20.6), while the supernatant fluid contains RT activity.

As shown in the flowchart (figure 20.4), supernatant fluids of cocultures are assayed for viral enzyme (RT) or viral antigen by EIA, and the cell pellets are assayed for viral antigen by IF and examined by EM for visualization of virions.

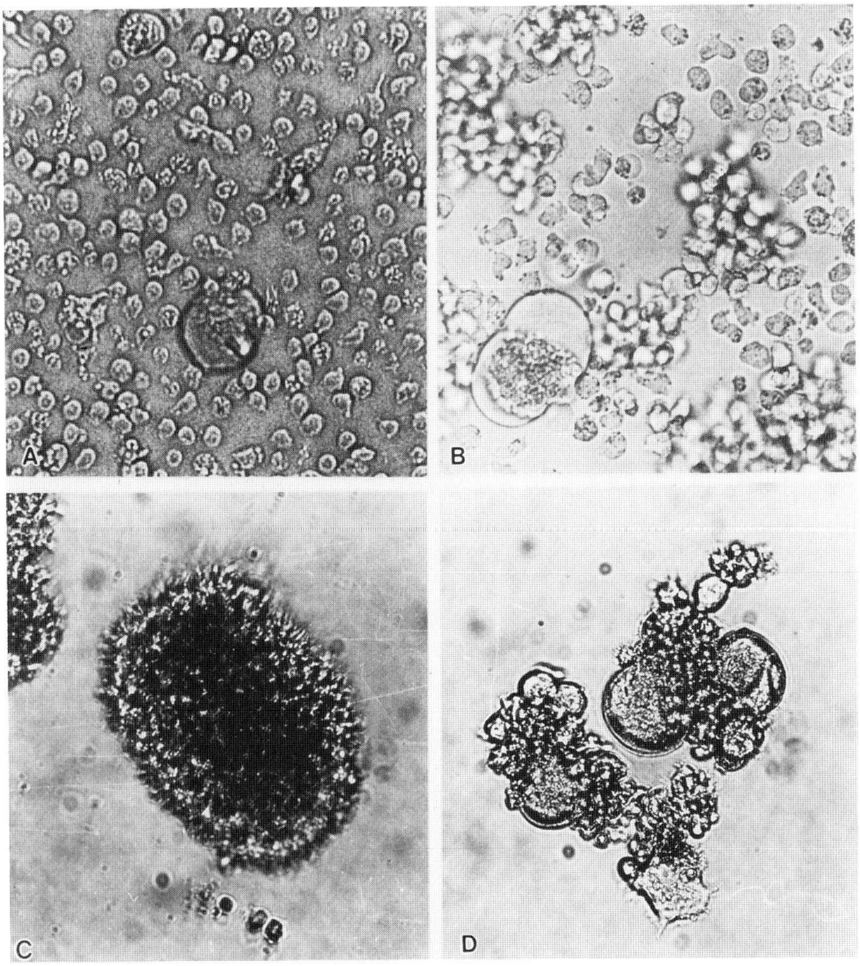

Figure 20.5 HIV-induced cytopathology in various cell lines: *(A)* HIV-1 infected PBMC cells showing syncytial cell formation (200×); *(B)* HIV-1 infected H9 cells showing syncytial and balloon cell formation (400 ×); *(C)* MT-2 cells control showing clumping of cells (200×); *(D)* MT-2 cells infected with HIV-1 showing multinucleated giant cells (200×). *(A, C,* and *D* courtesy D. Medina; *B* modified from B. P. Griffith, *Yale J. Biol. Med.* 60:579 [1987])

HUMAN IMMUNODEFICIENCY VIRUSES

Cytopathic Retroviruses

Laboratory strains of HIV-1 and HIV-2 produce characteristic syncytial cell formation in PBMC (figure 20.5*A*) and several cell lines, including H9 (figure 20.5*B*). MT-2 cells transformed by HTLV-I usually clump together (figure 20.5*C*) but disintegrate once infected with HIV-1, and multinucleated giant

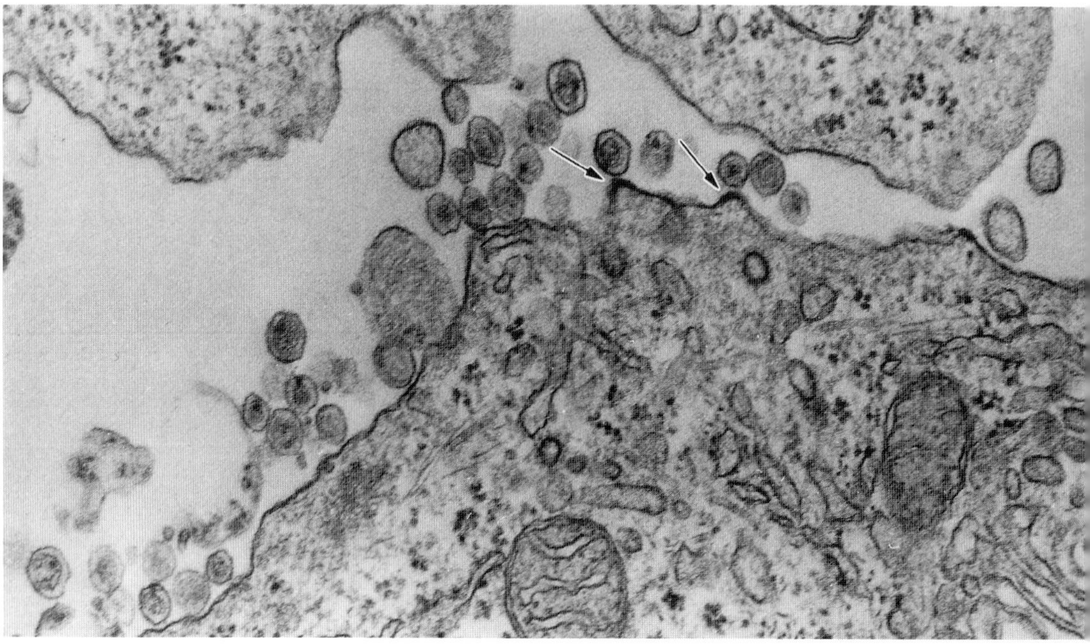

Figure 20.6 Electron micrograph showing HTLV-I extracellular virions and budding process (arrows) in MT-2 cells (60,000×). (Courtesy D. Medina)

cells are noted (figure 20.5*D*). In the MT-2 cells infected with HIV-1, large aggregates of mixed virions (both HIV-1 and HTLV-I) are commonly seen under EM examination (figure 20.7). Several cell lines have been used for HIV assays, but for the primary isolation of HIV-1 from patients, freshly separated PBMC from donor's blood that have been stimulated with PHA for one to three days are preferable, as described below. The following procedure is modified from the standard HIV coculture procedures published by F. B. Hollinger and coworkers (*J. Clin. Microbiol.* [1992]).

Coculture Procedure for HIV Isolation

Step 1: Processing of Donor-Patient PBMC

a. Separate the plasma from the anticoagulated blood within 12 hours of collection by centrifugation at 200 × *g* for 10 minutes at 20°–24°C. Freeze the plasma in aliquots.
b. Add 1 part diluent (PBS) to 1 part of the packed blood cells.
c. Overlay 4 parts of the diluted blood cells on 3 parts of Ficoll-Hypaque (Pharmacia); centrifuge at 400 × *g* for 30 minutes at 20°–24°C.

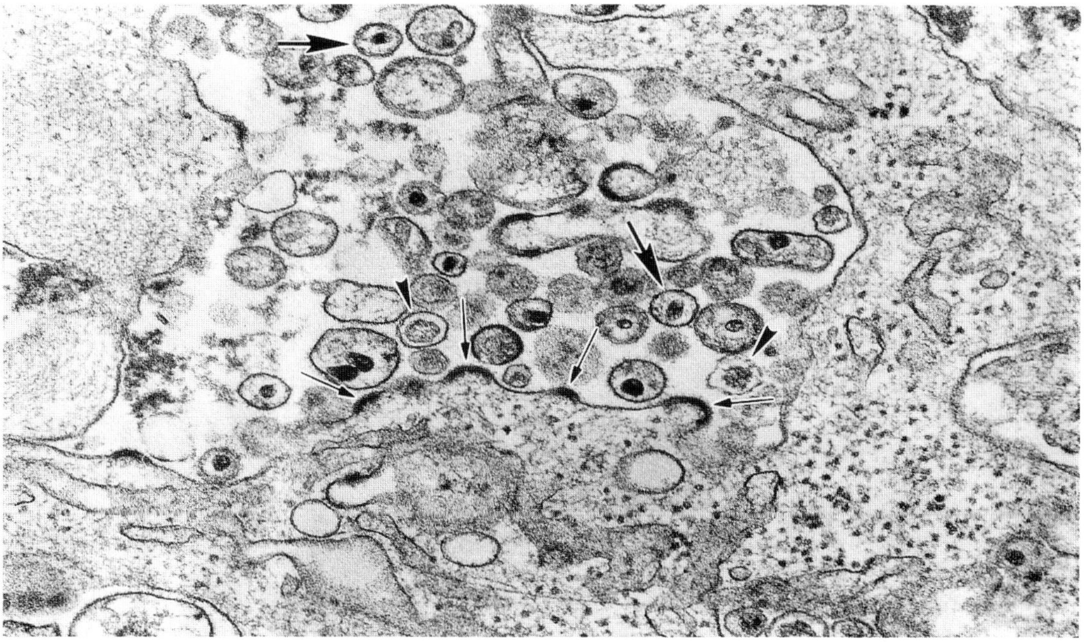

Figure 20.7 Electron micrograph showing mixed infection with HIV-1 and HTLV-I in MT-2 cells. Large arrows show HIV-1, arrow heads show HTLV-I, and small arrows show budding process (60,000×). (Courtesy D. Medina)

d. Remove PBMC and wash twice with PBS or Hanks' BSS.
e. Adjust the cell count to 2×10^6 cells/ml after enumeration.

Step 2: Stimulation of Donor Cells

Stimulate donor PBMC in culture (2×10^6/ml) in the stimulation medium with PHA (for formula, see below) within 12 hours of blood collection, 1–3 days prior to coculture. Avoid the use of frozen PBMC; use only anti-HIV negative donors' blood.

Step 3: HIV Coculture Procedure

Coculture patient's PBMC within 8 hours of blood collection as follows:

a. Sediment donor's PBMC (1- to 3-day-old PHA-stimulated) and resuspend in coculture medium with IL-2 (for formula, see below).
b. Enumerate donor's PBMC and adjust cell concentration to 2×10^6/ml.
c. Add an equal concentration of stimulated donor PBMC (10^6 cells/0.5 ml) and patient PBMC (10^6/0.5 ml) to give a final concentration of 2×10^6 cells/ml. Maintain the cell concentration at 2×10^6/ml.
d. Harvest 50% of the coculture medium every 3–4 days (keep

the harvested culture medium at −30° to 80°C for HIV p24 antigen determination) and replace with an equal volume of fresh coculture medium.

e. Add 5 × 10^5 freshly PHA-stimulated donor PBMC every week to the coculture for 4–8 weeks.

Step 4: Assay for HIV p24 Antigen

Test freshly harvested or frozen samples of supernatant fluids obtained from HIV cocultures by ELISA. ELISA kits for HIV p24 antigen detection are available commercially. In general, procedures for each kit are performed according to manufacturer specifications.

Virus Identification and Assay

SUPERNATANT FROM COCULTURED SAMPLES

Supernatant from cocultures (figure 20.4) can be tested for viral enzyme by RT assay and for viral antigen by the EIA method. When the supernatant fluid shows a positive RT assay, it indicates only the presence of a retrovirus. It does not indicate a specific virus type, which is determined by the detection of viral antigen with specific antibody.

REVERSE TRANSCRIPTASE (RT) ASSAY

Step 1: Polyethylene Glycol (PEG) Precipitation of HIV

a. Add 1 ml of PEG (30% PEG in 0.4 M NaCl) to 2 ml of cell-free supernatant. Mix well and let stand overnight at 4°C.
b. Centrifuge at 800 × *g* at 4°C for 45 minutes and discard the supernatant.
c. Add 100 μl of virus lysis buffer (for formula, see below) to the virus pellet and vortex.
d. Freeze samples at −20°C if not assayed immediately.

Step 2: RT Assay

a. Add 5 μl of the viral lysate to 45 μl of the assay reaction mixture (1X) (for formula, see below). Incubate mixture at 37°C.
b. After 60 minutes, add 40 μl of the reaction mixture onto a GFA (Whatman) Glass Fiber filter disc and drop disc immediately into cold 5% TCA containing 10 μM sodium pyrophosphate.
c. Wash filter disc three times for 5 minutes each in 5% TCA (use 5–10 ml TCA/filter).
d. Wash filter disc two times for 2 minutes each with 95% ethanol.

e. Air dry filter disc and count in a scintillation counter (expressed as CPM/ml).
f. Each test should include a known positive and negative control.

P24 ANTIGEN DETECTION FOR HIV-1

The detection of specific viral antigens allows the identification of each type of retrovirus. Antigen capture ELISA kits for the detection of HIV-1 antigen (p24) are commercially available and have been used for the detection of HIV-1 replication in cultured cells or in the blood of infected patients. Note that direct antigen detection in blood is not as sensitive as in culture, but antigen detection is most sensitive to detect the replication of the virus in cultured cells. For each test kit, usually the manufacturer provides specific procedures.

Cell Pellets from Cocultured Samples

IMMUNOFLUORESCENCE ASSAY

The presence of HIV-1 antigen in the cells can be detected using anti-HIV-1 antibody by indirect immunofluorescent staining test. In addition, the IF test is also being used as a confirmatory antibody test using known HIV-1-infected T-cell lines (see section below).

ELECTRON MICROSCOPY

When an EM facility is available, the examination of cell pellets fixed in 2% glutaraldehyde in PBS is the most reliable method for the identification of HIV and HTLV. EM shows the typical structure of each virion (figure 20.7).

Serodiagnosis

Because isolation of HIV is an elaborate procedure and requires restricted laboratory facilities, serodiagnosis plays an important role in the diagnosis of HIV infections. The methods used are ELISA or EIA, Western blot (WB), and IF. There are several ELISA kits for HIV-1 and HTLV-I that are licensed in the United States and available commercially.

Enzyme-linked Immunosorbent Assay (ELISA)

The most commonly used method for the detection of the HIV antibody is ELISA, a test that is easy to perform and appropriate for handling large numbers of specimens. Since the test is designed to be oversensitive, every ELISA positive specimen should be retested in duplicate. A confirmatory test is necessary, which is done either by WB or IF.

In the ELISA test, HIV antigens can be either purified whole viruses derived from cell cultures or recombinant antigens to capture any HIV specific antibodies. Following an incubation period, an enzyme-labeled probe is added to the antigen-antibody complexes and a color reaction is produced when a substrate is added to the labeled-probe-antigen-antibody complex (for details of the ELISA test, see chapter 8). False-positive responses detected by the use of ELISA kits may be due to the viral antigen prepared from T-lymphocytes carrying cellular antigens that are nonspecific to HIV infection.

Western Blot (WB) Test

WB is an immunoassay to identify antibodies to the different viral proteins (antigens, table 20.2). As a confirmatory test, WB is used to determine the presence of antibodies to the different HIV proteins, such as p24, gp41, and gp120 (figure 20.8). The presence of HIV antibodies in a given serum sample

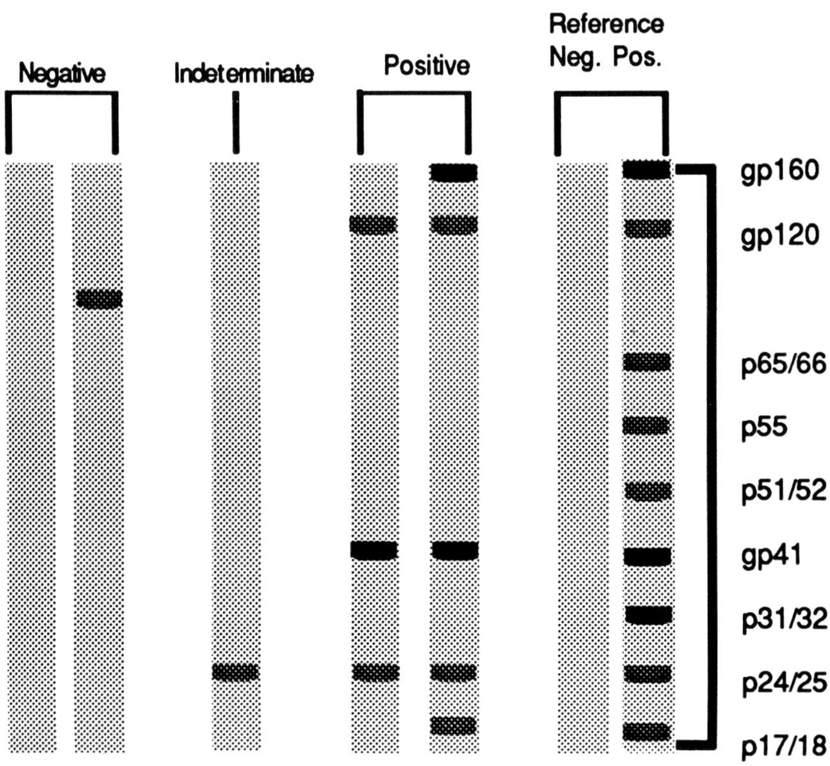

Figure 20.8 Western blot showing the presence of HIV-1 antibodies in discrete bands corresponding to the HIV-1 proteins (antigens) that have been electrophoretically separated by their molecular weights. Reference samples (known positive and negative sera) are included in each test. (Reproduced with permission, *AIDS 91 Summary*, Richmond, Va.: Philadelphia Sciences Group, 1992, 133)

can be determined by their reactivity with each of the HIV antigens that have been electrophoretically separated by their molecular weight. Commercially available WB kits contain strips of nontransferable paper on which electrophoresed HIV-1 or HTLV-I proteins have been immobilized. Specimens (patients' sera) are incubated with paper strips to allow antigen-antibody binding. If any antibodies are present, discrete band(s) corresponding to the HIV antigen(s) are visualized. Reference samples (known positive and negative sera) are included in each kit (examples are shown in figure 20.8 to illustrate the presence of HIV-1 antibodies in discrete bands).

Immunofluorescent (IF) Assay

The IF assay has been used to confirm the presence of HIV-1 antibodies. Various T-cell lines—including H9, CEM, HUT 78, and most recently, Pall cells (a T-cell leukemia cell line)—have been used for infections with HIV-1. IF employs antibodies to human IgG conjugated to a fluorescent dye. HIV-1 antigen in infected cells will bind any antibodies to HIV-1 in the test sample, which in turn will bind to the fluorescent-labeled antibodies to human IgG. Unbound fluorescent-labeled antibodies are removed by washing. Positive samples are identified by the fluorescent signal under fluorescence microscopy. An example is shown in figure 20.9.

Figure 20.9 Typical Fluorognost HIV-1 IF stained preparation showing HIV-1-induced antigen in infected Pall cells (400×). (Photo overexposed to increase cellular details; reproduced with permission, M. T. Sullivan et al., *J. Clin. Microbiol.* 30:2510 [1990])

Reagents for Reverse Transcriptase (RT)

Virus Lysis Buffer

Triton X-100 (10%)	5 ml
KCl 1 M	8 ml
Glycerol (100%)	30 ml
DTT (Dithiothreitol) 1 M	1 ml
Tris 1 M (pH 7.8)	5 ml
Distilled H_2O (sterile)	51 ml
Total volume	100 ml

Store at 4°C.

RT Reaction Mixture (5X)

Tris HCl 2 M (pH 8.5)	125 μl
DTT 1 M	50 μl
Triton X-100 (100%)	2.5 μl
Bovine serum albumin (50 mg/ml) (nuclease-free)	25 μl
$MgCl_2$ 1 M	30 μl
KCl 1 M	400 μl
Distilled H_2O	367.5 μl
Total volume	1,000 μl

RT Assay Reaction Mixture (1X)

RT reaction mixture (5X)	10 μl
Oligo-dT poly rA (5 u/ml) (template-primer)	5 μl
^3H TTP (1 μCi/ml)	5 μl
H_2O	5 μl
Total volume of reaction mixture	45 μl
Sample	5 μl
Total volume of assay mixture	50 μl

Medium for Cultivation of HIV

Stimulation Medium with PHA

RPMI-1640 with glutamine	740 ml
Fetal bovine serum (heat-inactivated)	200 ml
PHA-P (Difco or Sigma) 1%	10 ml
Delectinated IL-2 (natural delectinated is preferable)	50 ml

Antibiotics (penicillin 20 u/ml, streptomycin 200 μg/ml, or gentamyicin 50 μg/ml as needed)

Total volume 1,000 ml

(Note: Avoid adding amphotericin B, which may inhibit HIV replication.)

Coculture Medium with IL-2

RPMI-1640 with glutamine	750 ml
Fetal bovine serum (heat inactivated)	200 ml
Delectinated IL-2 (natural delectinated is preferable)	50 ml
Antibiotics (penicillin 200 u/ml, streptomycin 200 μg/ml as needed)	
Total volume	1,000 ml

SUPPLEMENTARY READING

AIDS 91 Summary. Diagnostic tests for HIV, chap. 4, pp. 122–51. Philadelphia Sciences Group, 1992.

Barre-Sinoussi, F.; Chermann, J. C.; Rey, F.; Nugeyre, M. Y.; Chamaret, S.; Gruest, J.; Dauguet, C.; Axler-Blin, C.; Vezinet-Brun, F.; Rouzioux, C.; Rozenbaum, W.; and Montagnier, L. Isolation of a T-lymphotropic retrovirus from a patient at risk for acquired immune deficiency syndrome (AIDS). *Science* 220:868–71 (1983).

Bayliss, G. J.; Jesson, W. J.; Mortimer, P.; McLean, K. A.; and Evans, B. A. Cultivation of human immunodeficiency virus from whole blood: Effect of anticoagulant and inoculum size on virus growth. *J. Med. Virol.* 31:161–64 (1990).

Brandt, C. D.; Rakusan, T. A.; Sison, A. V.; Josephs, S. H.; Saxena, E. S.; Herzog, K. D.; Parrott, R. H.; and Sever, J. L. Detection of human immunodeficiency virus type 1 infection in young pediatric patients by using polymerase chain reaction and biotinylated probes. *J. Clin. Microbiol.* 30:36–40 (1992).

Castro, B. A.; Weiss, C. D.; Wivicott, L. D.; and Levy, J. A. Optimal conditions for recovery of the human immunodeficiency virus from peripheral blood mononuclear cells. *J. Clin. Microbiol.* 26:2371–76 (1988).

Chrystie, I. L., and Almeida, J. C. Further studies of HIV morphocology by negative staining. *AIDS* 2:459–64 (1988).

Clavel, F.; Guetard, D.; Brun-Vezinet, F.; Chamaret, S.; Rey, M. A.; Santos-Ferreira, M. O.; Laurent, A. G.; Dauguet, C.; Katlama, C.; Rouzioux, C.; Klatman, D.; Champalimaud, J. L.; and Montagnier, L. Isolation of a new human retrovirus from West African patients with AIDS. *Science* 233:343–46 (1986).

Constantine, N. T.; Callahan, J. D.; and Watts, D. M. *Retroviral Testing: Essentials for Quality Control and Laboratory Diagnosis.* Boca Raton: CRC Press 1992.

Coombs, R. W.; Collier, A. C.; Allain, J. P.; Nikora, B.; Leuther, M.; Gjerset, G. F.; and Corey, L. Plasma viremia in human immunodeficiency virus infection. *N. Engl. J. Med.* 321:1626–31 (1989).

Davey, E. T., Jr., and Lane, H. C. Laboratory methods in the diagnosis and prognostic staging of infection with human immunodeficiency virus type 1. *Rev. Infect. Dis.* 12:912–30 (1990).

Dodd, R. Y., and Fang, C. T. The western immunoblot procedure for HIV antibodies and its interpretation. *Arch. Path. Lab. Med.* 114:240–45 (1990).

Fiscus, S.; Wallmark, E. B.; Folds, J. D.; Fryer, J.; and Van der Horst, C. M. Detection of infectious immune complexes in human immunodeficiency virus type 1 (HIV-1) infections; correlation with plasma viremia and CD4 cell counts. *J. Infect. Dis.* 164:765–69 (1991).

Gallo, R. C. Human retroviruses: A decade of discovery and link with human disease. *J. Infect. Dis.* 164:235–43 (1991).

Gallo, R. C.; Salahuddin, S. Z.; Popovic, M.; Shearer, G. M.; Kaplan, M.; Haynes, B. F.; Palker, T. J.; Redfield, R.; Oleske, J.; Safai, B.; White, G.; Foster, P.; and

Markham, P. D. Frequent detection and isolation of cytopathic retroviruses (HTLV-III) from patients with AIDS and at risk for AIDS. *Science* 224:500–503 (1984).

George, J. R., and Schochetman, G. Serologic tests for the detection of human immunodeficiency virus infection. In Schochetman, G., and George, J. R., eds., AIDS Testing: Methodology and Management Issues, pp. 48–69. New York: Springer-Verlag, 1992.

Griffith, B. P. Differential diagnosis of human retrovirus infections in the laboratory. *Yale J. Biol. Med.* 62:103–14 (1989).

Harry, D. J.; Jennings, M. B.; Yee, J.; and Carlson, J. R. Antigen detection for human immunodeficiency virus. *Clin. Microbiol. Rev.* 2:241–49 (1989).

Ho, D. D.; Moudgil, M. S.; and Alam, M. Quantitation of human immunodeficiency virus type 1 in the blood of infected persons. *N. Engl. J. Med.* 321:1621–25 (1989).

Hollinger, F. B.; Bremer, J. W.; Myers, L. E.; Gold, J. W. M.; McQuay, L.; and the NIH/NIAID/DIADS/ACTG Virology Laboratories. Standardization of sensitive human immunodeficiency virus coculture procedures and establishment of a multicenter quality assurance program for the AIDS clinical trials group. *J. Clin. Microbiol.* 30:1787–94 (1992).

Hsiung, G. D. Perspectives on retroviruses and the etiologic agents of AIDS. *Yale J. Biol. Med.* 60:505–14 (1987).

Jackson, J. B. Human immunodeficiency virus type 1 antigen and culture assays. *Arch. Path. Lab. Med.* 114:249–53 (1990).

Jackson, J. B., and Balfour, H. H. Practical diagnostic testing for human immunodeficiency virus. *Clin. Microbiol. Rev.* 1:124–38 (1988).

Jackson, J. B.; Coombs, R. W.; Sannerud, K.; Rhame, P. S.; and Balfour, H. H., Jr. Rapid and sensitive viral culture method for human immunodeficiency virus type 1. *J. Clin. Microbiol.* 26:1416–18 (1988).

Kalyanaraman, V. S.; Sarngadharan, M. G.; Robert-Guroff, M.; Miyoshi, I.; Blayney, D.; Golde, D.; and Gallo, R. C. A new subtype of human T-cell leukemia virus (HTLV-II) associated with a T-cell variant of hairy cell leukemia. *Science* 218:571–73 (1982).

Kanki, P. J.; Barin, F.; M'Boup, S.; Allan, J. S.; Romet-Limonne, J. L.; Marlink, R.; McLane, M. F.; Lee, T. H.; Arbeille, B.; and Denis, F. New human T-lymphotropic retroviruses related to simian T-lymphotropic virus type III (STLV-III AGM. *Science* 232:238–43 (1986).

Manzari, V.; Gismondi, A.; Barillari, G.; Morrone, S.; Modesti, A.; Albonici, L.; DeMarchis, L.; Fazio, V.; Gradilone, A.; Zani, M.; Frati, L.; and Santoni, A. HTLV-V: A new human retrovirus isolated in a Tac-negative T-cell lymphoma/leukemia. *Science* 238:1581–83 (1987).

Michalski, F. J. The diagnosis of human immunodeficiency virus infection: Progress in less than five years. *Yale J. Biol. Med.* 62:93–102 (1989).

Poiesz, B. J.; Ruscetti, F. W.; Gazdar, A. F.; Bunn, P. A.; Minna, J. D.; and Gallo, R. C. Detection and isolation of type C retrovirus particles from fresh and cultured lymphocytes of a patient with cutaneous T-cell lymphoma. *Proc. Natl. Acad. Sci.* (USA) 77:7415–19 (1980).

Popovic, M.; Sarngadharan, M. G.; Read, E.; and Gallo, R. C. Detection, isolation and continuous production of cytopathic retroviruses (HTLV-III) from patients with AIDS and pre-AIDS. *Science* 224:497–500 (1984).

Sison, A. V., and Campos, J. M. Laboratory methods for early detection of human immunodeficiency virus type 1 in newborns and infants. *Clin. Microbiol. Rev.* 5:238–47 (1992).

Sninsky, J. J., and Kwok, S. Detection of human immunodeficiency viruses by the polymerase chain reaction. *Arch. Path. Lab. Med.* 114:259–62 (1990).

Sullivan, M. T.; Mucke, H.; Kadey, S. D.; Fang, C. T.; and Williams, A. E. Evaluation of an indirect assay for confirmation of human immunodeficiency virus type I antibody in U. S. blood donor sera. *J. Clin. Microbiol.* 30:2509–10 (1992).

PART FOUR

DNA VIRUSES: RECOGNITION AND CHARACTERIZATION OF VIRUSES INFECTING HUMANS

VIRUS MORPHOLOGY AND VIRUS-INDUCED CELLULAR CHANGES

21

ADENOVIRIDAE

C. K. Y. FONG

ADENOVIRUSES

The initial isolation of an adenovirus was made in the early 1950s from a surgical specimen of human adenoid tissue, which when grown in cell culture after prolonged incubation underwent spontaneous degeneration. Similar agents were subsequently recovered from patients with acute respiratory diseases (ARD). It became apparent that adenoviruses are associated with a variety of clinical syndromes—commonly with pharyngoconjunctival fever, acute febrile pharyngitis, epidemic keratoconjunctivitis, atypical pneumonia, and acute hemorrhagic conjunctivitis (table 21.1). In immunosuppressed individuals, such as organ transplant patients, adenovirus types 34 and 35, as well as some other types, have been associated with diarrhea, pneumonia, hepatitis, and disseminated infections.

Electron microscopic examination of negatively stained preparations reveals that adenoviruses are nonenveloped, 70–90 nm in diameter (figure 21.1, inset). Each virion consists of 252 capsomeres arranged in icosahedral symmetry: 240 hexons and 12 pentons (or base-vertex capsomeres); each of the latter is attached to a fiber with a terminal knob. The length of the fiber is variable according to the serotype. Hexons are responsible for the group-reactive antigen and also for the type-specific neutralizing antigen. Pentons have cell-detaching activity and are responsible for the type-specific antigens, while the fiber accounts for the subgroup-reactive antigen. Terminal knobs of fibers are for the attachment of the virus to cells and are responsible for hemagglutinating activity. Physicochemical analysis indicates that adenovirus virions contain double-stranded DNA of $(20\text{–}30) \times 10^6$ molecular

ISBN 0-300-05845-4.

Table 21.1 Adenovirus Serotypes Associated with Clinical Diseases

Clinical Diseases	Adenovirus Serotypes Involved	
	Common	Less Common
Acute respiratory disease	4,7	3,14,21,35
Upper respiratory illness	1–3,5,7	4,6,11,18,35
Lower respiratory illness	3,4,7,21	1,2,5,35
Pertussis-like syndrome	5	1–3
Pharyngoconjunctival fever	3,4,7	1,11,14,16,19,37
Epidemic keratoconjunctivitis	8,19,37	3,4,7,11,21
Acute hemorrhagic conjunctivitis	11	2–8,14,15,19,37
Hemorrhagic cystitis	11	7,21,34,35
Diarrheal disease in infants	31,40,41	1,2,12–17,21,25,26,29
Meningitis, encephalitis	7	3,32
Ulcerative genital lesions	2,19,37	1,5,7,11,18,31
Diseases in immunosuppressed hosts	11,34,35	1,2,5,7,21,29,31,37–39

Modified from J. C. Hierholzer in *Diagnostic Procedures for Viral, Rickettsial, and Chlamydial Infections*, Schmidt, N. T., and Emmons, R. W., eds., 6th ed., Washington, D.C.: American Public Health Association, 1989, 224.

weight that codes for at least ten structural polypeptides. The virions are resistant to lipid solvents. In adenovirus infected cells, virions are assembled and accumulated in the nucleus and appear either singly (figure 21.1) or in crystalline arrays (figure 21.2).

Many adenovirus types have the ability to agglutinate erythrocytes of the rat, monkey, human type O, or other animal species. Based on their hemagglutinating property, DNA homology, and oncogenic property, human adenoviruses are divided into six subgenera (table 21.2).

To date, 47 human adenovirus serotypes have been characterized. Adenovirus types 12, 18, and 31 are able to induce tumors in experimental animals and are highly oncogenic. Many types are able to transform cells in vitro. In humans, however, the association of adenovirus with cancer has not been documented. Adenoviruses have been recovered from many other animal species, including chimpanzees, monkeys, dogs, cattle, and chickens. Each adenovirus type replicates best in the tissues of the host species from which the virus was isolated. All adenoviruses isolated from mammalian species belong to the genus *Mastadenovirus* and share a common group antigen. Adenoviruses isolated from fowl belong to the genus *Aviadenovirus*.

Specimen Source and Inoculation

Throat, eye, and rectal swabs, stool, urine, and tissue biopsies are good sources for human adenovirus isolation. These specimens can be inoculated

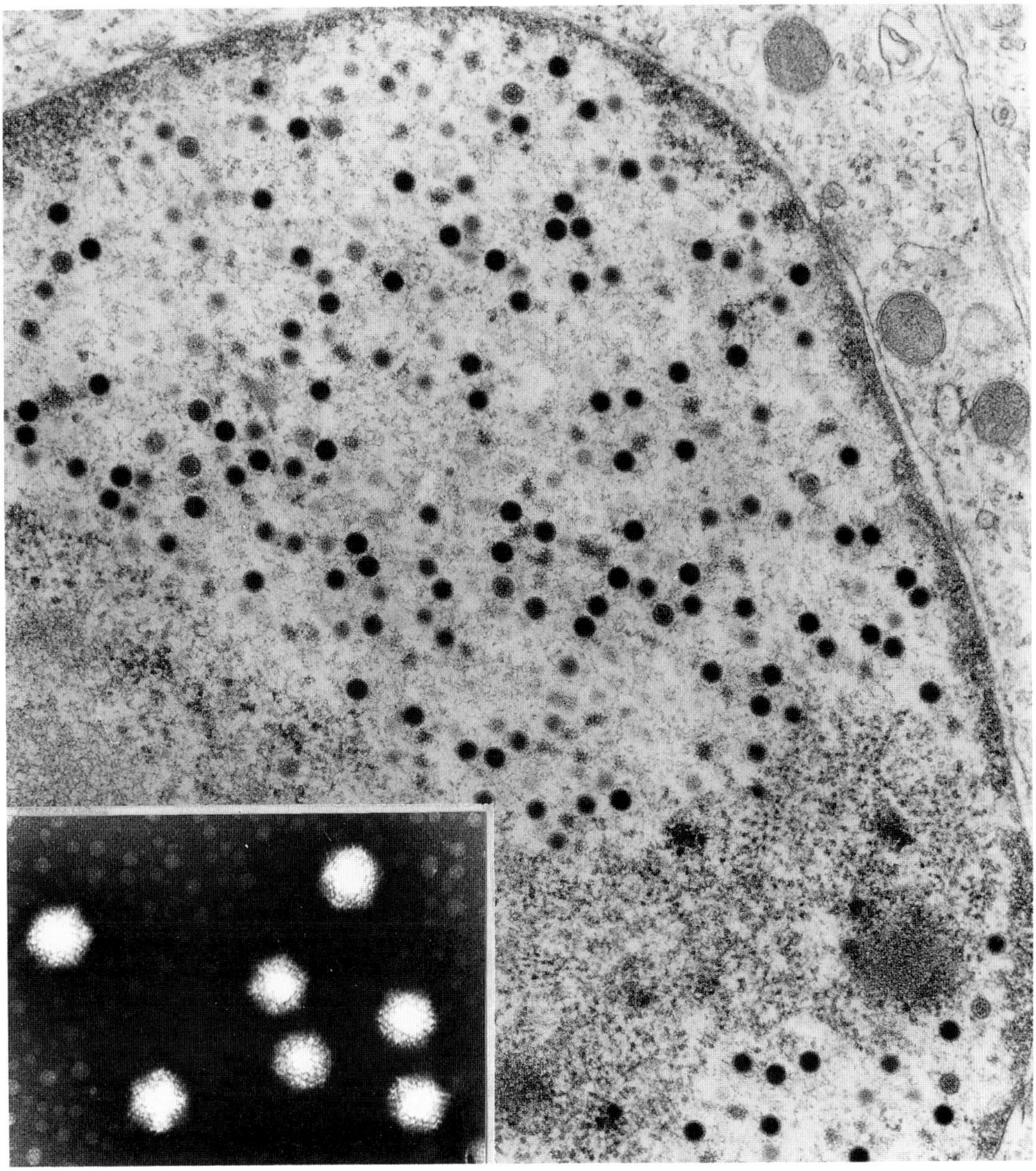

Figure 21.1 Electron micrograph of adenovirus type 2 virion in an infected HEK cell, showing virions scattered in the nucleus (38,000×). Inset: Adenovirus virions negatively stained with phosphotungstic acid (100,800×)

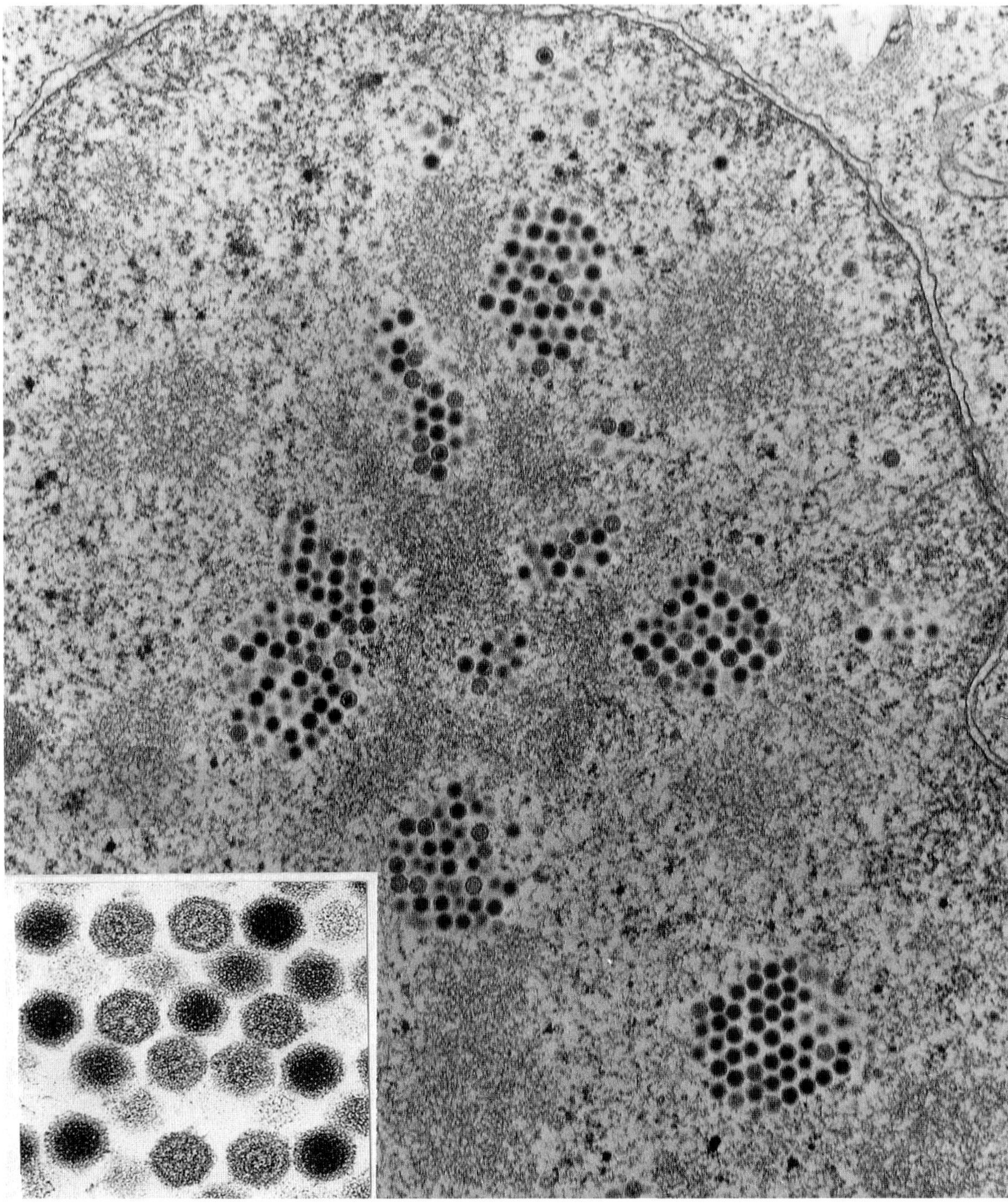

Figure 21.2 Electron micrograph of adenovirus type 7 virions in an infected HEK cell, showing virus aggregates in crystalline arrays in the nucleus (30,000×). Inset: Higher magnification of adenovirus virions in thin section (100,800×)

Table 21.2 Subgenera of Human Adenoviruses: DNA Homology, Oncogenicity, and Hemagglutinating Properties*

Subgenus	Serotypes	% DNA Homology		Oncogenic Property	Hemagglutination Pattern[a]
		Intrageneric	Intergeneric		
A	12,18,31	48–69	8–20	High	III
B[b]	3,7,11,16,34,35	89–94	9–20	Weak	I
	14,21				
C	1,2,5,6	99–100	10–16	Negative	III
D[c]	8,9,37	95–99	4–17	Negative	II
	10,19,26,27,36,38,39				
	13,43				
	15,22,23,30				
	17,24,32,33,42				
	20,25,28,29				
	44,45,46,47				
E	4		4–23	Negative	III
F	40,41	62–69	15–22	Negative	III

*This table is adapted from Wadell (1988), Hierholzer (1973, 1989), and Hierholzer et al. (1991).

a. I = complete HA of monkey RBC; II = complete HA of rat RBC; III = incomplete HA of rat RBC and the HA activity can be enhanced in the presence of heterotypic antiserum.

b. In subgenus B, serotypes 3, 7, 11, 16, 34, and 35 agglutinate monkey RBC with high HA titers (range 128–8,192); serotypes 14 and 21 agglutinate monkey RBC with lower HA titers (range 16–1,024).

c. All members of subgenus D agglutinate rat RBC. In addition, certain serotypes also agglutinate human or mouse RBC. Serotypes 20, 25, 28, and 29 agglutinate rat RBC only. Serotypes 8, 9, and 37 agglutinate human, mouse, guinea pig, and dog RBC.

into a variety of cell cultures of human origin (see virus isolation section below).

Virus Isolation, Propagation, and Assay

SELECTION OF CELL CULTURE

Primary human embryonic kidney (HEK) cells are the most sensitive cells for human adenoviruses but are expensive and not always readily available. Human amnion cells, human diploid fibroblast cells, and several passaged cell lines derived from human cancer tissues—including A549, Hep-2, HeLa, and KB cells—are also susceptible to adenovirus infection. The cell line most commonly used in clinical laboratories for adenovirus isolation is A549 since it is almost as sensitive as HEK cells. Primary monkey kidney cells are relatively resistant for adenovirus replication. Types 2 and 5 have a broader host spectrum. They can replicate in cell cultures of other animal species, such as primary guinea pig embryo cells. Adenovirus types 40 and 41, previously classified as noncultivable in cell cultures, can be grown and produce low titers in Graham's 292 cells, a cell line derived from HEK, which was transformed by adenovirus type 5.

Recently, centrifugation culture has been applied to a variety of viruses for rapid detection. After a brief period of cultivation in a susceptible cell culture, virus detection is performed by immunofluorescence or immunoperoxidase staining using monoclonal or polyclonal antibody to the hexon antigen of adenoviruses. The early detection of adenovirus antigen by centrifugation culture has been reported in two days' postinoculation, while a conventional culture would take three to five days or longer for CPE to become visible.

CYTOPATHIC EFFECT IN CELL CULTURE WITH FLUID MEDIUM

Most adenovirus serotypes produce a characteristic cytopathic effect consisting of marked rounding and clumping in cell cultures, especially in continuous cell lines. The typical CPE—large grapelike clusters of infected cells—is recognized readily in HEK cells (figure 21.3*B*). In passaged cell lines inoculated with clinical specimens, CPE may be delayed and usually requires three to five days or longer. The length of time required to produce CPE depends on the virus type, the cell culture used, and the amount of virus present in the specimen. It may take two to seven days or longer (figures 21.4, 21.5).

CYTOPATHOLOGY OF INFECTED CELLS IN STAINED PREPARATIONS

Adenoviruses in infected tissues or cell cultures produce typical intranuclear inclusions, which can be recognized by hematoxylin-eosin staining or the Feulgen reaction method (see chapter 11). Adenovirus-induced inclusions

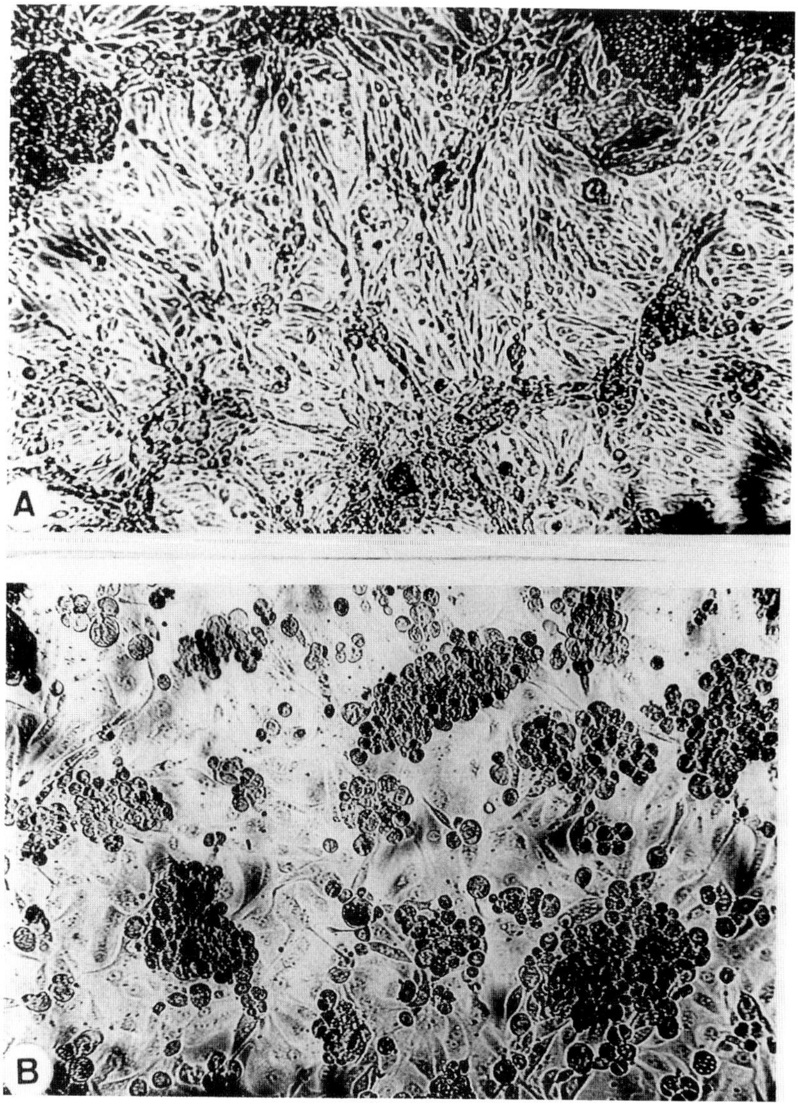

Figure 21.3 Cytopathic effect induced by an adenovirus in HEK cell culture (100×): *(A)* Uninfected HEK cultures; *(B)* CPE induced by adenovirus type 3

are basophilic and Feulgen-positive (containing DNA), distinctly recognizable from those produced by other viruses. Inclusions induced by certain adenovirus serotypes show some level of dissimilarity. For example, serotypes 1, 2, 5, and 6 produce multiple, well-defined eosinophilic intranuclear inclusions at an early stage of infection. Later, these inclusions become basophilic (Feulgen-positive), granular clusters spaced throughout the nuclei (figure 21.6*B*). Others, like serotypes 3, 4, and 7, produce large basophilic inclusions that occupy the entire nucleus.

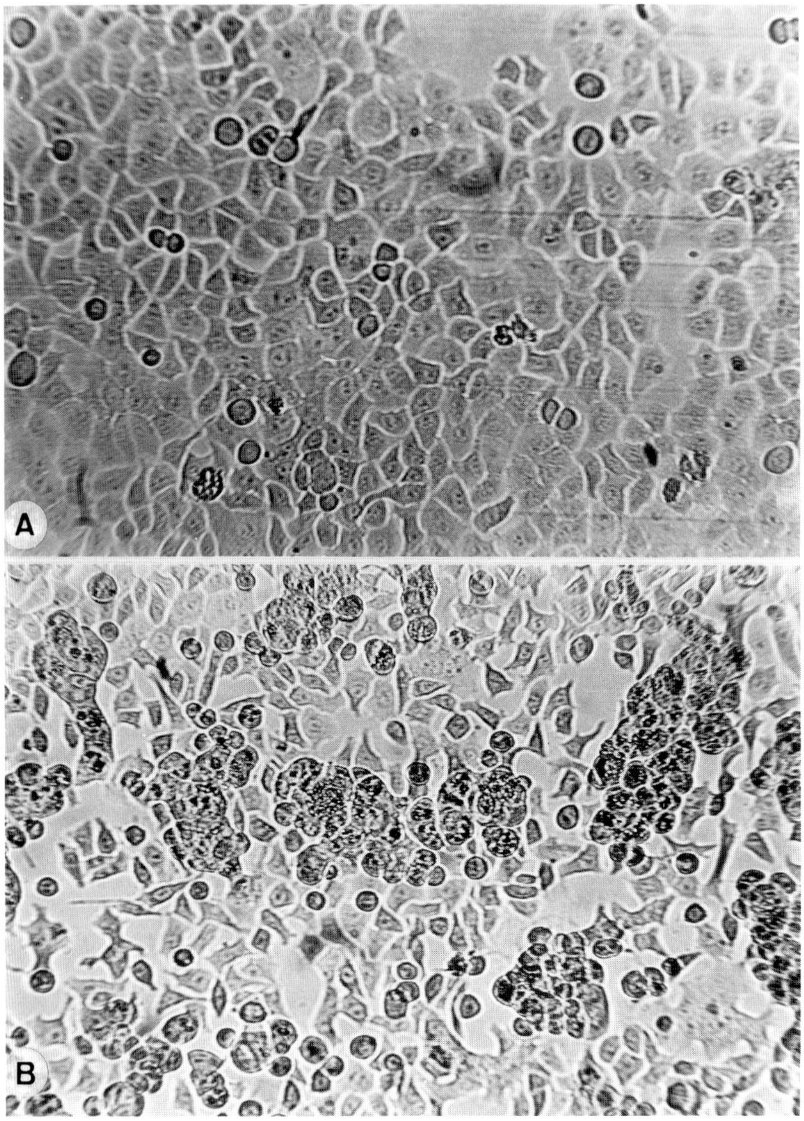

Figure 21.4 Cytopathic effect induced by an adenovirus in Hep-2 cell culture (100×): *(A)* Uninfected Hep-2 culture; *(B)* CPE induced by adenovirus type 3

Identification as Adenovirus Group

IMMUNOFLUORESCENT AND IMMUNOPEROXIDASE STAINING

All adenovirus-infected cells contain group-specific hexon antigens, which react with antibody against the common hexon antigen. Thus, a simple and rapid identification of all human adenovirus types can be done using monoclonal or polyclonal antibody to the common hexon antigen. Indirect IF

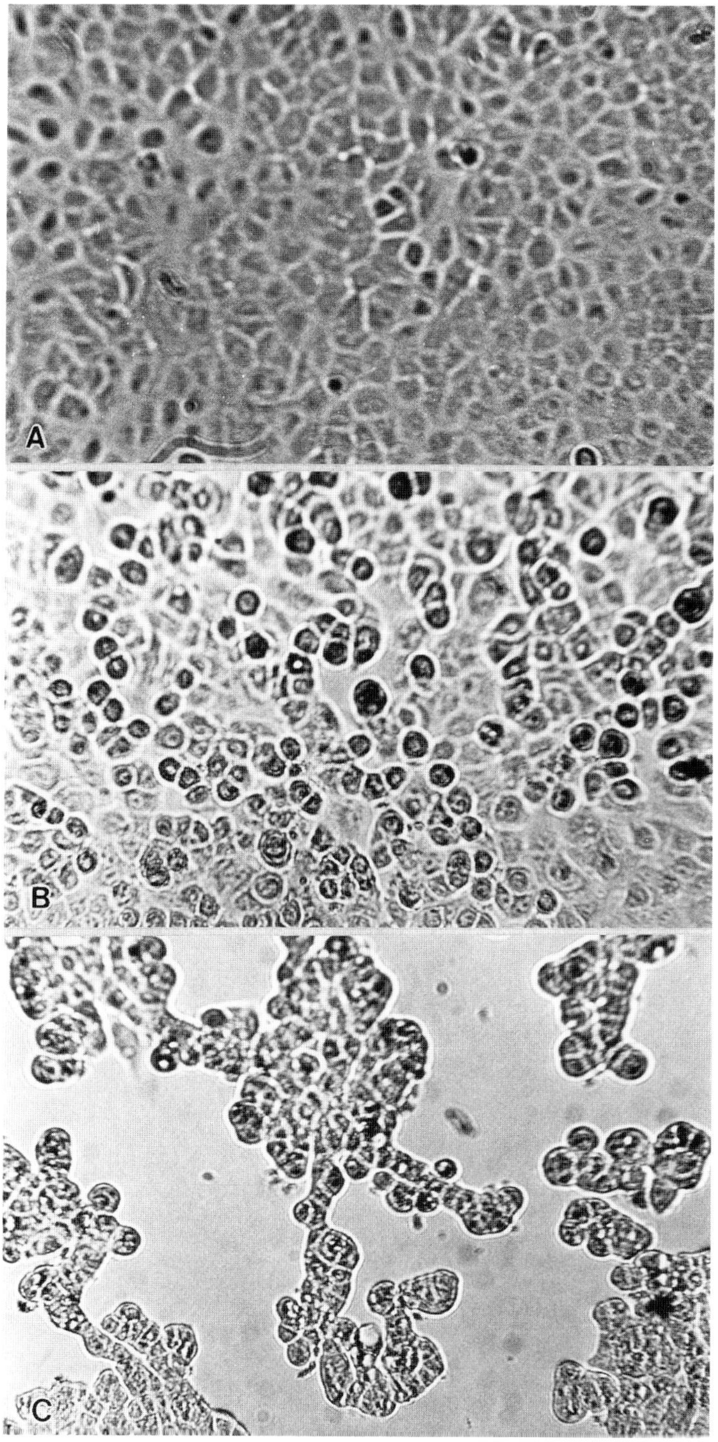

Figure 21.5 Cytopathic effect induced by an adenovirus in A549 cell culture (100×): *(A)* Uninfected A549 cell culture; *(B)* CPE induced by adenovirus type 7, 24 hours after infection; *(C)* CPE induced by adenovirus type 7, 48 hours after infection

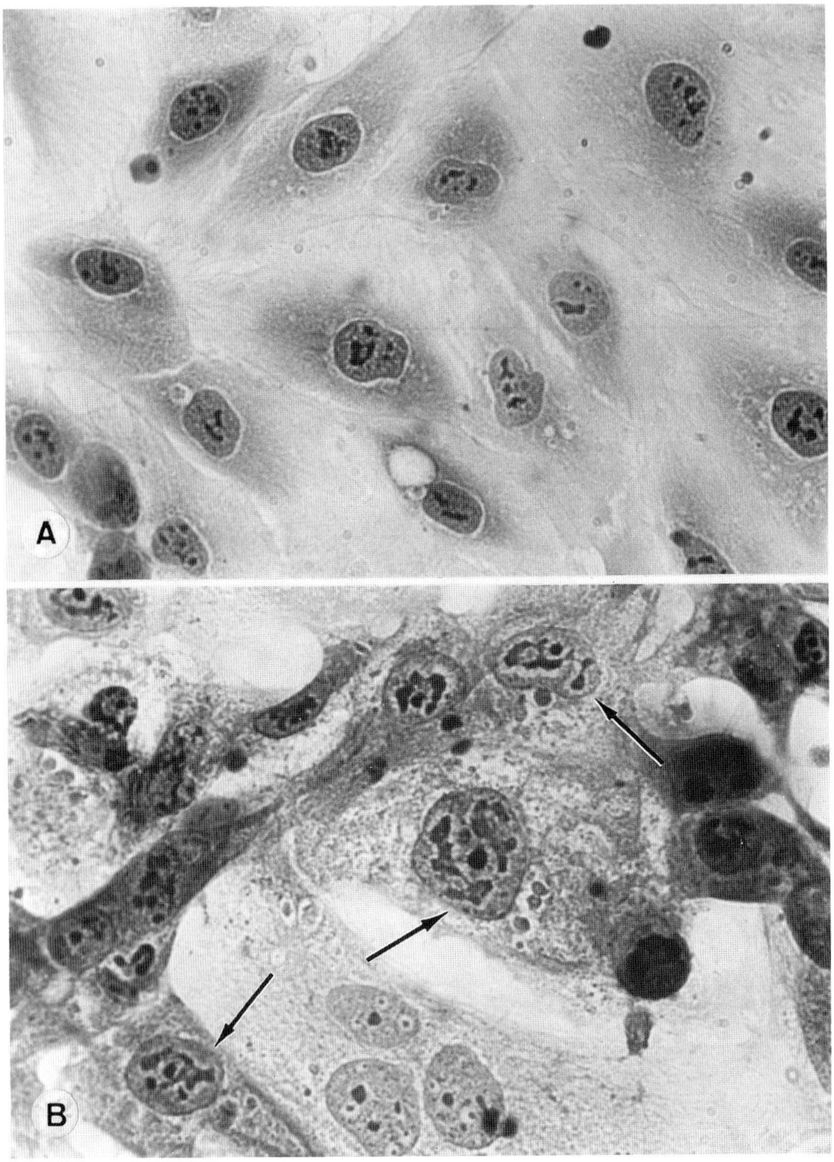

Figure 21.6 Intranuclear inclusions induced by an adenovirus in HEK cells (H&E, 400×): *(A)* Uninfected HEK cells; *(B)* Basophilic intranuclear inclusions (arrows) induced by adenovirus type 2

staining is the most common method currently used for adenovirus identification. Monoclonal antibody to adenovirus hexon antigen is commercially available. The conjugate, fluorescein isothiocyanate (FITC)–anti-mouse IgG or horseradish peroxidase–anti-mouse IgG can be purchased from commercial sources. Infected cells suspected of adenovirus infection are detached by trypsin treatment or by scraping. A drop of cell suspension is placed on a slide; after air drying, the cell spots are fixed with acetone and stained either by IF (color plate 6) or IP procedure (see chapter 7).

ENZYME IMMUNOASSAY (EIA) FOR ANTIGEN DETECTION

Several EIA formats, direct or indirect assay systems for adenoviruses, have been described; kits are commercially available. Manufacturer instructions should be followed in each case. EIA's are particularly useful for identifying enteric adenoviruses in diarrheic stools because they are difficult to grow in cell cultures.

ELECTRON MICROSCOPY

Adenovirus has a distinct morphology and can be readily recognized under electron microscopic examination, but EM does not identify the specific virus type. Culture fluid from adenovirus-infected cells can be applied directly to an electron microscope grid, stained with 2% phosphotungstic acid (PTA), and examined with EM. Adenoviruses, especially types 40 and 41, can be detected directly from stool specimens using EM procedure (for procedures, see chapter 13).

COMPLEMENT-FIXATION TEST

The CF test has been used for many years as a reliable method of adenovirus genus identification, but it is not type-specific. Since the CF test is a time-consuming and labor-intensive method, many clinical laboratories may not have this test available (for procedures, see chapter 10).

DNA-HYBRIDIZATION METHODS

DNA in situ hybridization using a DNA probe for the common hexon genes has been applied in adenovirus identification. In situ hybridization is also useful in tissue sections since adenovirus inclusions can be confused with inclusions induced by cytomegalovirus. The DNA probe is biotinylated by nick translation to incorporate biotin-labeled dUTP. The major steps in the procedure are as follows:

1. Samples of clinical specimens, including tissue sections or spots of infected cultured cells, are fixed in 4% buffered paraformaldehyde and treated briefly with an enzyme, trypsin.
2. Following the addition of the hybridization mixture containing the denaturing agent and biotinylated DNA probe, the sample is covered with a coverslip, sealed, and steam-heated at 80°C to denature the DNA in the sample. Then hybridization reaction follows at 37°C.
3. The coverslips are removed; the unbound DNA probe is washed off with buffered solution. The amount of hybridized DNA in the sample

is detected by incubating the slide with streptavidin-biotinylated peroxidase at 37°C.

4. After washing, the slide is incubated with a substrate, diaminobenzidine-hydrogen peroxide, washed with water, and counterstained with hematoxylin.
5. Positive cells are seen with their nucleus colored brown.

Identification of Adenovirus Serotypes

CONVENTIONAL NEUTRALIZATION TEST

For a type-specific identification of adenovirus, a serum neutralization test (NT) is commonly performed.

1. Mix equal volumes (0.35 ml) of virus (100 $TCID_{50}$/0.1 ml) and 2-fold serial dilutions of heat-inactivated type-specific antiserum. Incubate for 1 hour at 22°C.
2. Inoculate the serum-virus mixtures into triplicate tubes of A549 cell culture with medium removed, 0.2 ml per tube. Incubate for 1 hour at 37°C to enhance virus adsorption and refeed with maintenance medium.
3. Culture tubes are then incubated at 37°C for 7 days and examined daily for CPE. The virus type-specific antiserum that inhibits CPE identifies the adenovirus type.

MODIFIED THREE-DAY NEUTRALIZATION TEST

This modified NT is performed in primary rhesus monkey kidney cells or passaged monkey kidney cell lines such as Vero or $LLC\text{-}MK_2$. The test does not measure the inhibition of viral infectivity, but rather the inhibition of toxicity induced by the penton capsomeres since human adenoviruses replicate poorly or not at all in these cell systems in the absence of the helper SV_{40}. The following procedure is as described according to D. A. Stevens and coworkers (1967).

1. Determine the virus toxicity titer by making 2-fold serial dilutions; each dilution is inoculated into triplicate tubes of cell cultures, 0.1 ml each.
2. After 3 days, the cultures are examined for morphologic changes caused by the toxicity of the virus. The highest dilution of virus producing 25%–50% cell rounding is considered the endpoint, which is equal to 1 unit.
3. Prepare the virus challenge dose to contain 2 units. For NT, mix equal volumes of each of the 2-fold serial dilutions of adenovirus type-specific antiserum (0.35 ml) and 2 units of virus dilution. Incubate virus-antiserum mixtures for 1 hour at 22°C.
4. Inoculate three culture tubes for each virus-antiserum mixture, 0.2

ml of virus-serum mixture per tube. Uninfected cells or 0.1 ml of virus-alone inoculated cells are used as controls.

5. Incubate all inoculated cultures for 1 hour at 37°C, refeed with maintenance medium, and incubate at 37°C for 3 days.
6. Examine culture tubes microscopically. The virus type is indicated by inhibition of toxicity by the type-specific antiserum.

MICRONEUTRALIZATION TEST

The microneutralization test was described using rhesus monkey kidney or Vero cells to measure toxicity, and Hep-2 cells to measure infectivity. The procedure in Vero cells described by J. C. Hierholzer and P. G. Bingham (1978) is as follows: Virus titrations and neutralization tests are performed in 96-well flat bottom cell culture plates.

1. Make 2-fold serial dilutions of virus suspension in growth medium (0.05 ml) with a multiple channel micropipetter in a 96-well plate (6 replicates).
2. Add 0.05 ml Vero cell suspension containing 175,000 cells/ml to each well, including cell control wells.
3. After gentle mixing, cover the plate, and incubate in a CO_2 incubator at 37°C for 5 days.
4. Stain with 1.3% crystal violet–5% formaldehyde solution for ease in reading the plate.
5. The endpoint of the virus titer is the highest dilution showing 75%–100% cell degeneration (= 1 unit).
6. For neutralization test, 2 units of virus is incubated with type-specific antiserum in cell culture plates, followed by adding Vero cells or Hep-2, as described above.
7. Cell control and virus back-titration should be included.
8. Incubate cell culture plates as above, stain, and read.
9. The virus type is indicated by inhibition of CPE in Hep-2 cells or inhibition of toxicity in Vero cells by the homologous antiserum.

HEMAGGLUTINATION (HA) AND HEMAGGLUTINATION-INHIBITION (HI) TESTS

For the type-specific identification of adenovirus isolates, ideally the HA test should be performed first to determine the subgenus, followed by HI or NT to determine specific serotype. Fresh red blood cells from the rhesus monkey, rat, human type O, mouse, guinea pig, and dog are used. HA and HI are performed in a microtiter plate according to the procedure described by Hierholzer (1973). Table 21.2 shows the hemagglutinating properties of the different adenovirus subgenera. Some have incomplete HA patterns of rat RBC that can be enhanced in the presence of heterotypic antiserum (see below).

Hemagglutination Test. For the adenovirus subgenus giving an incomplete HA pattern, heterologous type 4 or 6 antiserum must be added in order to render a complete HA reaction for certain serotypes, as described below. (Before using them, heterologous types 4 and 6 antisera should be heat-inactivated and adsorbed with rat RBC.)

1. Make four sets (each set in duplicate) of serial 2-fold dilutions of the virus isolate (1:2 to 1:4,096) in 0.05 ml amounts in a 96-well microtiter V plate using different diluents as follows: To sets 1 and 2, use plain PBS; to set 3, use PBS with 1% type 4 antiserum (for subgenus A, types 12, 18, 31; subgenus C types 1, 2, 5, 6; and subgenus F types 40 and 41); to set 4 use PBS with 1% type 6 antiserum (for subgenus E, type 4).
2. Add an equal volume of 0.4% RBC to each well as follows: To set 1, add rhesus monkey RBC; to sets 2, 3, and 4, add rat RBC.
3. Include RBC controls consisting of equal volumes of each of the different diluents and RBC types.
4. Cover plates, mix gently, and incubate at 37°C in a water bath for 1 hour.
5. Read for HA. RBC control wells should show a complete sedimentation of RBC at the center of the bottom of the well. The HA titer is defined as the reciprocal of the highest dilution of virus that gives complete HA. This dilution endpoint is equivalent to 1 HA unit per 0.05 ml. The patterns of HA titers with monkey and rat RBC in plain PBS and in the presence of two heterologous antiserum diluent systems—that is, types 4 and 6 antisera—determine the subgenus of the adenovirus isolate (see table 21.2).

Hemagglutination-Inhibition Test. Once the subgenus of the adenovirus isolate has been determined by HA, the HI test is performed using specific antisera against all types within the same subgenus. (The antisera are heat-inactivated and adsorbed with the appropriate RBC.)

1. Dilute the antisera, 2-fold serially, with PBS in 0.025 ml amounts in a 96-well microtiter V plate. For subgenera A, C, and E, isolates giving incomplete HA pattern with rat RBC, antisera dilutions are prepared in PBS with the appropriate heterotypic type 4 or 6 antiserum diluent, as stated above.
2. Add 4 HA units of virus in 0.025 ml to each well.
3. Incubate virus-antiserum mixtures at 22°C or 37°C for 1 hour followed by adding the appropriate RBC suspension, 0.05 ml per well.
4. Make a back-titration of the test antigen consisting of 2-fold serial dilutions containing 4, 2, 1, and 0.5 HA units in 0.05 ml amounts.

5. Make an RBC control consisting of 0.05 ml PBS and 0.05 ml of RBC suspension.
6. Set up serum controls consisting of 0.025 ml antiserum plus 0.025 ml PBS and 0.05 ml RBC suspension.
7. Incubate all plates at 37°C for 1 hour.
8. The adenovirus isolate is identified by the type-specific antiserum which inhibits HA completely.

Serodiagnosis

A significant rise in antibody titers in paired sera—collected during the acute and convalescent stages of infection—to the adenovirus genus-specific antigen or hexon, regardless of specific serotype, by the complement-fixation test is an indication of a recent adenovirus infection. Type-specific antibody titer rises are detected by means of HI or neutralization tests. A 4-fold or greater rise in antibody titer is of diagnostic significance. As a rule, the CF test is less sensitive than the HI or neutralization test in determining the antibody rise of paired sera.

SUPPLEMENTARY READING

Adam, E.; Lengyel, A.; Takacs, M.; Erdei, J.; Fachet, J.; and Nasz, I. Grouping of monoclonal antibodies to adenovirus hexons by their cross-reactivity. *Arch. Virol.* 87:61–71 (1986).

Albert, M. J. Enteric adenoviruses. *Arch. Virol.* 88:1–17 (1986).

Ambinder, R. F.; Burns, W.; Forman, M.; Charache, R.; Arthur, R.; Beschorner, W.; Santos, G.; and Saral, R. Hemorrhagic cystitis associated with adenovirus infection in bone marrow transplantation. *Arch. Intern. Med.* 146:1400–1 (1986).

Cepko, C. L.; Whetstone, C. A.; and Sharp, P. A. Adenovirus hexon monoclonal antibody that is group specific and potentially useful as a diagnostic reagent. *J. Clin. Microbiol.* 17:360–64 (1983).

Chandler, F. W., and Gorelkin, L. Immunofluorescence staining of adenovirus in fixed tissues pretreated with trypsin. *J. Clin. Microbiol.* 17:371–73 (1983).

Dawson, C. R.; Hanna, L.; Wood, T. R.; and Despain, R. Adenovirus type 8 keratoconjunctivitis in the United States. III, Epidemiologic, clinical and microbiologic features. *Am. J. Ophthalmol.* 69:473–80 (1970).

de Jong, J. C.; Wigand, R.; Kidd, A. H.; Wadell, G.; Kapsenberg, J. G.; Muzerie, C. J.; Wermenbol, A. G.; and Firtzlaff, R. G. Candidate adenoviruses 40 and 41: Fastidious adenoviruses from human infant stool. *J. Med. Virol.* 11:215–31 (1983).

Espy, M. J.; Hierholzer, J. C.; and Smith, T. F. Effect of centrifugation on the rapid detection of adenovirus in shell vials. *Am. J. Clin. Path.* 88:358–60 (1987).

Fong, C. K. Y.; Landry, M. L.; and Hsiung, G. D. Replication of human adenoviruses in guinea pig embryo cells: An ultrastructural study. *Arch. Virol.* 93:247–60 (1987).

———. Adenovirus infection in epidemic keratocon-junctivitis. *V.A. Practitioner* 9:102–7 (1992).

Furesz, J.; Armstrong, R.; and Nagler, F. Adenovirus serum neutralization in micro-tissue cultures of rhesus monkey kidney cells. *Can. J. Microbiol.* 18:1603–6 (1972).

Gary, G. W.; Hierholzer, J. C.; and Black, R. E. Characteristics of noncultivable adenoviruses associated with diarrhea in infants: A new subgroup of human adenoviruses. *J. Clin. Microbiol.* 10:96–103 (1979).

Ginsberg, H. S. Adenovirus Structural Proteins. In Fraenkel-Conrat, H., and Wagner, R. R., eds., *Comprehensive Virology*, vol. 13, pp. 409–57. New York: Plenum, 1979.

Gomes, S. A.; Nascimento, J. P.; Siqueira, M. M.; Krawczuk, M. M.; Pereira, H. G.; and Russell, W. C. In situ hybridization with biotinylated DNA probes: A rapid diagnostic test for adenovirus upper respiratory infections. *J. Virol. Meth.* 12:105–10 (1985).

Green, M.; Mackey, J. K.; Wold, W. S.; and Rigden, P. Thirty-one human adenovirus serotypes (Ad1-Ad31) form five groups (A–E) based upon DNA genome homologies. *Virology* 93:481–92 (1979).

Harmon, M. W.; Drake, S.; and Kasel, J. A. Detection of adenovirus by enzyme-linked immunosorbent assay. *J. Clin. Microbiol.* 9:342–46 (1979).

Harmon, M. W., and Pawlik, K. M. Enzyme immunoassay for direct detection of influenza type A and adenovirus antigens in clinical specimens. *J. Clin. Microbiol.* 15:5–11 (1982).

Hasler, P., and Wigand, R. The susceptibility of Vero cell cultures for human adenoviruses. *Med. Microbiol. Immun.* 164:267–76 (1978).

Hierholzer, J. C. Adenoviruses. In Schmidt, N. J., and Emmons, R. W., eds., *Diagnostic Procedures for Viral, Rickettsial and Chlamydial Infections*, 6th ed., pp. 219–64. Washington, D.C.: American Public Health Association (1989).

———. Further subgrouping of the human adenoviruses by differential hemagglutination. *J. Inf. Dis.* 128:541–50 (1973).

Hierholzer, J. C., and Bingham, P. G. Vero microcultures for adenovirus neutralization tests. J. Clin. Microbiol. 7:499–506 (1978).

Hierholzer, J. C.; Stone, Y. O.; and Broderson, J. R. Antigenic relationships among the forty-seven human adenoviruses determined in reference horse antisera. *Arch. Virol.* 121:179–97 (1991).

Hierholzer, J. C.; Suggs, M. T.; and Hall, E. C. Standardized viral hemagglutination and hemagglutination-inhibition tests. II, Description and statistical evaluation. *Appl. Microbiol.* 18:824–33 (1969).

Hierholzer, J. C.; Wigand, R.; Anderson, L. J.; Adrian, T.; and Gold, J. W. Adenoviruses from patients with AIDS: A plethora of serotypes and a description of five new serotypes of subgenus D (types 43–47). *J. Infect. Dis.* 158:804–13 (1988).

Hierholzer, J. C.; Wigand, R.; and de Jong, J. C. Evaluation of human adenoviruses 38, 39, 40 and 41 as new serotypes. *Intervirology* 29:1–10 (1988).

Hilleman, M. R., and Werner, J. H. Recovery of a new agent from patients with acute respiratory illness. *Proc. Soc. Exp. Biol. Med.* 85:183–88 (1954).

Huebner, R. J.; Rowe, W. P.; Ward, T. G.; Parrott, R. H.; and Bell, J. A. Adenoidal pharyngeal-conjunctival agents: A newly recognized group of common viruses of the respiratory system. *N. Engl. J. Med.* 251:1077–86 (1954).

Jawetz, E.; Kimura, S. J.; Nicholas, A. N.; Thygeson, P.; and Hanna, L. New type of APC virus from epidemic keratoconjunctivitis. *Science* 122:1190–91 (1955).

Johansson, M. E.; Uhnoo, I.; Svensson, L.; Pattersson, C. A.; and Wadell, G. Enzyme-linked immunosorbent assay for detection of enteric adenovirus 41. *J. Med. Virol.* 17:19–27 (1985).

Krisher, K. K., and Menegus, M. M. Evaluation of three types of cell culture for recovery of adenovirus from clinical specimens. *J. Clin. Microbiol.* 25:1323–24 (1987).

Landry, M. L.; Fong, C. K. Y.; Neddermann, K.; Solomon, L.; and Hsiung, G. D. Disseminated adenovirus infection in an immunocompromised host: Pitfalls in diagnosis. *Am. J. Med.* 83:555–59 (1987).

Mahafzah, A. M., and Landry, M. L. Evaluation of immunofluorescent reagents, centrifugation, and conventional cultures for diagnosis of adenovirus infection. *Diag. Microbiol. Infect. Dis.* 12:407–11 (1989).

Norrby, E. The structural and functional diversity of adenovirus capsid components. *J. Gen. Virol.* 5:221–36 (1969).

Pieniazek, D.; Pieniazek, N. J.; Macejak, D.; and Luftig, R. B. Enteric adenovirus 41 (Tak) requires low serum for growth in human primary cells. *Virology* 178:72–80 (1990).

Rowe, W. P.; Huebner, R. J.; Gilmore, L. K.; Parrott, R. H.; and Ward, T. G. Isolation of a cytopathogenic agent from human adenoids undergoing spontaneous degeneration in tissue culture. *Proc. Soc. Exp. Biol. Med.* 84:570–73 (1953).

Schmidt, N. J.; Lennette, E. H.; and King, C. J. Neutralizing, hemagglutination-inhibiting, and group complement-fixing antibody responses in human adenovirus. *J. Immun.* 96:64–74 (1966).

Stevens, D. A.; Schaeffer, M.; Fox, J. P.; Brandt, C. D.; and Romano, M. Standardization and certification of reference antigens and antisera for thirty human adenovirus serotypes. *Am. J. Epidemiol.* 86:617–33 (1967).

Takiff, H. E.; Straus, S. E.; and Garon, C. F. Propagation and in vitro studies of previously noncultivable enteric adenovirus in 293 cells. *Lancet* 2:832–34 (1981).

Wigand, R.; Adrian, T.; and Bricout, F. A new human adenovirus of subgenus D: Candidate adenovirus type 42. *Arch. Virol.* 94:283–86 (1987).

22

HERPESVIRIDAE

M. L. LANDRY AND G. D. HSIUNG

Herpesviruses are widespread in nature, infecting humans as well as many lower animal species. Six human herpesviruses have been recognized: herpes simplex virus 1 (HSV-1) and HSV-2; varicella-zoster virus (VZV); cytomegalovirus (CMV); Epstein-Barr virus (EBV); and human herpesvirus-6 (HHV-6) (table 22.1). In addition, human infections with a herpesvirus of monkeys (herpesvirus simiae or herpesvirus B) have been reported in individuals in contact with monkeys or monkey tissues.

The herpesviruses are icosahedral viruses with 162 capsomeres containing a double-stranded DNA genome. Mature virions range from 150 to 200 nm. The presence of a double membrane-bound lipoprotein envelope (figure 22.1) renders herpesviruses susceptible to inactivation by lipid solvents such as diethyl ether. While all herpesviruses are morphologically similar, they are biologically distinct and differ in the composition of their genomes, capsids, and envelope proteins. Virus multiplication occurs in the nucleus and Cowdry type A intranuclear inclusions, each surrounded by a halo, may be seen in H&E-stained preparations of infected cells.

Although each of the human herpesviruses is associated with characteristic clinical manifestations, infection is most often inapparent. After primary infection, herpesviruses remain latent within the host despite high levels of neutralizing antibody and may periodically reactivate to cause recurrent clinical disease or to result in virus shedding without symptoms.

Laboratory diagnosis of the human herpesviruses involves a spectrum of techniques, depending upon the virus sought. Some methodologies are routine in the diagnostic laboratory, and others are reserved for research settings.

ISBN 0-300-05845-4.

Table 22.1 Host Cell Spectrum and Pathogenicity of Human Herpesviruses

Virus	Common Specimens	Cytopathic Effect in Cell Culture Systems					Pathogenicity in	
		MK	HDF	Hep-2	RK	CE	Embryonated Eggs	Newborn Mice
Herpes simplex virus type 1 (HSV-1)	Vesicle fluid, lesion swab, throat swab	+/−	+	+	++	−	+[a]	+
Herpes simplex virus type 2 (HSV-2)	Vesicle fluid, lesion swab	+/−	+	+	++	+	+[b]	+
Varicella-zoster virus (VZV)	Vesicle fluid, lesion swab	+/−	++	−	−	−	−	−
Cytomegalovirus (CMV)	Urine, saliva, blood leukocytes, bronchial lavage	−	++	−	−	−	−	−
Epstein-Barr virus (EBV)	Throat swab,[c] blood leukocytes	−	−	−	−	−	−	−
Human herpesvirus type 6 (HHV-6)	Blood leukocytes[c]	−	−	−	−	−	−	−

\+ virus-induced change; − no change

a. Pinpoint pocks

b. Large clear pocks

c. Virus isolated by cocultivation of clinical specimens with umbilical cord blood leukocytes

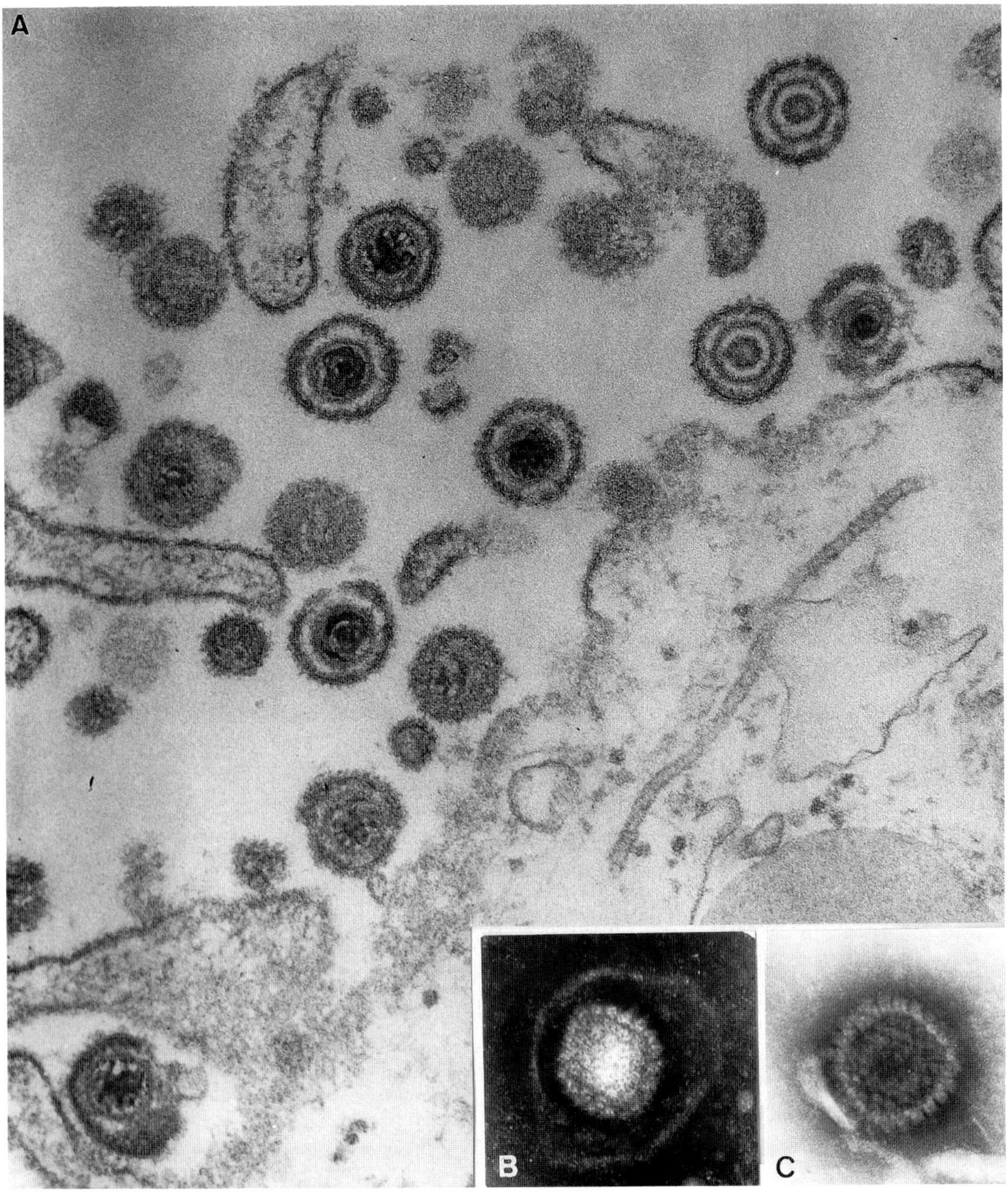

Figure 22.1 Electron micrograph of HSV-1 in a thin section and PTA-stained preparations: *(A)* Thin section of an infected rabbit kidney cell culture showing many extracellular enveloped virions (100,800×); *(B)* Enveloped herpesvirus virion stained with PTA (165,000×); *(C)* Naked nucleocapsid stained with PTA (169,000×)

HERPES SIMPLEX VIRUS

Specimen Source and Inoculation

Vesicle fluid, lesion, throat or eye swabs, and tissues such as brain, liver, and lung are all potential sources for virus isolation in the appropriate clinical setting. Collection of specimens early in illness when virus titers are high provides best results. Both HSV-1 and HSV-2 have a wide host spectrum, growing rapidly in a variety of cell cultures of both human and nonhuman origin. All specimens can be inoculated into a variety of cell culture systems, including primary rabbit kidney (RK), guinea pig embryo (GPE), and mink lung (ML) cells, which are more sensitive than commonly used human diploid fibroblast, Hep-2, A549, and Vero cell lines. The use of more sensitive cell systems will enhance the isolation rate from specimens containing low titers of virus, as well as shorten the time for detection of CPE.

Virus Isolation, Propagation, and Assay

CYTOPATHIC EFFECT IN CELL CULTURES WITH FLUID MEDIUM

HSV induces rapid cellular degeneration in many cell types, as described above. Passage of HSV can be accomplished by inoculation of supernate or infected cells. Clusters of degenerated cells can be readily observed in infected RK (figure 22.2*B*) and A549 cell cultures (figure 22.2*D*) but are less distinct in rhesus monkey kidney cells. Although CPE progresses more slowly in human fibroblasts, cell rounding occurs in two to four days (figure 22.2*F*). HSV-2 can be distinguished from HSV-1 by its propensity to induce syncytia formation in infected cells (figure 22.3). When the virus titer in the inoculum is very high, generalized cell rounding may be seen instead of distinct foci. CPE can be detected within 24–48 hours in the majority of specimens when highly sensitive cell systems are inoculated but may take several days longer when less sensitive systems are used or when the virus titer in the inoculum is low. Virtually all isolates are obtained by five to seven days, unless the patient is on antiviral therapy, which may delay the detection of HSV.

CYTOPATHOLOGY IN STAINED PREPARATIONS

In hematoxylin-eosin-stained preparations, intranuclear inclusions are basophilic in nature (figure 22.4) and Feulgen-positive. As the infection progresses, the virions move from the nucleus into the cytoplasm and inclusions are converted into an eosinophilic, Feulgen-negative inclusion body. Thus, when eosinophilic inclusions are observed, there are fewer identifiable virions present. In cell scrapings from skin lesions stained with Wright's or

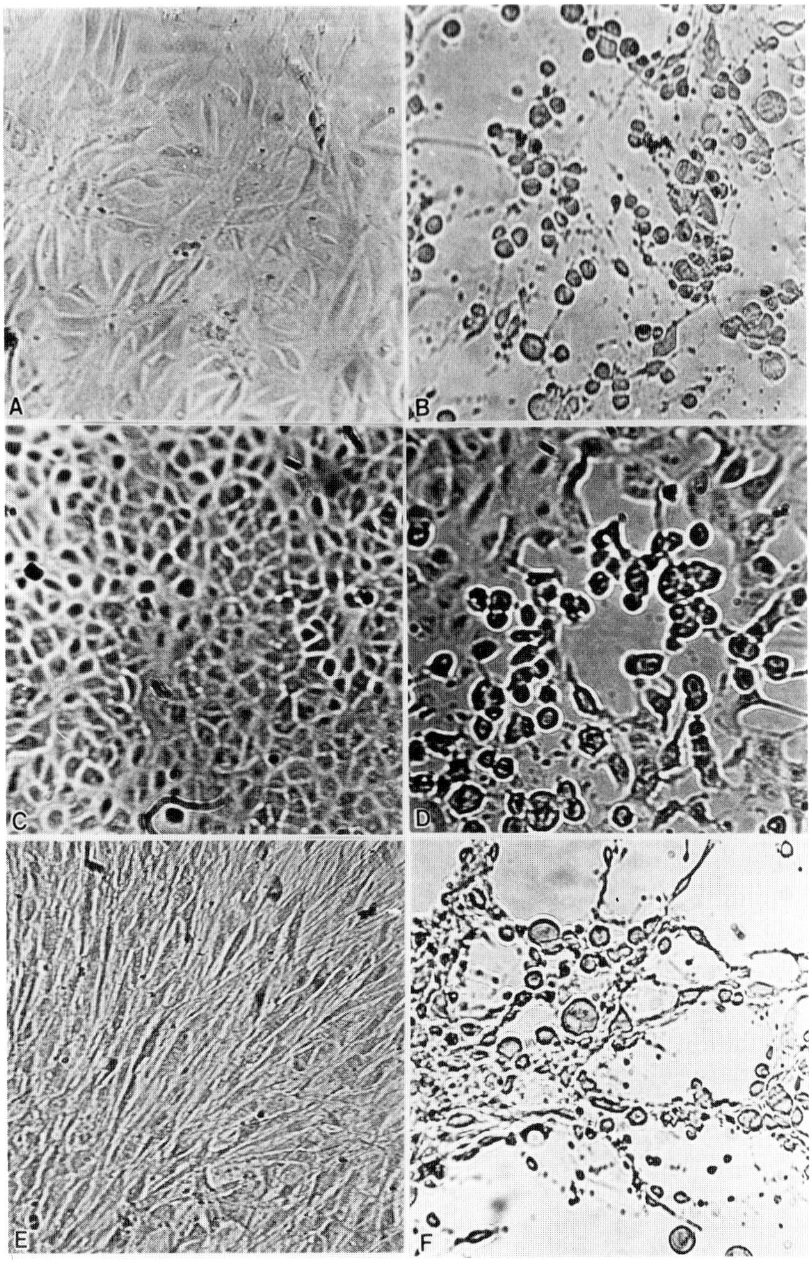

Figure 22.2 Cytopathic effects induced by HSV-1 in different cell cultures (100×): *(A)* Uninfected rabbit kidney (RK) culture; *(B)* HSV-1-infected RK cells, 24 hours after inoculation; *(C)* Uninfected A549 culture; *(D)* HSV-1-infected A549 cells, 48 hours after inoculation; *(E)* Uninfected HDF culture; *(F)* HSV-1-infected HDF cells, 72 hours after inoculation

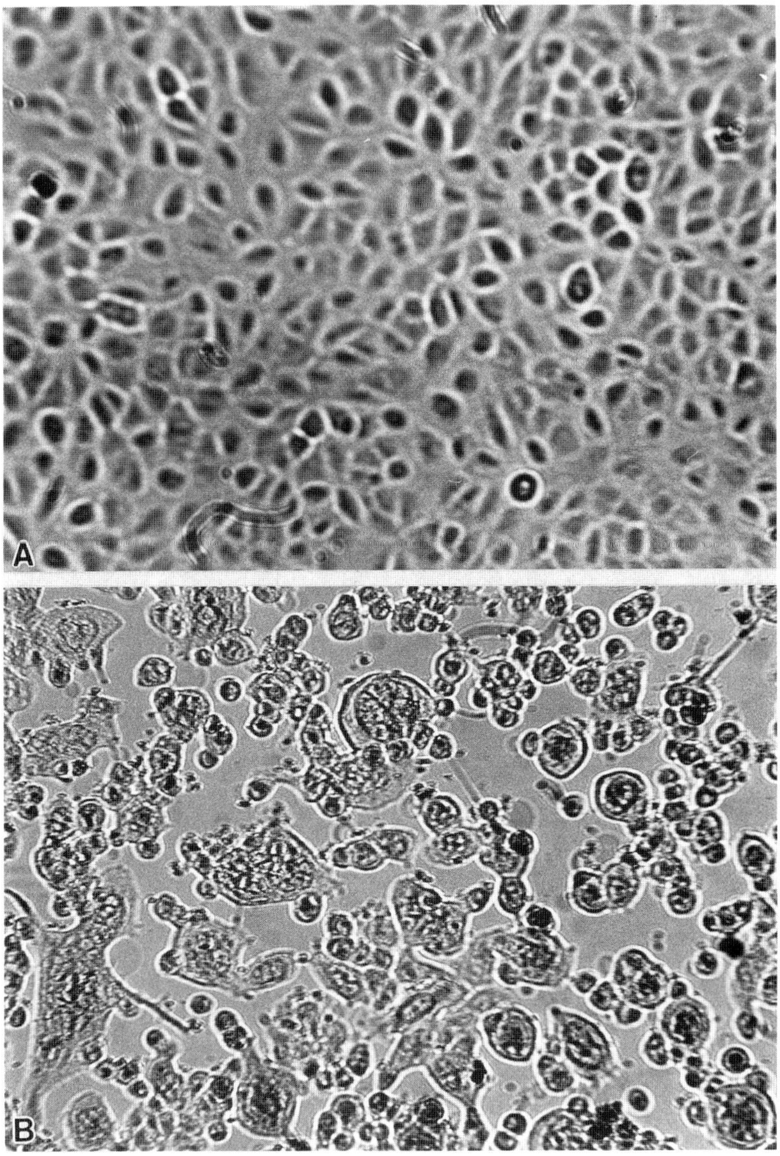

Figure 22.3 Cytopathic effects induced by HSV-2 in A549 cell culture (100×): *(A)* Uninfected A549 culture; *(B)* Syncytia formation induced by HSV-2 in A549 cells

Giemsa stain (Tzanck preparation), multinucleated giant cells are seen in both HSV and VZV infection (color plate 17; multinucleated cells are also seen in pap smears [see color plate 16]). Staining of skin lesion smears with HSV specific antibody can provide a rapid diagnosis but is less sensitive than virus isolation.

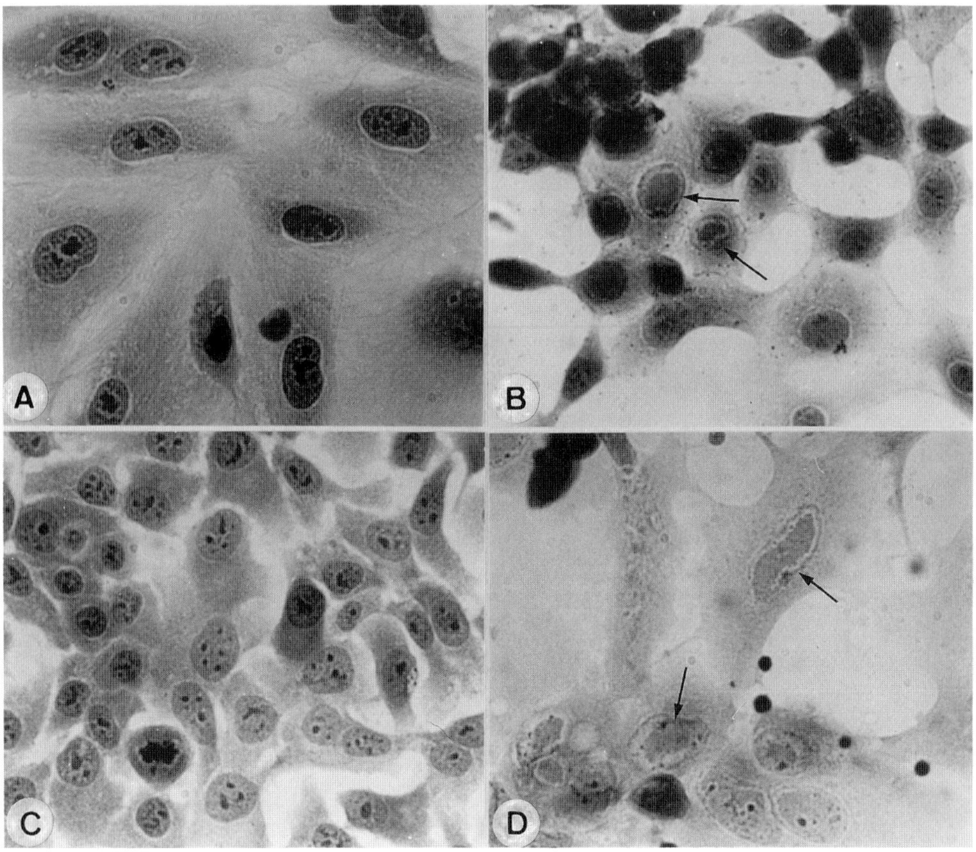

Figure 22.4 Intranuclear inclusions induced by HSV-1 in HEK and Hep-2 cell cultures, 24 hours after infection (H&E, 400×): *(A)* Uninfected HEK culture; *(B)* Basophilic and eosinophilic inclusions (arrows) induced by HSV in HEK cells; *(C)* Uninfected Hep-2 culture; *(D)* Basophilic and eosinophilic inclusions (arrows) induced by HSV in Hep-2 cells

MOUSE PATHOGENICITY

Although newborn mice are not commonly used for the isolation of herpes simplex virus in the laboratory, after intracerebral or intraperitoneal inoculation they develop a characteristic encephalitis. Cowdry type A inclusions are usually found in the hematoxylin-eosin-stained sections of infected mouse brain tissue. Mice have been used commonly as animal models for herpes simplex virus infections.

PLAQUE FORMATION

Plaque formation has been used for assaying HSV either in flask cultures under an agar overlay medium or in plastic plate cultures under methylcellulose overlay (chapter 6). Distinct plaques appear two to three days after inoculation and are easily enumerated. This technique is particularly useful for the antiviral susceptibility assay.

Identification of Isolates

PRESUMPTIVE IDENTIFICATION BY CYTOPATHIC EFFECT

Typical cytopathic change in a wide variety of cell cultures, rapid onset, and progression of CPE all support a presumptive diagnosis of HSV. Adenovirus and high-titered CMV inocula can, however, sometimes induce similar CPE in selected cell systems.

SPECIFIC IDENTIFICATION

IF or IP staining of virus-infected cells with monoclonal antibody is the most rapid and convenient method for virus identification (chapter 7, color plates 1, 2, 15). In situ hybridization with biotinylated probes has also been used.

Neutralization with HSV polyclonal antiserum in cell culture, either by the inhibition of virus-induced CPE or plaque reduction, can be used to identify HSV but does not readily distinguish type 1 from type 2 owing to extensive antigenic cross between the two viruses (see figure 6.3).

DIFFERENTIATION OF HERPES SIMPLEX VIRUS 1 AND 2

With the availability of type-specific monoclonal antibodies, HSV-1 and HSV-2 can be readily differentiated by IF or IP staining of infected cells. As stated above, HSV-2 induces extensive syncytial clusters in A549 cells (figure 22.3), but not HSV-1 (figure 22.2). Previously, advantage was taken of biologic differences in cell selectivity between the two viruses; HSV-2 has the ability to produce plaques in CE cells, whereas HSV-1 does not. Similarly, chorioallantoic membrane (CAM) inoculation was used for differentiation of the two types; HSV-1 produces pinpoint pocks on the CAM of embryonated eggs, whereas HSV-2 produces large clear pocks.

ELECTRON MICROSCOPY

Direct visualization of HSV virions in thin sections of infected cells or in negatively stained preparations provides an important method for rapid diagnosis. Distinct nucleocapsids, either in groups or scattered, are easily recognizable in infected cells obtained from brain biopsy (figure 22.5) or in infected cultured cells (figure 22.6). In addition, HSV-2 often induces fibers (figure 22.7B), a characteristic that appears to be a distinct morphologic characteristic that is lacking in HSV-1 infected cells.

Serodiagnosis

To determine immune status to HSV, a sensitive assay such as an ELISA, IF, or latex agglutination test may be used. The complement-fixation test, as well as

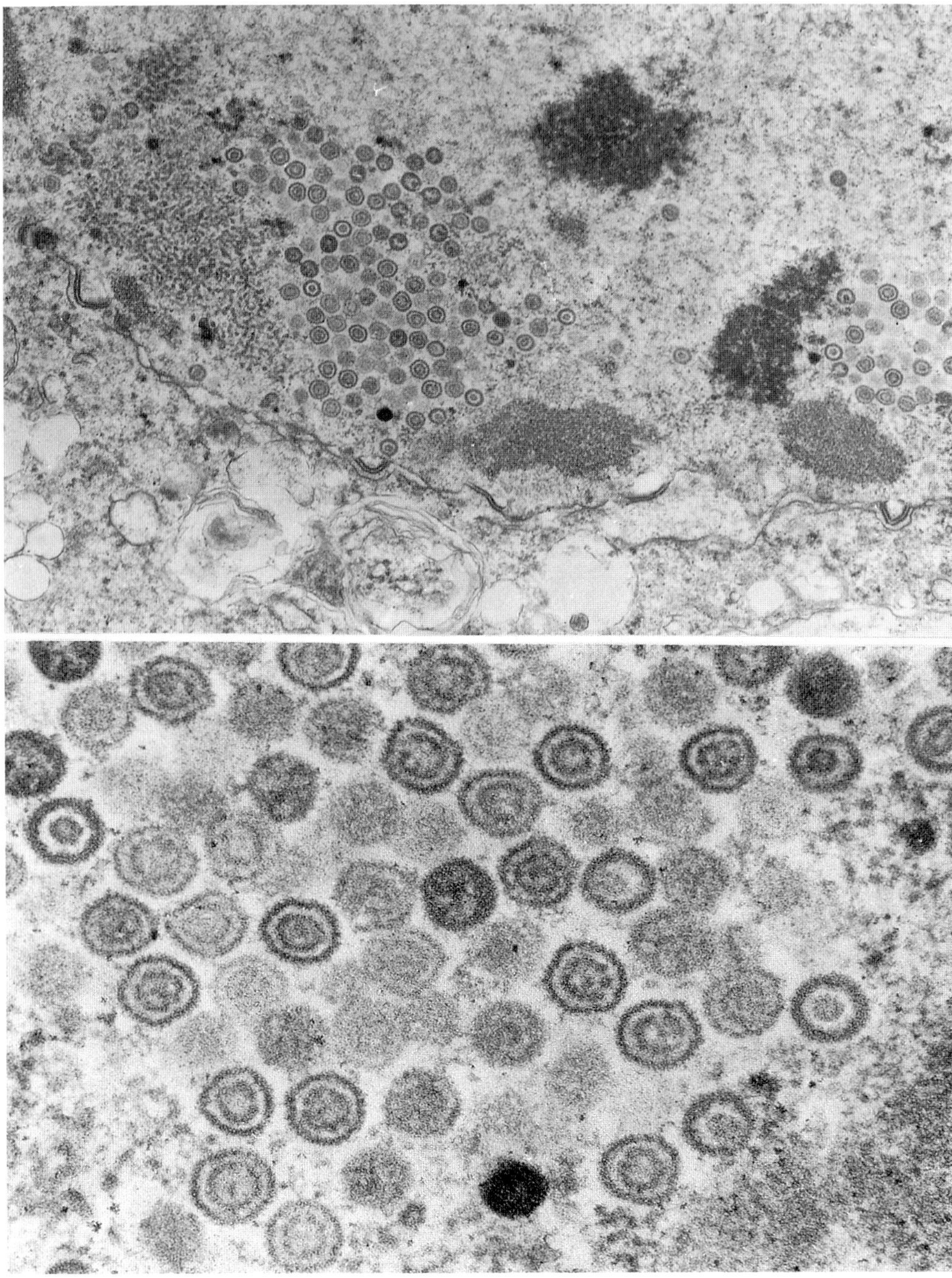

Figure 22.5 Electron micrograph of HSV-1 in an infected neuron from a human brain biopsy (courtesy M. J. August). *(A)* Aggregates of nucleocapsids in the nucleus (24,000×); *(B)* Higher magnification of nucleocapsids in *(A)* (100,800×) (modified from C. K. Y. Fong and G. D. Hsiung, *Virology and Rickettsiology*, 1978, vol. 1, pt. 1, 94)

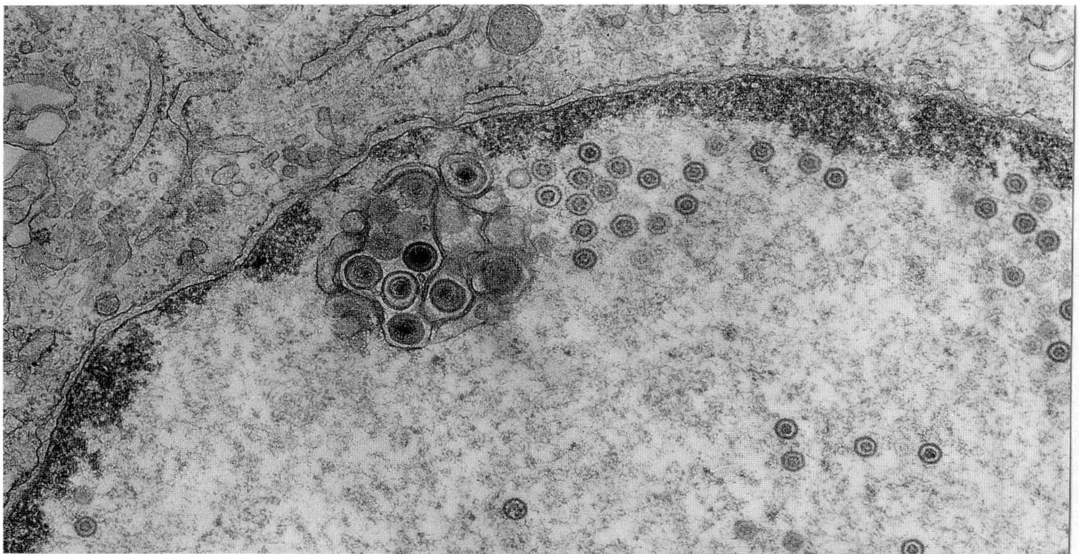

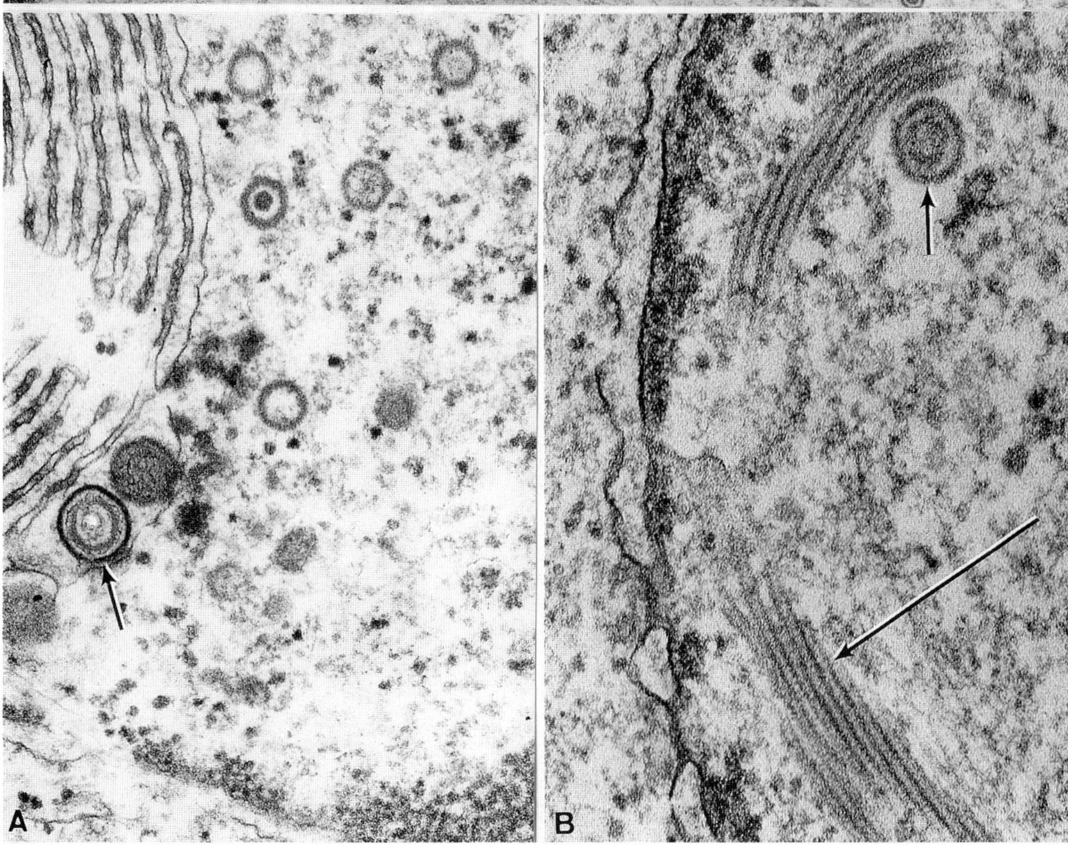

Figure 22.6 (*top*) Electron micrograph of HSV-1 in an infected rabbit kidney cell showing nucleocapsids and enveloped virions in the nucleus (30,000×)

Figure 22.7 (*bottom*) Electron micrographs of HSV-2 in an infected rabbit kidney cell. (Modified from C. K. Y. Fong and G. D. Hsiung, *Virology and Rickettsiology,* 1978, vol. 1, pt. 1, 93.) *(A)* Portion of a nucleus containing nucleocapsids, enveloped virions (arrow), and membrane duplication (64,800×); *(B)* Portion of nucleus showing filamentous structures (long arrow) and a virion (short arrow) induced by HSV-2 (100,800×)

the assays above, can be used to demonstrate antibody rise in primary HSV infection. In recurrent HSV disease, however, serology has limited value since antibody titers often show no rise. Commonly used serologic tests also do not reliably distinguish between HSV-1 and HSV-2. For this purpose, a western blot or an ELISA using recombinant type-specific proteins has been performed to detect antibody to type-specific viral proteins.

VARICELLA-ZOSTER VIRUS

Specimen Source and Inoculation

Vesicle fluid and lesion swabs are good sources for VZV isolation. Involved tissues and CSF can also yield virus isolates. Collecting specimens early in the disease aids virus recovery, as does the prompt inoculation of cultures. Human diploid fibroblasts such as WI-38 and MRC-5 are sensitive to VZV. VZV also replicates in HEK and to a lesser degree in continuous human heteroploid cell lines such as A549.

Virus Isolation, Propagation, and Assay

In contrast to HSV, VZV does not cause disease in newborn mice or chick embryos.

CYTOPATHIC EFFECT IN CELL CULTURE WITH FLUID MEDIUM

VZV isolation generally requires three to seven days of cultivation, though up to 14 days may be required when a low titer of virus is present in the sample. CPE consists of foci of rounded cells that progress slowly to involve other areas of the monolayer (figure 22.8). This is in marked contrast to the rapid CPE progression of HSV as well as the propensity of HSV to induce CPE in nonprimate cell cultures. Since VZV is highly cell-associated in culture, serial propagation requires the transfer of intact infected cells. To enhance the rapidity and sensitivity of VZV isolation, inoculation of shell vial cultures followed by IF staining with monoclonal antibody at two to five days can be performed.

CYTOPATHOLOGY IN STAINED PREPARATIONS

VZV induces intranuclear inclusions in infected cells. Typical Cowdry type A intranuclear inclusions can be seen in hematoxylin-eosin-stained preparations. In Tzanck smears made from cell scrapings from the base of skin lesions, multinucleated giant cells are seen and are indistinguishable from HSV infection (color plate 17).

Figure 22.8 Cytopathic effects induced by VZV in HDF cell culture (100×): *(A)* Uninfected HDF cell culture; *(B)* Focus of VZV-induced CPE three days postinoculation; *(C)* VZV-induced CPE six days postinoculation

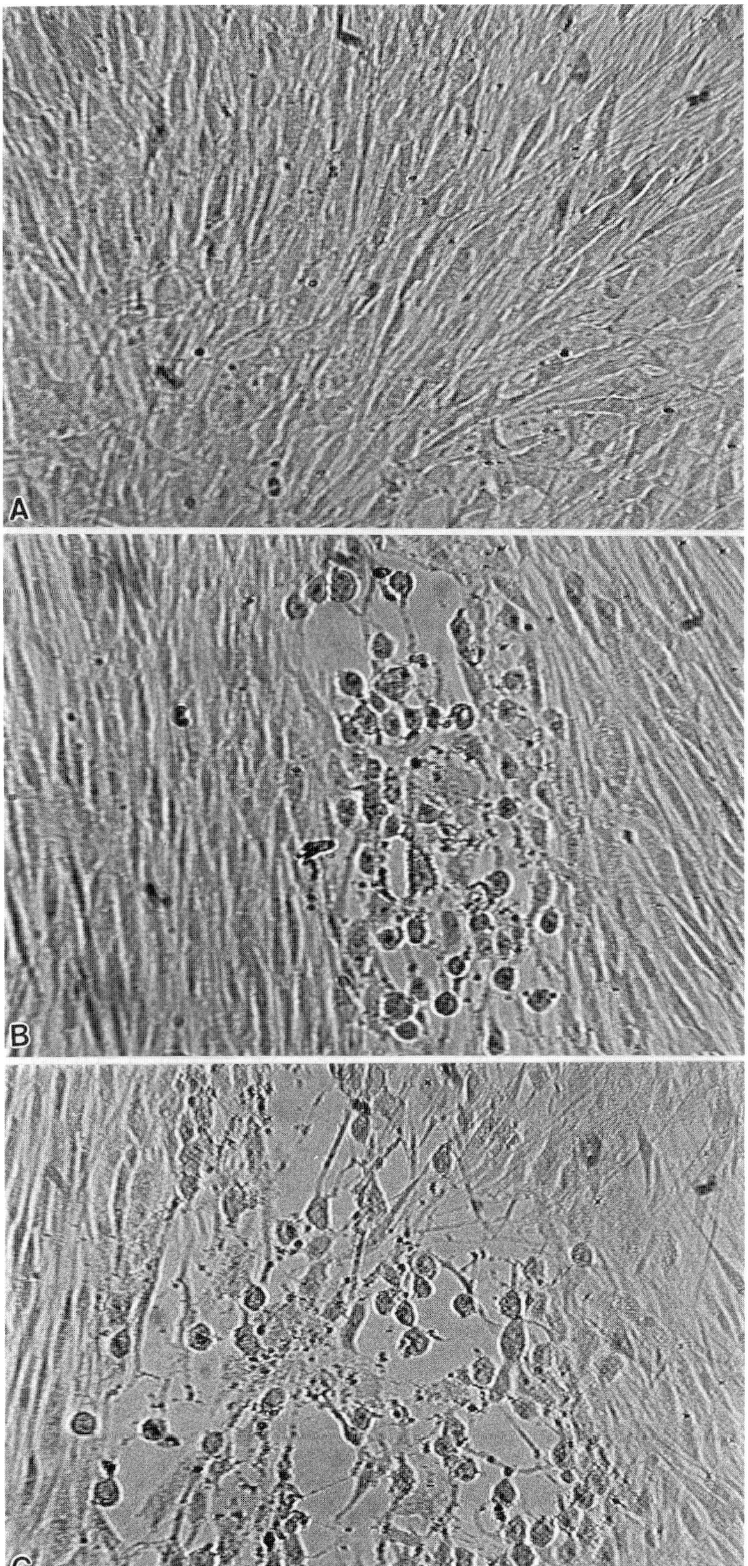
A
B
C

The most sensitive and rapid method for the diagnosis of VZV is direct IF staining of cell scrapings from skin lesions with monoclonal antibody to VZV, provided an adequate number of cells are examined by an experienced observer.

Identification of Isolates

Most VZV isolates are obtained from skin lesions, which can be similar in appearance to HSV-induced lesions. Characteristic focal CPE with slow progression and limited host cell range distinguish VZV from HSV isolates in the laboratory. CMV also induces foci of CPE in HDF cells with slow progression. CMV-induced foci, however, tend to be more refractile and do not exhibit the granular degeneration typical of VZV (compare figures 22.8 and 22.11). In addition, the usual specimen sources for CMV such as urine, saliva, and blood are unlikely to yield VZV.

Specific identification of isolates is obtained by transferring infected cells from cell cultures to welled slides followed by IF staining with VZV-specific antibody. Owing to the slow progression of CPE, there is often a delay until adequate CPE has developed to allow identification by IF. If a more rapid result is desired, infected cells can be passaged into shell vial cultures and the coverslips fixed and stained at 48 hours after inoculation (color plate 5).

ELECTRON MICROSCOPY

Virions in VZV-infected cell cultures examined by thin-section are often distorted (figure 22.9) and can be distinguished from HSV and CMV (figures 22.1, 22.14). Distinct herpesvirus virions are easily found in vesicle fluids obtained from patients with chicken pox (figure 22.10*A*) and examined by the negative staining technique.

Serodiagnosis

For immune status testing, a sensitive assay such as the fluorescent antibody to membrane antigen (FAMA), ELISA, or latex agglutination test can be used. To determine an antibody rise, a complement-fixation test is standard, although the above tests can also be used. A heterologous rise in antibody titer, however, can be seen between HSV and VZV, leading to diagnostic confusion.

CYTOMEGALOVIRUS

Specimen Source and Inoculation

Specimens for isolation of CMV include urine, saliva, cervical secretions, semen, blood leukocytes, bronchoalveolar lavage fluid, and involved tissues such as lung, liver, and gastrointestinal biopsies. Human CMV has a narrow host range, replicating only in human fibroblast cells such as WI-38, MRC-5,

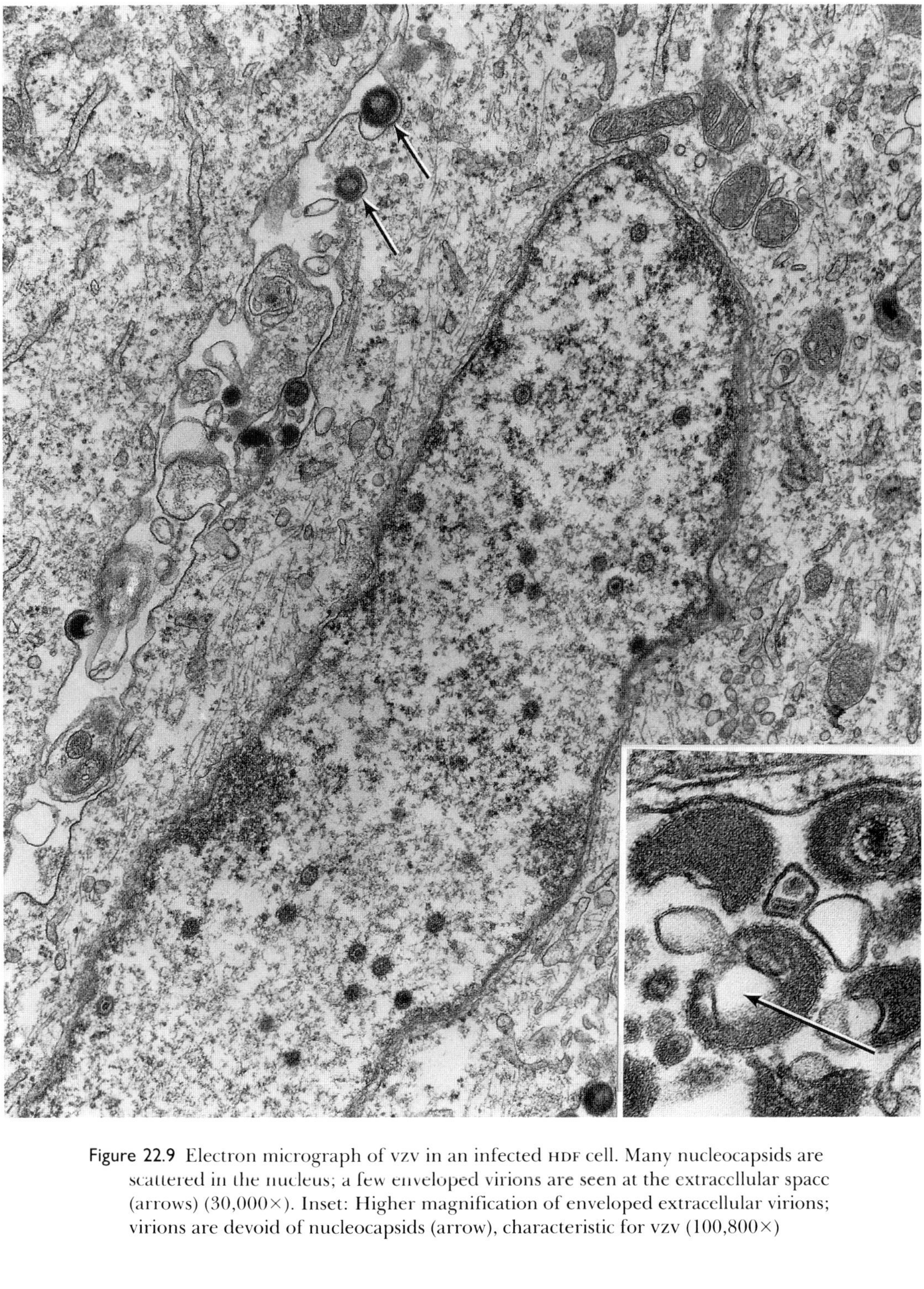

Figure 22.9 Electron micrograph of VZV in an infected HDF cell. Many nucleocapsids are scattered in the nucleus; a few enveloped virions are seen at the extracellular space (arrows) (30,000×). Inset: Higher magnification of enveloped extracellular virions; virions are devoid of nucleocapsids (arrow), characteristic for VZV (100,800×)

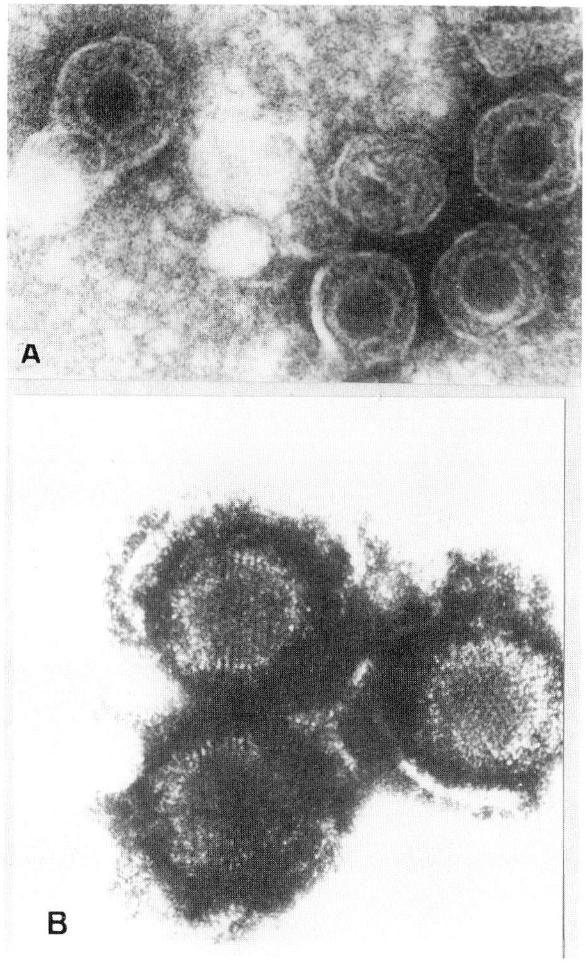

Figure 22.10 Negative staining of VZV and CMV virions in clinical specimens: *(A)* VZV virions seen in vesicular lesions, stained with PTA (65,000×) (F. W. Doane and N. Anderson, *Comparative Diagnosis of Viral Diseases*, E. Kurstak and C. Kurstak, eds., vol. 2. New York: Academic Press, 1977, 512); *(B)* CMV virions seen in a urine sample stained with PTA (150,000×) (F. K. Lee et al., *New Engl. J. Med.* 299:1266 [1978])

and human newborn foreskin. Toxicity of specimens can be a significant problem, leading to monolayer degeneration before CPE can be detected. Mixing urine in an equal volume of viral transport medium prior to inoculation, rinsing monolayers after virus adsorption, and changing media 24 hours after inoculation all help to lessen toxic effects.

Virus Isolation, Propagation, and Assay

Human CMV is host-specific and replicates only in human fibroblast cells in cultures freshly prepared. Use of subconfluent monolayer enhances the sensitivity of isolation.

CYTOPATHIC EFFECT IN CELL CULTURES WITH FLUID MEDIUM

CMV-induced CPE consists of foci of enlarged, rounded refractile cells, appearing 3–28 days after inoculation (figure 22.11). The mean onset of CPE is seven to eight days, but when a very high virus titer is present in the inoculum, as in specimens from congenitally infected infants, generalized rounding similar to HSV can be seen one day after inoculation. Although virions are released into the supernatant to a slight extent, passage of CMV to fresh cultures requires transfer of intact cells. Furthermore, when the number of infectious virions in the inoculum is low, and only one or two foci are present, these foci may occasionally disappear. Therefore, if CPE does not progress over a period of a few days, infected cultures should be passaged. If only a few foci are detected, it is important to passage all the cells in the culture, rather than an aliquot, to ensure the transfer of infected cells.

RAPID DIAGNOSIS

A more rapid diagnosis of CMV can be obtained by using the centrifugation of shell vial culture technique. Infected cells are fixed and stained with monoclonal antibody to CMV-induced early antigen 24–48 hours postinoculation, and positive results are obtained before the development of CPE (color plate 3). Since early antigens are produced prior to DNA synthesis in infected cells, detection of DNA by hybridization requires a longer incubation period in cell culture. For optimum sensitivity and to obtain a broader spectrum of virus isolates, the conventional culture should be performed in parallel. Recently, the CMV antigenemia assay provided same-day detection of CMV infection in leukocytes (see below).

CYTOPATHOLOGY IN STAINED PREPARATIONS

Characteristic large *cytomegalic* cells with intranuclear inclusions are commonly seen in CMV-infected cultured cells (figure 22.12) and infected tissues (color plate 23). Occasionally, cytoplasmic inclusions can be seen in addition to nuclear inclusions in tissue sections of biopsy or autopsy materials. Inclusions, however, may not be detected in the presence of active infection and can be confused with other viruses, including adenovirus. Confirmatory immunostaining or hybridization tests are therefore recommended (color plate 24). Examination of exfoliated cells in fresh urine specimens from congenitally infected infants may reveal cytomegalic cells with prominent inclusions in 25%–50% of cases.

CMV-infected leukocytes can be detected and enumerated by IP or IF staining of peripheral blood leukocyte preparations using pooled monoclonal antibodies to CMV lower-matrix phosphoprotein (pp65) (figure 22.13). The number of CMV antigen-positive cells is generally directly proportional to the severity of disease. The sensitivity of the CMV antigenemia assay is dependent on the antibodies used, the fixative, minimal use of endogenous

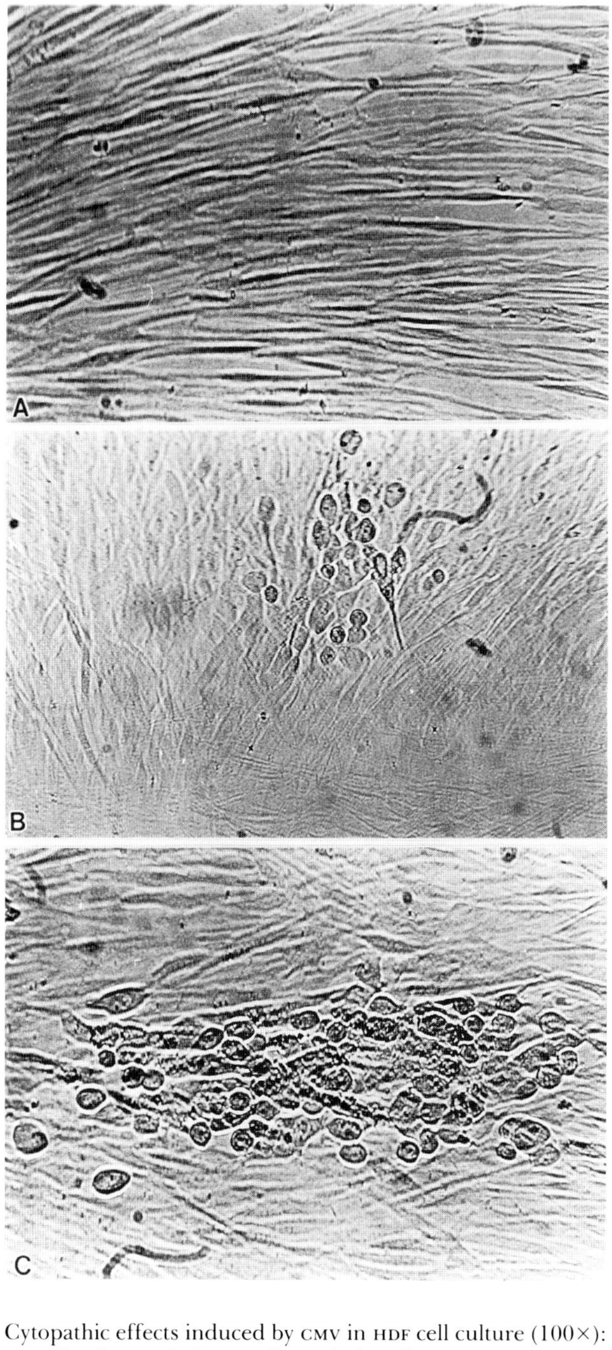

Figure 22.11 Cytopathic effects induced by CMV in HDF cell culture (100×): *(A)* Uninfected HDF cell culture; *(B)* Focus of CMV-induced CPE three days postinoculation; *(C)* Focus of CMV-induced CPE five days postinoculation

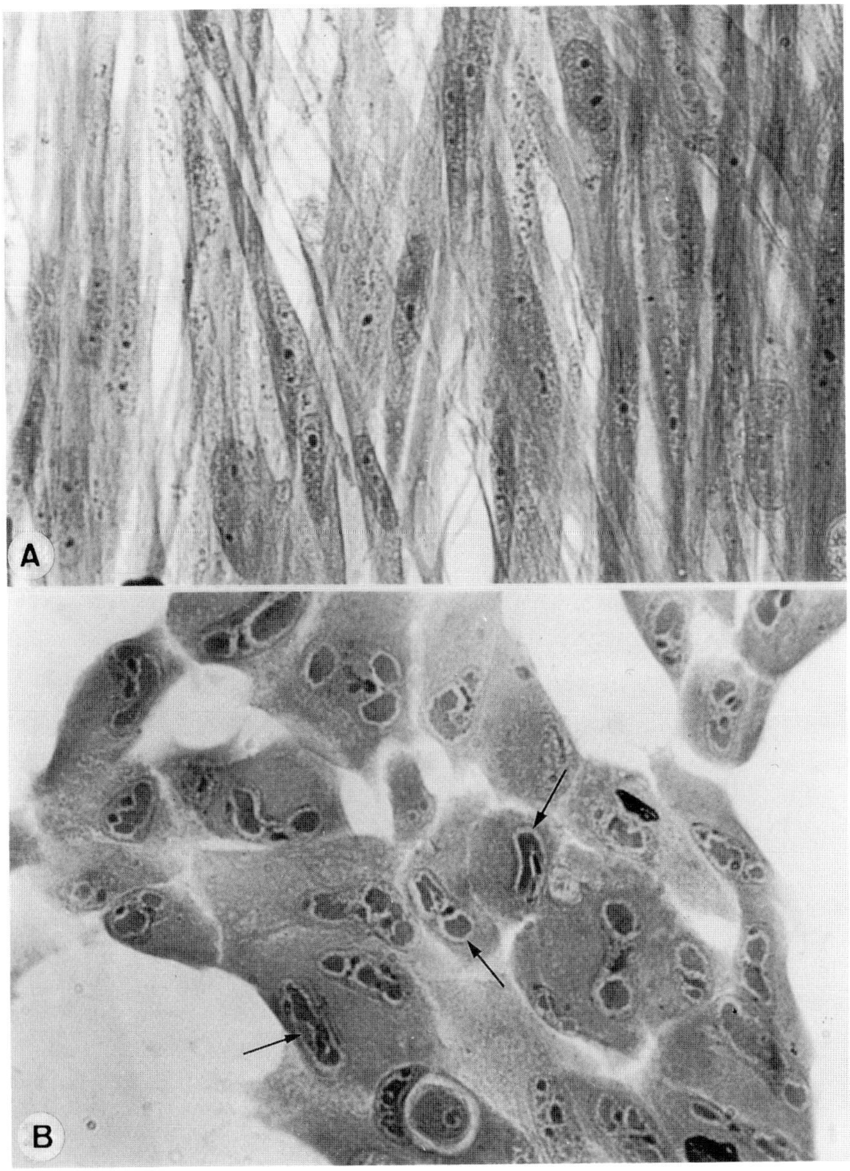

Figure 22.12 Intranuclear inclusions induced by human CMV in HDF cell culture (H&E, 400×): *(A)* Uninfected HDF; *(B)* Eosinophilic intranuclear inclusions (arrows) in HDF cells

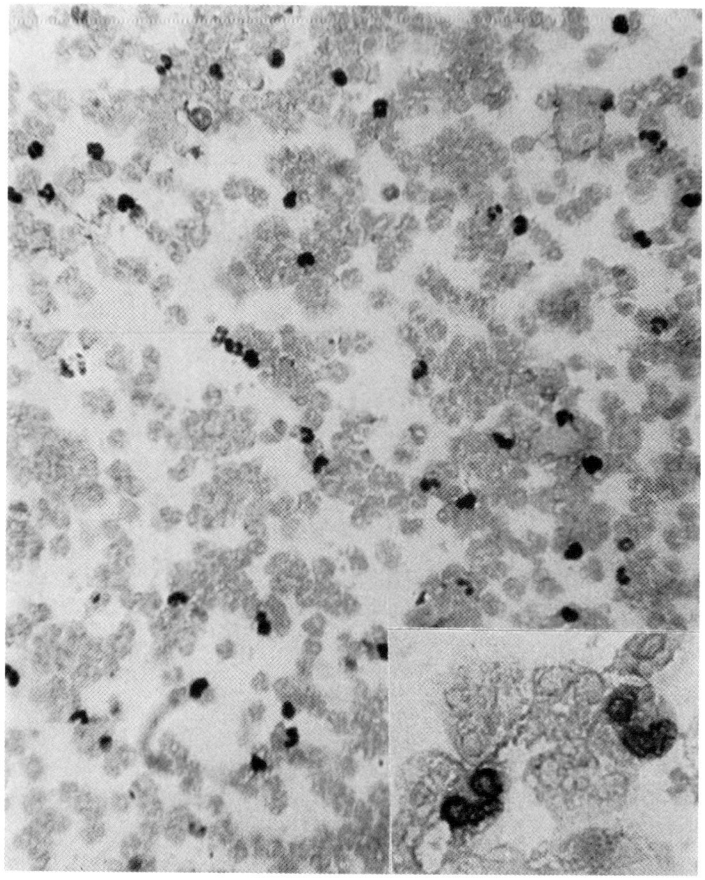

Figure 22.13 Immunoperoxidase staining of CMV-induced early antigens in peripheral blood leukocytes (100×). Inset: CMV-infected polymorphonuclear cells showing dark nuclear staining (400×)

peroxidase inhibitors, and the processing of samples promptly after blood collection. Detailed procedures have been published, and antibodies to CMV pp65 are commercially available.

Identification of Isolates

Presumptive identification of CMV isolates is based on their cytopathology, slow growth rate, and narrow host range in cell culture. Adenovirus can, however, cause similar CPE in human fibroblasts and can be isolated from the same specimens as CMV. For specific identification, once 50% of the cell monolayer is involved, infected cells can be transferred to a welled-slide and stained with CMV-specific antibody (color plate 4 compared with color plate 3). Since CPE progresses slowly, a more rapid identification can be obtained by the passage of infected cultures with only a few foci into shell vial cell cultures, followed by IF staining at 24–48 hours using monoclonal antibody to CMV-induced early antigen.

ELECTRON MICROSCOPY

A rapid diagnosis of CMV in congenitally infected infants can be made by negative staining of urine sediment. The distinct herpesvirus morphology is easily identified (figure 22.10*B*). Still, CMV cannot be distinguished from HSV, which can be found occasionally in urine. Examination of infected tissues or cell cultures by thin-section can distinguish CMV from other herpesviruses because CMV has a different ultrastructure within the infected cell (figure 22.14). In cultured cells, dense bodies (DB) with or without nucleocapsids can be seen in both the cytoplasm and the extracellular space.

Serodiagnosis

For immune status testing, a sensitive assay such as latex agglutination (LA), ELISA, IFA, or indirect hemagglutination can be used. IgM antibody to CMV can be detected in both primary and reactivation infection either by IFA or ELISA. To determine a significant antibody titer rise, the CF test is standard, although LA, IFA, IHA, and quantitative ELISA can also be used. Unfortunately, anti-CMV antibody titer may rise in the absence of clinical disease or may fail to rise in the presence of severe illness, especially in immunocompromised hosts.

EPSTEIN-BARR VIRUS

Diagnosis of EBV infection is generally made by serology. The techniques of virus isolation and detection of EBV antigen or nucleic acid in infected tissues are not widely available and are largely confined to research laboratories.

Specimen Source and Inoculation

For isolation of EBV, 5–10 ml of throat gargle in viral transport medium, 10 ml of heparinized blood, or biopsies of affected tissues such as lymph nodes are acceptable specimens. In EBV-associated tumors, virus may be latent and not able to be isolated in culture.

Virus Isolation, Propagation, and Assay

EBV cannot be grown in monolayer cultures routinely employed in a diagnostic laboratory. Isolation of EBV requires the cocultivation of clinical specimens with B lymphocytes derived from umbilical cord blood. If EBV is present, the cord lymphocytes will be transformed into a continuous cell line containing

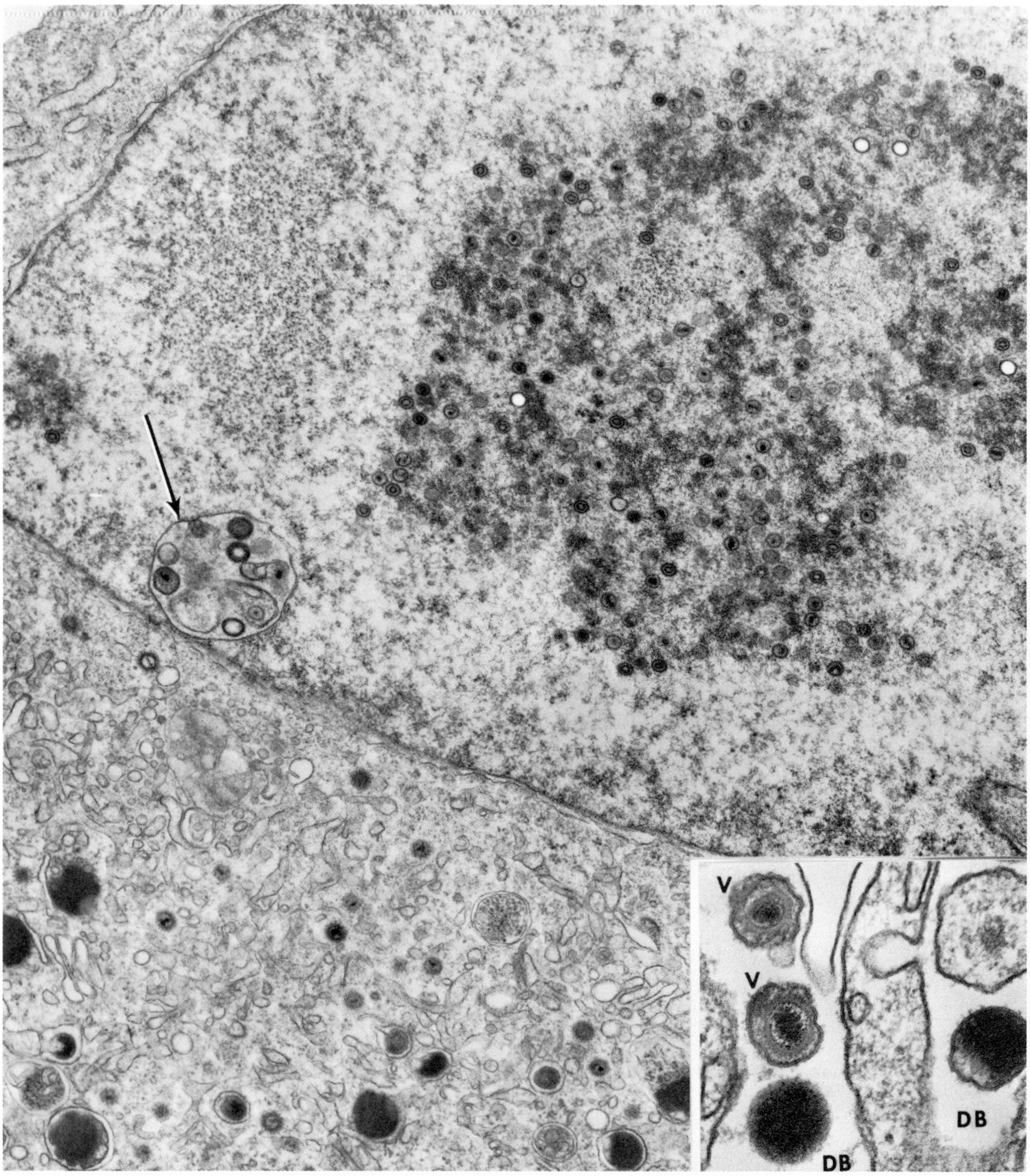

Figure 22.14 Electron micrograph of a human CMV-infected HDF cell showing intranuclear inclusion containing scattered nucleocapsids. A few enveloped virions inside a vacuole are indicated by an arrow (19,680×). Inset: Enveloped mature virions *(V)* and dense bodies *(DB)* (64,800×)

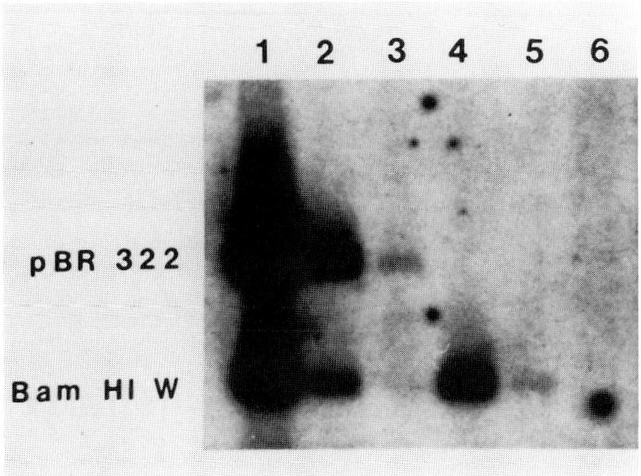

Figure 22.15 Detection of EBV DNA in EBV-infected cell lines by Southern blot analysis. (Modified with permission, M. L. Landry and C. K. Y. Fong, *Clinics Lab. Med.* 5:513–29 [1985].) DNA samples in all lanes were digested with the restriction enzyme Bam HI, fractionated by gel electrophoresis, transferred to nitrocellulose, then probed with ^{32}P-labeled recombinant plasmid pBR322 containing Bam HI W fragment of the EBV genome: lanes 1–3, titration of the probe (100, 10, and 1 pg); lanes 4–5, EBV positive cell line (two concentrations of DNA) showing hybridization with EBV probe (Bam HI W fragment); lane 6, uninfected human placenta cell DNA. (Original courtesy Dr. Warren Andiman)

the EBV genome. Evidence of transformation is indicated by an abrupt increase in the total number of cells, the rapid production of acid, growth of cells in clumps, and the capacity to be subcultured indefinitely. Transformation usually occurs 30–90 days after initiation of the culture.

Identification of Isolates

Several EBV-induced antigens may be expressed in transformed cord blood lymphocytes. Only EBV nuclear antigen (EBNA), however, is present in all EBV-infected cells and can be detected by anticomplement immunofluorescence (ACIF).

Viral Antigen or Genome Detection

The presence of EBV in infected cells can be demonstrated by the detection of EBNA by ACIF or by the detection of EBV DNA, using either in situ, spot, or Southern blot hybridization techniques (figure 22.15). These techniques are essential in establishing the etiology of EBV-associated tumors but are not used in a routine clinical virology laboratory.

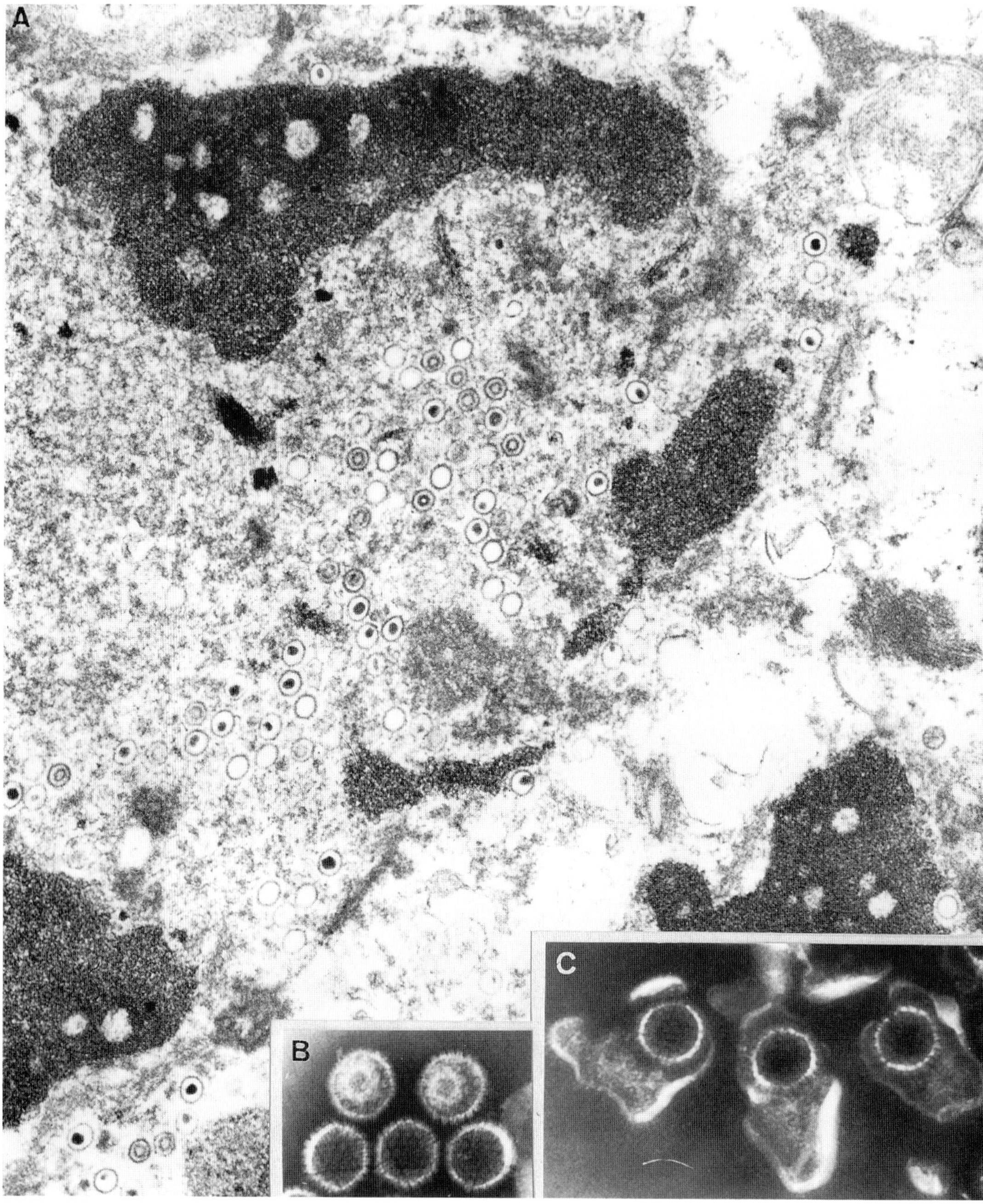

Figure 22.16 Electron micrographs of EBV and EBV-infected lymphoblast producing virions. (Modified from G. Miller, *J. Inf. Dis.* 130:187 [1974].) *(A)* Thin section of an EBV producer cell showing many nucleocapsids in the nucleus at various stages of development (43,560×); *(B)* Purified EBV nucleocapsids stained with PTA (101,000×); *(C)* Enveloped EBV virions stained with PTA (101,000×)

ELECTRON MICROSCOPY

Detection of EBV virions in lymphoid tissues is usually not feasible, since infection is often latent and not productive. High concentrations of mature virions have been demonstrated, however, in tongue lesion biopsies from patients with oral leukoplakia and in EBV-infected lymphoblasts in culture (figure 22.16).

Serodiagnosis

Isolation of EBV is not commonly used because of the specialized techniques and long incubation period required and because of the frequency of the shedding of EBV in persons with no symptoms. Thus, serology is the method of choice in primary infections. In patients with infectious mononucleosis (IM), a positive Paul-Bunnell heterophile antibody test is considered diagnostic. Heterophile antibodies are present during the first month of illness and decrease rapidly thereafter. In general, 10% 15% of patients and up to 50% of children may not have detectable heterophile antibodies. For these individuals, EBV-specific antibody is determined. Antibodies to three antigen complexes are commonly measured for diagnostic purposes: viral capsid antigen (VCA), early antigen (EA), and EBNA (figure 22.17, *A* and *B*).

At the time of clinical presentation of IM, IgG and IgM antibodies to VCA are present in high titers in the serum (figure 22.18). VCA-IgM usually disappears after four to eight weeks, whereas VCA-IgG declines but persists for life. Antibodies to EA transiently rise during acute infection. EA antibodies may reappear during reactivation. Antibodies to EBNA increase during convalescence and persist for life. Thus, with a single serum specimen, acute or convalescent, past infection with EBV can be determined. Commercially prepared slides are now available to detect EBV-specific antibodies using an IF or IP technique.

HUMAN HERPESVIRUS-6

HHV-6 is a recently recognized member of the Herpesviridae family and has been associated with roseola infantum (exanthem subitum), an infectious disease of infancy. HHV-6 was initially named human B-lymphotropic virus from the cell type in which it was originally discovered. It was subsequently found to grow in several other cell types and was thus renamed HHV-6. Like other herpesviruses, infection is prevalent and often inapparent.

Specimen Source and Inoculation

HHV-6 has been isolated from peripheral blood samples. Ten (10) ml of heparinized blood should be collected and mononuclear cells separated by Ficoll-Hypaque gradient.

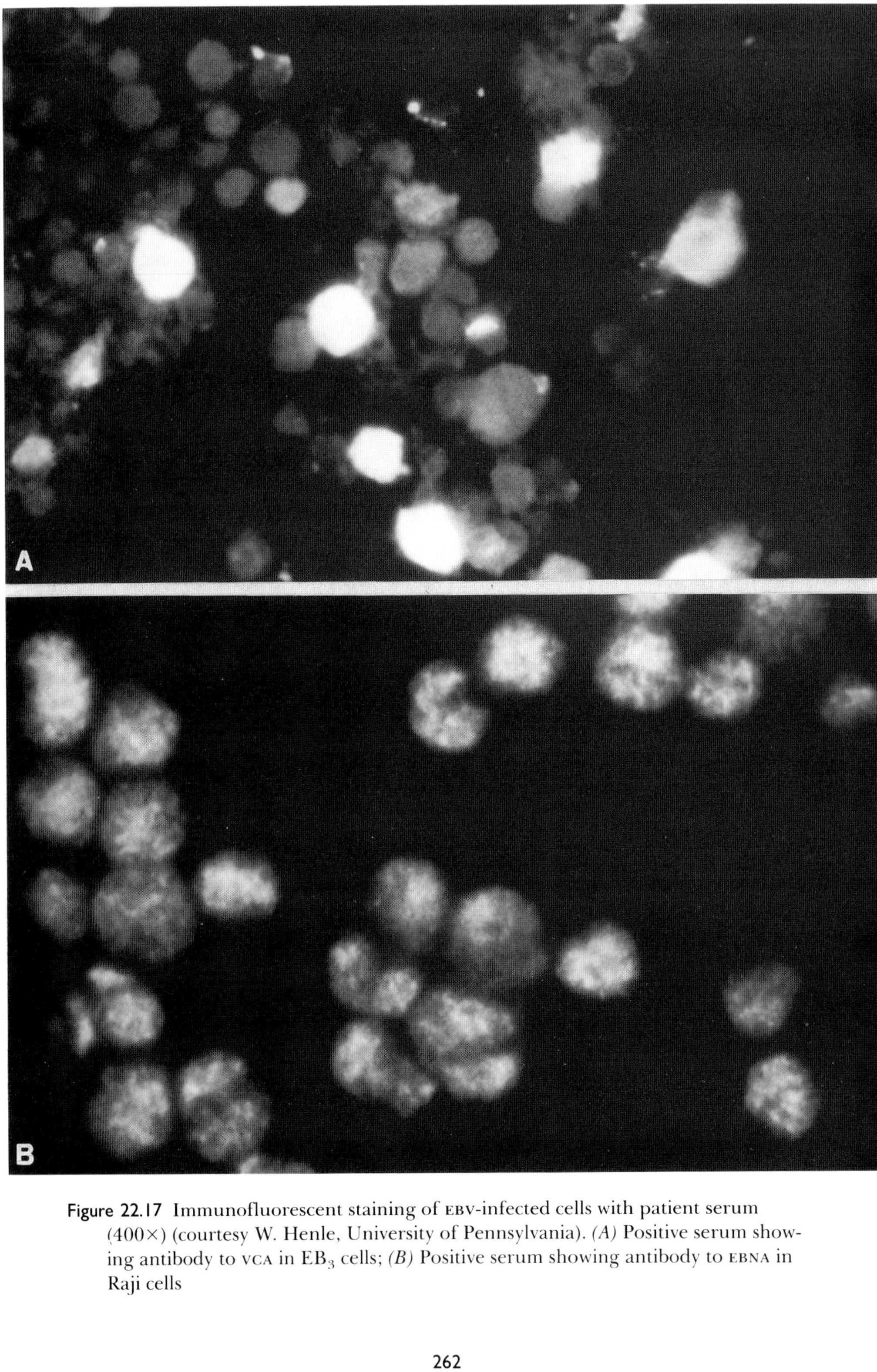

Figure 22.17 Immunofluorescent staining of EBV-infected cells with patient serum (400×) (courtesy W. Henle, University of Pennsylvania). *(A)* Positive serum showing antibody to VCA in EB_3 cells; *(B)* Positive serum showing antibody to EBNA in Raji cells

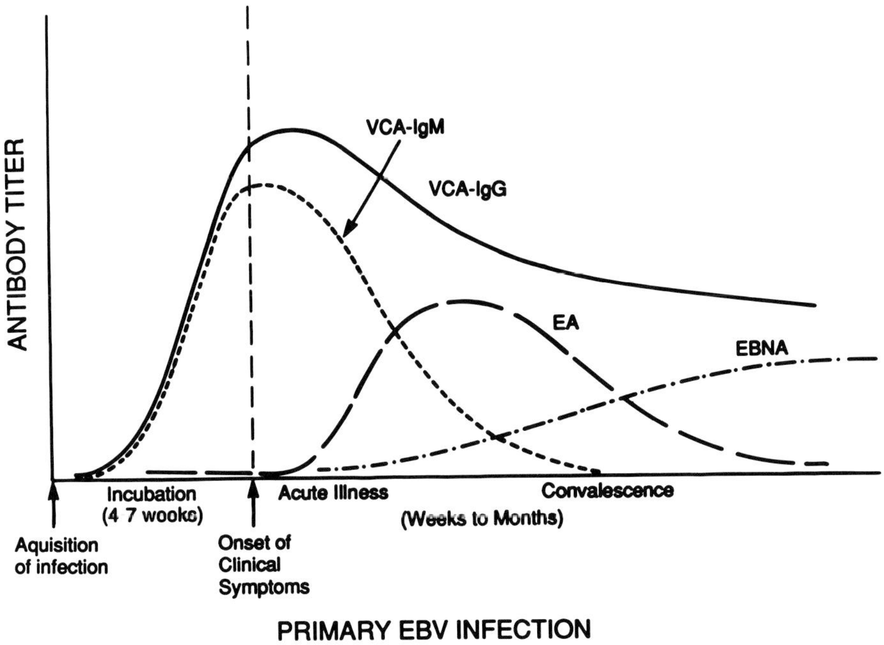

Figure 22.18 The appearance, rise and fall of antibodies to EBV-induced antigens during primary infection. VCA, viral capsid antigen; EA, early antigen; EBNA, EB nuclear antigen. (Modified from W. A. Andiman, *Diagnostic Procedures for Viral, Rickettsial and Chlamydial Infections,* 6th ed., N. J. Schmidt and R. W. Emmons, eds., 1989, 432)

Virus Isolation, Propagation, and Assay

HHV-6 can be isolated by cocultivation of mononuclear cells from patient's peripheral blood with phytohemagglutinin and interleukin 2–stimulated umbilical cord blood or adult donor lymphocytes (HHV-6 negative). Seven to ten days after cocultivation, cells are tested for HHV-6 antigen by ACIF.

Identification of Isolates

CPE may be detected as large balloonlike cells (figure 22.19), however, such cells can occur in the absence of infection. Thus, it is essential to confirm the presence of HHV-6 by IFA or preferably ACIF.

ELECTRON MICROSCOPY

Typical herpesvirus virions can be seen in infected cells in culture. In fact, HHV-6 was first discovered by EM examination of infected peripheral blood lymphocytes (figure 22.20).

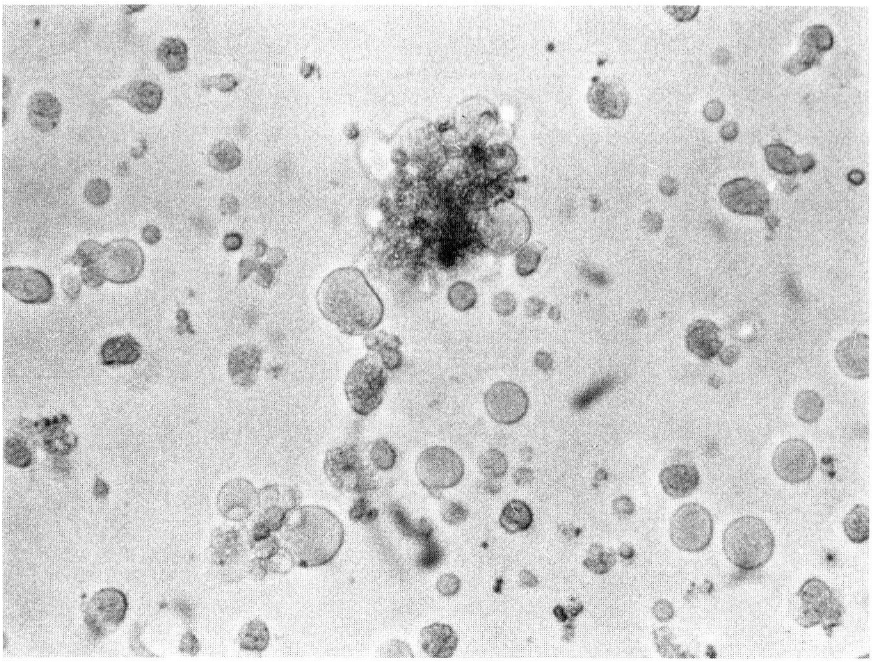

Figure 22.19 Syncytial cells induced by HHV-6 in infected mononuclear cells cocultivated with cord blood leukocytes. (Courtesy Mingxu Guan and F. Diaz-Mitonia, Children's Hospital in Eastern Ontario, Ottawa, Canada, 1992)

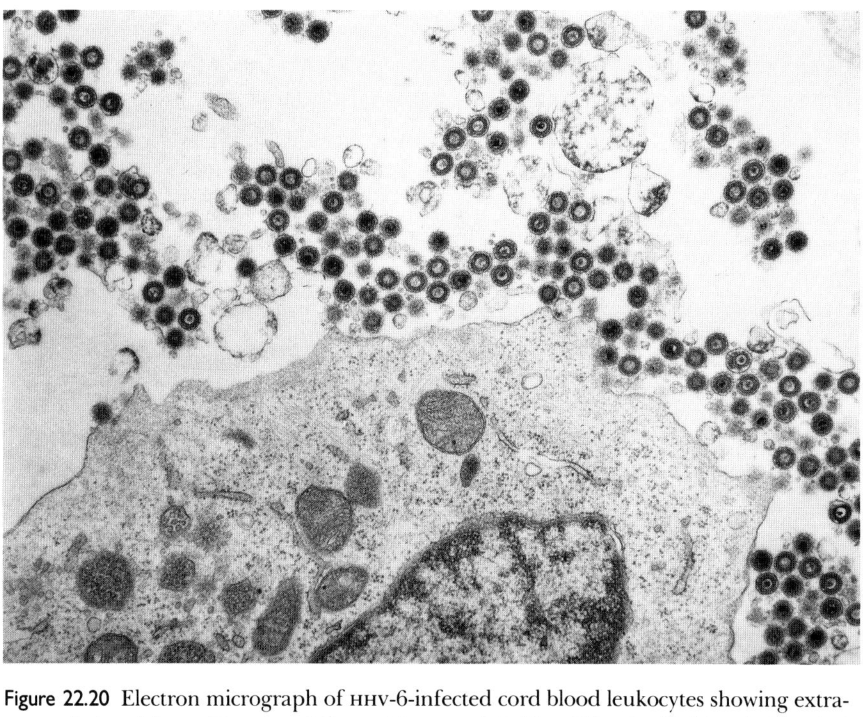

Figure 22.20 Electron micrograph of HHV-6-infected cord blood leukocytes showing extracellular virions. (Courtesy Mingxu Guan and F. Diaz-Mitonia, Children's Hospital of Eastern Ontario, Ottawa, Canada, 1992)

Viral Genome Detection

PCR has been used to detect HHV-6 DNA in the peripheral blood of patients with AIDS and lymphoproliferative disorders. Owing to the prevalence of HHV-6 infection in the general population, the significance of these findings is unclear.

Serodiagnosis

Serologic assays have only recently been developed and have been troubled with both specificity and sensitivity problems. Antibodies to HHV-6 can be determined by IFA, ACIF, or ELISA but are still performed primarily in the research setting.

SUPPLEMENTARY READING FOR HERPES SIMPLEX VIRUS

Balkovic, E. B., and Hsiung, G. D. Comparison of immunofluorescence with commercial monoclonal antibodies to biochemical and biological techniques for typing clinical herpes simplex virus isolates. *J. Clin. Microbiol.* 22:870–72 (1985).

Corey, L., and Spear, P. G. Infections with herpes simplex viruses. *N. Engl. J. Med.* 314:686–91; 749–57 (1986).

Fayram, L.; Aarnaes, S. L.; Peterson, E. M.; and de la Maza, L. M. Evaluation of five cell types for the isolation of herpes simplex virus. *Diagn. Microbiol. Infect. Dis.* 5:127–33 (1986).

Landry, M. L., and Ferguson, D. Comparison of a latex agglutination test with enzyme-linked immunosorbent assay for detection of antibody to varicella-zoster virus. *J. Clin. Microbiol.* 31:3031–33 (1993).

Lee, F. K.; Coleman, R. M.; Pereira, L.; Bailey, P. D.; Tatsumo M.; and Nahmias, A. J. Detection of herpes simplex virus type 2-specific antibody with glycoprotein G. *J. Clin. Microbiol.* 22:641–44 (1985).

Lycke, E., and Jeansson, S. Herpesviridae: Herpes simplex virus. In Lennette, E. H.; Halonen, P.; and Murphy, F. A., eds., *Laboratory Diagnosis of Infectious Diseases.* Vol. 2, *Viral, Rickettsial and Chlamydial Diseases,* pp. 211–29. New York: Springer-Verlag, 1988.

Moseley, R. C.; Corey, L.; Benjamin, D.; Winter, C.; and Remington, M. L. Comparison of viral isolation, direct immunofluorescence, and indirect immunoperoxidase techniques for detection of genital herpes simplex virus infection. *J. Clin. Microbiol.* 13:913–18 (1981).

Schmidt, N. J.; Dennis, J.; Devlin, J.; Devlin, V.; Gallo, D.; and Mills, J. Comparison of direct immunofluorescence and direct immunoperoxidase procedures for detection of herpes simplex virus antigen in lesion specimens. *J. Clin. Microbiol.* 18:445–48 (1983).

Straus, S. E.; Rooney, J. F.; Sever, J. L.; Seidlin, M.; Nusinoff-Lehrman, S.; and Cremer, K. Herpes simplex virus infection: Biology, treatment and prevention. *Ann. Intern. Med.* 103:404–41 (1985).

Whitley, R. J. Herpes simplex virus. In Fields, B. N., and Knipe, D. M., eds., *Fields Virology,* pp. 1843–88. New York: Raven, 1990.

Woods, G. L., and Young, A. Use of A-549 cells in a clinical virology laboratory. *J. Clin. Microbiol.* 26:1026–28 (1988).

Zhao, L. M.; Landry, M. L.; Balkovic, E. S.; and Hsiung, G. D. Impact of cell culture sensitivity and virus concentration on rapid detection of herpes simplex virus by cytopathic effects and immunoperoxidase staining. *J. Clin. Microbiol.* 25:1401–5 (1987).

SUPPLEMENTARY READING FOR VARICELLA-ZOSTER VIRUS

Drew, W. L., and Mintz, L. Rapid diagnosis of varicella-zoster virus infection by direct immunofluorescence. *Am. J. Clin. Path.* 73:699–701 (1980).

Gershon, A. A., ed. The First International Conference on the Varicella-Zoster Virus. *J. Infect. Dis.* 166:S1–68 (1992).

Hyman, R. W., ed. *Natural History of Varicella-Zoster Virus.* Boca Raton: CRC Press, 1987.

Landry, M. L.; Cohen, S. D.; Mayo, D. R.; Fong, C. K. Y.; and Andiman, W. A. Comparison of fluorescent antibody to membrane antigen, indirect immunofluorescence, and a commercial enzyme linked immunosorbent assay for determination of immunity to varicella-zoster virus. *J. Clin. Microbiol.* 25:832–35 (1987).

Landry, M. L., and Ferguson, D. Comparison of a latex agglutination test with enzyme-linked immunosorbent assay for detection of antibody to varicella-zoster virus. *J. Clin. Microbiol.* 31:3031–33 (1993).

Schirm, J.; Meulenberg, J. J.; Pastoor, G. W.; van Voorst-Vader, P. C.; and Schroder, F. P. Rapid detection of varicella-zoster virus in clinical specimens using monoclonal antibodies on shell vials and smears. *J. Med. Virol.* 28:1–6 (1989).

Shehab, Z., and Brunell, P. A. Enzyme-linked immunosorbent assay for susceptibility to varicella. *J. Infect. Dis.* 148:472–76 (1983).

Straus, S. E. Varicella-zoster virus infections: Biology, natural history, treatment and prevention. *Ann. Intern. Med.* 108:221–37 (1988).

Takahashi, M. Herpesviridae: Varicella-zoster virus. In Lennette, E. H.; Halonen, P.; and Murphy, F. A., eds., *Laboratory Diagnosis of Infectious Diseases.* Vol. 2, *Viral, Rickettsial and Chlamydial Diseases.* pp. 261–83. New York: Springer-Verlag, 1988.

Weller, T. H. Varicella and herpes zoster. Changing concepts of a not-so-benign virus. *N. Engl. J. Med.* 309:1362–68; 1434–40 (1983).

Williams, V.; Gershon, A. A.; and Brunell, P. A. Serological response to varicella-zoster membrane antigens measured by immunofluorescence. *J. Infect. Dis.* 136:669–72 (1974).

Zaia, J. A., and Oxman, M. Antibody to varicella-zoster virus-induced membrane antigen: Immunofluorescence assay using monodisperse glutaraldehyde-fixed cells. *J. Infect. Dis.* 136:519–30 (1977).

SUPPLEMENTARY READING FOR CYTOMEGALOVIRUS

Drew, W. L. Herpesviridae: Cytomegalovirus. In Lennette, E. H.; Halonen, P.; and Murphy, F. A., eds., *Laboratory Diagnosis of Infectious Diseases.* Vol. 2, *Viral, Rickettsial and Chlamydial Diseases,* pp. 247–60. New York: Springer-Verlag, 1988.

Beckwith, D. G.; Halstead, D. C.; Alpaugh, K.; Schweder, A.; Blount-Fronefield, D. A.; and Toth, K. Comparison of a latex agglutination test with five other methods for determining the presence of antibody against cytomegalovirus. *J. Clin. Microbiol.* 21:328–31 (1985).

Gerna, G.; Revello, M. G.; Percivalle, E.; and Morini, F. Comparison of different immunostaining techniques and monoclonal antibodies to the lower matrix phosphoprotein (pp65) for optimal quantitation of human cytomegalovirus antigenemia. *J. Clin. Microbiol.* 30:1232–37 (1992).

Gleaves, C. A.; Hursh, D. A.; Rice, D. H.; and Myers, J. D. Detection of cytomegalovirus from clinical specimens in centrifugation culture by in situ hybridization and monoclonal antibody staining. *J. Clin. Microbiol.* 27:21–23 (1989).

Gleaves, C. A.; Smith, T. F.; Shuster, E. A.; and Pearson, G. R. Rapid detection of cytomegalovirus in MRC-5 cells inoculated with urine specimens by using low speed centrifugation and monoclonal antibody to an early antigen. *J. Clin. Microbiol.* 19:917–19 (1984).

Gregory, W. W., and Menegus, M. A. Effect of incubation temperature on isolation of cytomegalovirus from fresh clinical specimens. *J. Clin. Microbiol.* 18:1003–5 (1983).

Landry, M. L., and Ferguson, D. Comparison of quantitative cytomegalovirus antigenemia assay with culture methods and correlation with clinical disease. *J. Clin. Microbiol.* 31:2851–56 (1993).

Leland, D. A.; Barth, K. A.; Cunningham, E. B.; Jansen, J.; Tricot, G. J.; and French, M. L. V. Evaluation of four methods for cytomegalovirus antibody detection for use by a bone marrow transplantation service. *J. Clin. Microbiol.* 27:176–78 (1989).

Marcris, M. P.; Nahmias, A. J.; Bailey, P. D.; Lee, F. K.; Visintine, A. M.; and Braun, A. W. Electron microscopy in the routine screening of newborns with congenital cytomegalovirus infection. *J. Virol. Meth.* 2:315–20 (1981).

Mayo, D. R.; Brennan, T.; Sirpenski, S. P.; and Seymour, C. Cytomegalovirus antibody detection by the three commercially available assays and complement fixation. *Diagn. Microbiol. Infect. Dis.* 3:455–59 (1985).

Paya, C. V.; Wold, A. D.; Ilstrup, D. M.; and Smith, T. F. Evaluation of number of shell vial cell cultures per clinical specimen for rapid diagnosis of cytomegalovirus infection. *J. Clin. Microbiol.* 26:198–200 (1988).

Rubin, R. H., ed. Cytomegalovirus infections: Epidemiology, diagnosis, and treatment strategies. *Rev. Infect. Dis.* 12(7):S691–860 (1990).

Stagno, S.; Britt, W. J.; and Pass, R. F. Cytomegalovirus. In Schmidt, N. J., and Emmons, R. W., eds., *Diagnostic Procedures for Viral, Rickettsial and Chlamydial Infections*, 6th ed., pp. 321–78. Washington, D.C.: American Public Health Association, 1989.

van der Bij, W.; Torensma, R.; van Son, W. J.; Anema, J.; Schirm, J.; Tegzess, A. M.; and The, T. H. Rapid immunodiagnosis of active cytomegalovirus infection by monoclonal antibody staining of blood leukocytes. *J. Med. Virol.* 25:179–88 (1988).

SUPPLEMENTARY READING FOR EPSTEIN-BARR VIRUS

Andiman, W. A. Epstein-Barr Virus. In Schmidt, N. J., and Emmons, R. W., eds., *Diagnostic Procedures for Viral, Rickettsial and Chlamydial Infections*, 6th ed., pp. 401–52. Washington, D.C.: American Public Health Association, 1989.

Andiman, W. A.; Eastman, R.; Martin, K.; Katz, B. Z.; Rubinstein, A.; Pitt, J.; Pahwa, S.; and Miller, G. Opportunistic lymphoproliferations associated with Epstein-Barr viral DNA in infants and children with AIDS. *Lancet* 2:1390–93 (1985).

Henle, W.; Henle, G.; and Horwitz, C. A. Epstein-Barr virus specific diagnostic tests in infectious mononucleosis. *Hum. Path.* 5:551–64 (1974).

Horwitz, C. A.; Henle, W.; Henle, G.; Rudnick, H.; and Latts, E. Long term serological follow-up of patients for Epstein-Barr virus after recovery from infectious mononucleosis. *J. Infect. Dis.* 151:1150–53 (1985).

Lennette, E. T. Herpesviridae: Epstein-Barr Virus. In Lennette, E. H.; Halonen, P.; and Murphy, F. A., eds., *Laboratory Diagnosis of Infectious Diseases* Vol. 2, *Viral, Rickettsial and Chlamydial Diseases,* pp. 230–46. New York: Springer-Verlag, 1988.

Miller, G. Epstein Barr Virus. In Fields, B. N., and Knipe, D. M., eds., *Fields Virology,* pp. 1921–58. New York: Raven, 1990.

Paul, J. R., and Bunnell, W. W. The presence of heterophile antibodies in infectious mononucleosis. *Am. J. Med. Sci.* 183:80–104 (1932).

Schooley, R. T.; Carey, R. W.; Miller, G.; Henle, W.; Eastman, R.; Mark, E. J.; Kenyon, K.; Wheeler, E. O.; and Rubin, R. Chronic Epstein-Barr virus infection associated with fever and interstitial pneumonitis. Clinical and serological features and response to antiviral chemotherapy. *Ann. Intern. Med.* 104:636–43 (1986).

Wreghitt, T. G.; Sargaison, M.; Sutehall, G.; Woodward, C. G.; Scott, J.; English, T. A.; and Wallwork, J. A study of Epstein-Barr virus infections in heart and heart and lung transplant recipients. *Transplantation Proc.* 21:2502–3 (1989).

zur Hauzen, H.; Schulte-Holthausen, H.; Klein, G.; Henle, W.; Henle, G.; Clifford, P.; and Santesson, L. EBV DNA in biopsies of Burkitt's tumors and anaplastic carcinomas of the nasopharynx. *Nature* (London) 228:1056–58 (1970).

SUPPLEMENTARY READING FOR HUMAN HERPESVIRUS TYPE 6

Ablashi, D. V.; Josephs, S. F.; Buchbinder, A.; Hellman, K.; Nakamura, S.; Llana, T.; Lusso, P.; Kaplan, M.; Dahlberg, J.; Memon, S.; Imam, F.; Ablashi, K. L.; Markham, P. D.; Kramarsky, B.; Krueger, G. R. F.; Biberfeld, P.; Wong-Staal, F.; Salahuddin, S. Z.; and Gallo, R. C. Human B-lymphotropic virus (human herpesvirus-6). *J. Virol. Meth.* 21:29–48 (1988).

Asano, Y.; Yoshikawa, T.; Suga, S.; Yazaki, T.; Hatz, T.; Nagai, T.; Kajita, Y.; Ozaki, T.; and Yoshida, S. Viremia and neutralizing antibody response in infants with exanthem subitum. *J. Pediatr.* 141:535–39 (1989).

Lopez, C.; Pellett, P.; Stewart, J.; Goldsmith, C.; Sanderlin, K.; Black, J.; Warfield, D.; and Feorino, P. Characteristics of human herpesvirus-6. *J. Infect. Dis.* 157:1271–73 (1988).

Okuno, T.; Takahashi, K.; Balachandra, K.; Shiraki, K.; Yamanishi, K.; Takahashi, M.; and Baba, K. Seroepidemiology of human herpesvirus 6 infection in normal children and adults. *J. Clin. Microbiol.* 27:651–53 (1989).

Oren, I., and Sobel, J. D. Human herpesvirus type 6: Review. *Clin. Infect. Dis.* 14:741–46 (1992).

Salahuddin, S. Z.; Ablashi, D. V.; Markham, P. D.; Josephs, S. F.; Sturzenegger, S.; Kaplan, M.; Halligan, G.; Biberfeld, P.; Wong-Staal, F.; Kramarsky, B.; and Gallo, R. C. Isolation of a new virus, HBLV, in patients with lymphoproliferative disorders. *Science* 234, 596–601 (1986).

23

PAPOVAVIRIDAE

F. J. MICHALSKI

Papovaviridae consists of two distinct subfamilies of viruses, Papillomavirinae and Polyomavirinae (table 23.1). The name papova virus was originally derived from the first two letters of three separate groups of viruses that are morphologically similar and are classified together. These are rabbit *pa*pillomavirus, mouse *po*lyomavirus, and simian *va*cuolating virus (SV_{40}). After this classification was made, these viruses, although morphologically similar, were shown to be unrelated genetically and immunologically.

HUMAN PAPILLOMAVIRUSES

Morphologically, the papillomaviruses (figure 23.1) are 50–55 nm in diameter. They are small nonenveloped viruses with 72 capsomeres and double-stranded, circular DNA that multiply in the cell nucleus.

The viral etiology of human warts (papillomas) was established in 1907 by the transmission of warts in a human-to-human inoculation of a cell-free extract of wart tissue. To date, human papillomaviruses (HPV) have not been grown in routine cultures or laboratory animals. Thus, until the 1970s, only animal papillomaviruses were studied in the search for the etiology of tumors and to elucidate the molecular biology of cancer. With the availability of sensitive methods for the detection of viral nucleic acid in recent years, diagnosis of papillomavirus is now possible. This is a timely development since the incidence of genital warts has increased and genital HPV has been linked to cervical cancer. It is now recognized that there are more than 60 types of human papillomaviruses (HPV) (table 23.2). If two papillomavirus

ISBN 0-300-05845-4.

Table 23.1 Representative Papovaviruses of Humans and Animals

Virus Subfamily	Common Virus Types	Source of Virus Specimen	Methods of Diagnosis
Papillomavirinae			
Human papillomavirus	More than 60 types	Warts, tumors, skin	1. Not yet cultivable 2. Detection by DNA hybridization 3. EM
Animal papillomavirus	Rabbit, bovine, canine, hamster	Warts, tumors, skin	1. Not yet cultivable 2. Detection by DNA hybridization 3. EM
Polyomavirinae			
Human polyomavirus	BK JC	Urine Brain tissue of PML patient	1. Cell culture in research 2. Antigen detection by ELISA 3. DNA hybridization 4. Antibody detection by ELISA
Animal polyomavirus	Mouse polyomavirus Simian vacuolating virus (SV_{40})	Mouse cell culture Monkey kidney cell culture	Cell culture and serodiagnosis Cell culture isolation and serodiagnosis

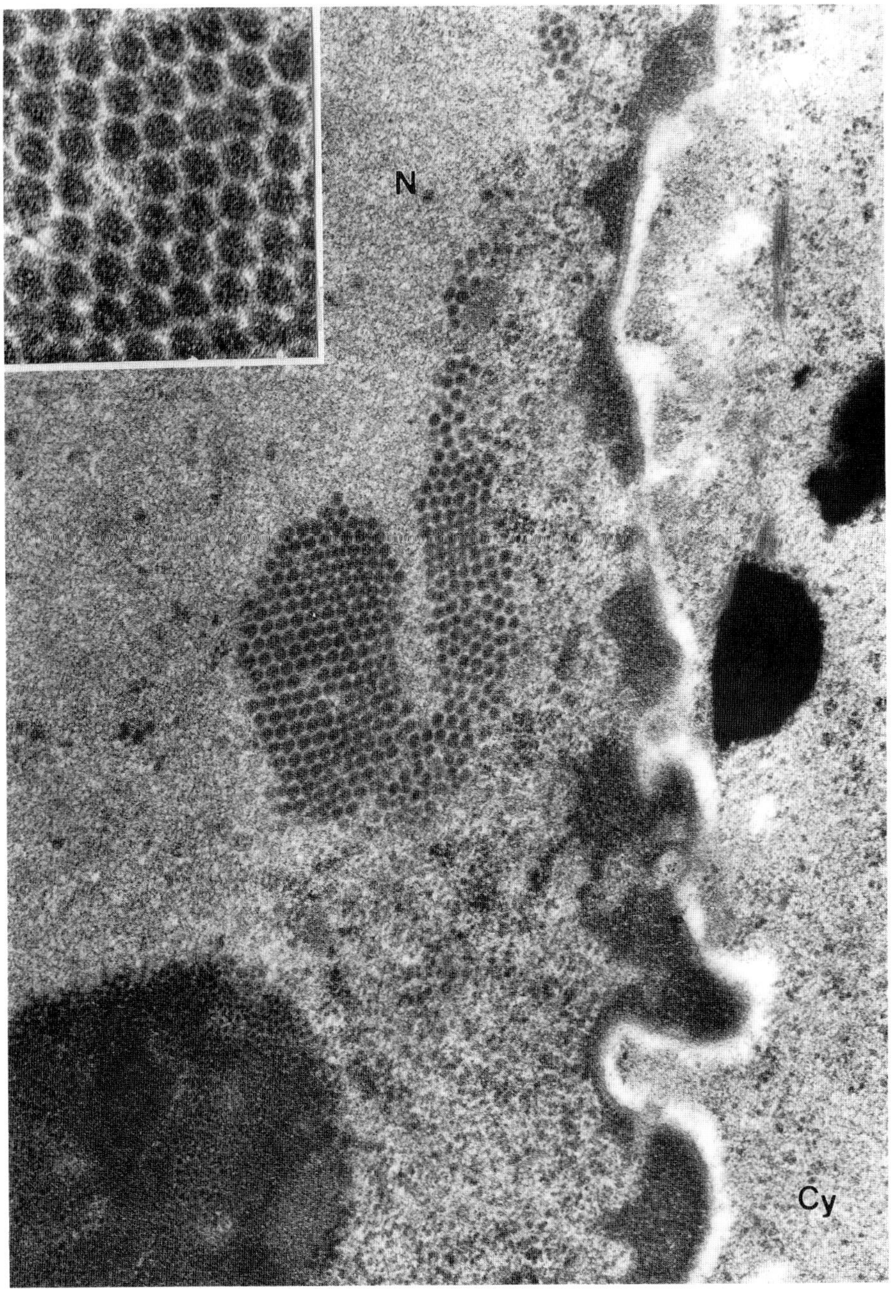

Figure 23.1 Electron micrograph of human papillomavirus virions in the nucleus of a cell from a wart (38,400×). Inset: Higher magnification of the viral aggregate (108,900×). (Courtesy S. Klaus)

Table 23.2 Distribution of Human Papillomavirus Types and Clinical Features

Location	Clinical Features	HPV Types
Skin	Cutaneous warts	1, 2, 3, 4, 7, 10, 26, 27, 28, 29, 37, 38, 41, 46, 48, 49
	Associated with epidermodysplasia verruciformis	5, 8, 9, 12, 14, 15, 17, 19, 20, 21, 22, 23, 24, 25, 36, 47, 50
Genital tract mucosa	Condyloma acuminatum, cervical carcinoma, etc.	6, 11, 16, 18, 31, 33, 34, 35, 39, 40, 42, 43, 44, 45, 51, 53, 54, 55, 56, 58
Oral cavity mucosa	Focal epithelial hyperplasia	13, 32
Other sites		30, 57

Adapted from K. V. Shah, Papillomaviruses. In E. H. Lennette, ed., *Laboratory Diagnosis of Viral Infection,* 2nd ed. New York: Marcel Dekker, 1992.

DNAS show a less than 50% homology as determined by duplex formation in a liquid reassociation reaction, they are regarded as different types. HPV-1 is predominant in plantar warts, HPV-2 and HPV-4 in common warts, and HPV-3, HPV-10, HPV-28, and HPV-41 in flat cutaneous warts. Epidemiological studies have shown that genital warts caused by HPV are the most common form of sexually transmitted disease today. Approximately 20 HPV types have a predilection for genital infection. Certain types of these HPV have been shown to be associated with cervical carcinoma and with a high, intermediate, and low risk of progression to malignancy (see table 23.3).

Table 23.3 Three Biological Categories of Risks of Progression to Malignancy among Human Papillomavirus Types

Risk of Progression to Malignancy	HPV Types	Source of Original Virus Isolation
High	16, 18	Cervical carcinoma
	45, 56	Cervical intraepithelial neoplasia
Intermediate	31, 51, 52, 58	Cervical intraepithelial neoplasia
	33	Cervical carcinoma
	35	Cervical adenocarcinoma
Low	6	Condyloma acuminatum
	11	Laryngeal papilloma
	42	Vulvar papilloma
	43	Vulvar hyperplasia
	44	Vulvar condyloma

Specimen Source for Virus Detection

Cell scrapings or swabs from female genital tracts and tissue biopsy from warts or cancer cells are the specimens of choice. Specimens are sent to special virus laboratory in sampling tubes containing 1 ml of Vira Pap transport medium supplied by Digene Diagnostics, Inc. If in situ diagnosis is attempted, smears from the female genital tract or fixed tissues are required.

Virus Isolation, Antigen Detection, and Assay

ISOLATION OF INFECTIOUS VIRUS IN CELL CULTURE

Human papillomaviruses have so far not been grown in cell culture or laboratory animals, since virus replication is limited to terminally differentiated squamous epithelial cells.

VIRAL ANTIGEN OR GENOME DETECTION

As stated above, human papillomaviruses have not been grown in cell cultures. Consequently, few reagents for the immunological detection of antigens are available. Capsid antigen has been detected in tissue sections or cell spots by an immunoperoxidase (IP) test using a broadly cross-reactive, genus-specific immune serum prepared in rabbits immunized with bovine papillomavirus.

Human papillomaviruses are best diagnosed today by detection of viral nucleic acid in tissues or exfoliated cells. This can be done before or after DNA amplification by PCR. Methods of detection include Southern blot hybridization, dot blot hybridization, and tissue in situ hybridization. In 1988, the FDA licensed a modified dot blot RNA-DNA hybridization kit that could detect seven types of human genital HPV—such as types 6, 11, 16, 18, 31, 33, and 35. It consists of two separate tests, Vira Pap (figure 23.2) and Vira Type. If found positive by Vira Pap, then the HPV typing can be performed by Vira Type. Both tests, in their present configuration, were found to be complicated (39 manipulations), time-consuming, limited to only seven viruses, and subjective in interpretation. Many variables make the performance of the test difficult. Figure 23.3 shows an example of the results of Vira Pap test. This test is not intended for use as a screening test for the general population but rather as a way to identify those patients requiring closer surveillance for oncogenic potential after an equivocal or low-grade result of a Papanicolaou (PAP) smear screening. Several studies have indicated that conventional cytology (PAP testing and physical examination) with HPV DNA testing can increase diagnostic sensitivity and specificity greatly. An improved version of this HPV DNA test using chemiluminescent rather than radioisotope labeling to distinguish low-risk HPV types (6, 11, 42, 43, and 44) from intermediate (31, 33, 35, 51, 52, and 58) and high-risk HPV types (16, 18, 45, and 56) is being evaluated clinically (table 23.3).

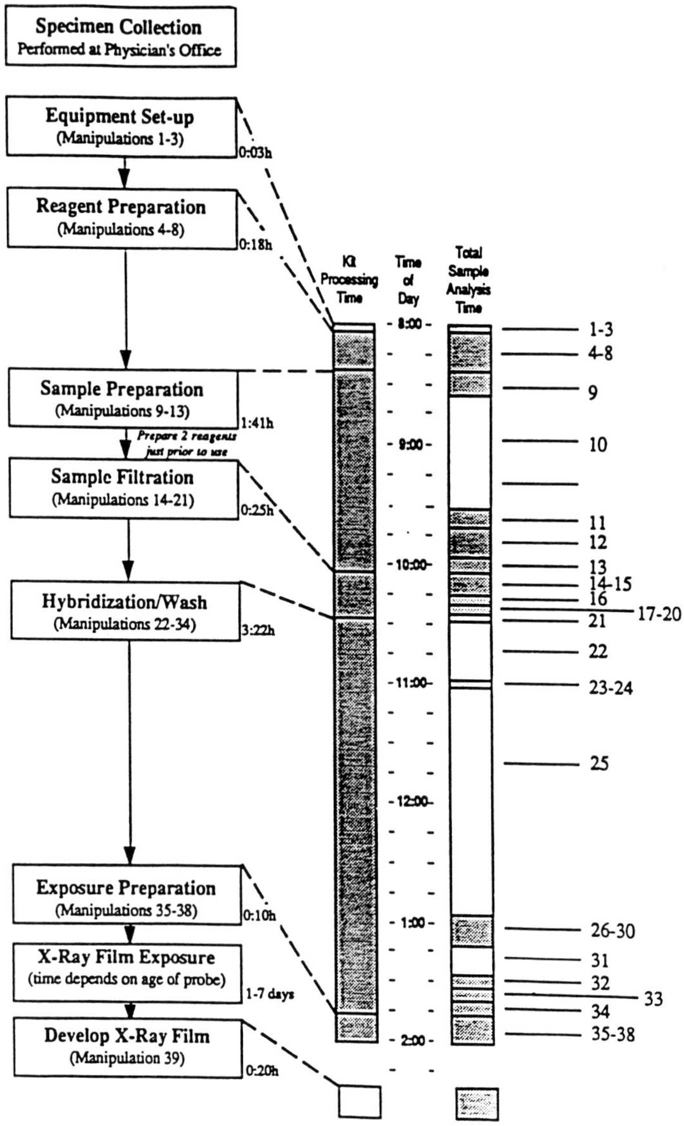

Figure 23.2 Vira Pap test procedures. Left-hand bar: Shaded areas represent time spent using the kit; right-hand bar: Shaded areas represent hands-on time. Clear areas represent incubation time; total analysis time 6:20 h + 1–7 days; kit processing time 6:00 h; total hands-on time 2:35 h; kit hands-on time 2:15 h. (Courtesy J. D. Love and A. B. Jenson, "Clinical Test Facts: Comparative Report on HPV Tests," *Bioline* 1990)

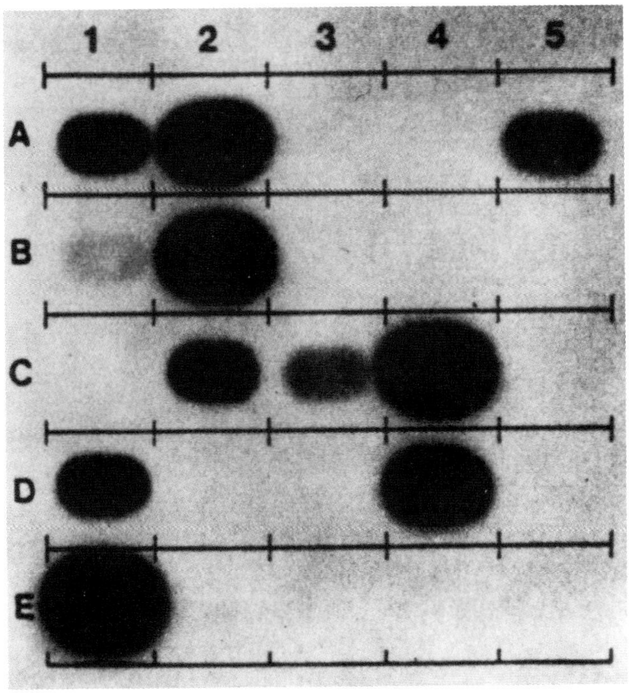

Figure 23.3 Sample results of Vira Pap test. Autoradiogram of dot blot hybridization by Digene Vira Pap screen test for human papillomaviruses 6, 11, 16, 18, 31, 33, 35. Controls are shown: high and low positive in squares A1 and B1, respectively, and negative in square C1. Positive specimens indicate presence of one or more of the seven types of HPV DNA as observed in squares *D1, E1, A2, B2, C2, C3, C4, D4,* and *A5*. (Photograph courtesy of Digene)

ELECTRON MICROSCOPIC EXAMINATION

In addition to the clinical diagnosis of papillomavirus infection, direct visualization of specific virions in clinical specimens by either negative staining or thin-sectioning techniques are convincing evidence of HPV infection. Figure 23.1 illustrates a thin section of a human wart showing HPV virions in aggregate in an infected nucleus.

Serodiagnosis

Infections due to human papillomaviruses cannot be diagnosed by serology, since reagents for antibody detection are not available. Research is under way to develop recombinant viral proteins that might possibly be used to develop a serological test.

HUMAN POLYOMAVIRUSES

Morphologically, the polyomaviruses are 40–45 nm in diameter. Like the papillomaviruses but slightly smaller, their virions are small and nonenveloped, with 72 capsomeres and double-stranded circular DNA that replicate in the nucleus. In a thin section of an infected cell, numerous virions can be seen scattered throughout the nucleus (figures 23.4, 23.5*A*).

In 1965, papovavirus-like virions in the brain tissue of a patient with a rare chronic demyelinating disease, progressive multifocal leukoencephalopathy (PML), were noted. In 1971, two human polyomaviruses were isolated: BK virus from the urine of an immunosuppressed patient and JC virus from the brain of a PML patient. Subsequently, serological studies have shown that BK and JC are common infections of humans.

Specimen Source for Virus Detection

Urine for BKV or brain tissue for JCV are submitted for virus isolation attempts, which are still performed mainly for research purposes.

Virus Isolation, Antigen Detection, and Assay

ISOLATION OF INFECTIOUS VIRUS IN CELL CULTURE

Human polyomaviruses JC and BK can be grown in cell culture, but it is very difficult and time-consuming. In primary isolation, CPE takes a long time—6–31 days—to develop, and may require several blind passages. JC is best recovered from primary human fetal glial cells (PHFG). BK is slightly easier to grow and has been isolated in PHFG, primary epithelial cells derived from human tissues, primary human embryonic kidney and lung cells, human foreskin fibroblasts, and in monkey cell lines such as Vero and CV-1 cells. Rapid detection of BK virus is done using shell vial cell culture tested with cross-reactive monoclonal antibody to the T antigen of SV_{40} by indirect immunofluorescence. Nuclear fluorescence can be seen in BK virus–infected cells as early as 16 hours postinoculation; 6–28 times more foci were seen at 36 hours postinoculation. Supernates of BK virus–infected cultures hemagglutinate human O or guinea pig RBC, although some isolates lack HA activity.

VIRAL ANTIGEN OR GENOME DETECTION

IF and IP staining methods can be used for BK and JC polyomavirus antigens, but reagents are not commercially available. An ELISA test has been developed for the detection of JC or BK viral antigens in urine for research purposes.

Human polyomaviruses can be diagnosed by Southern blot, dot blot,

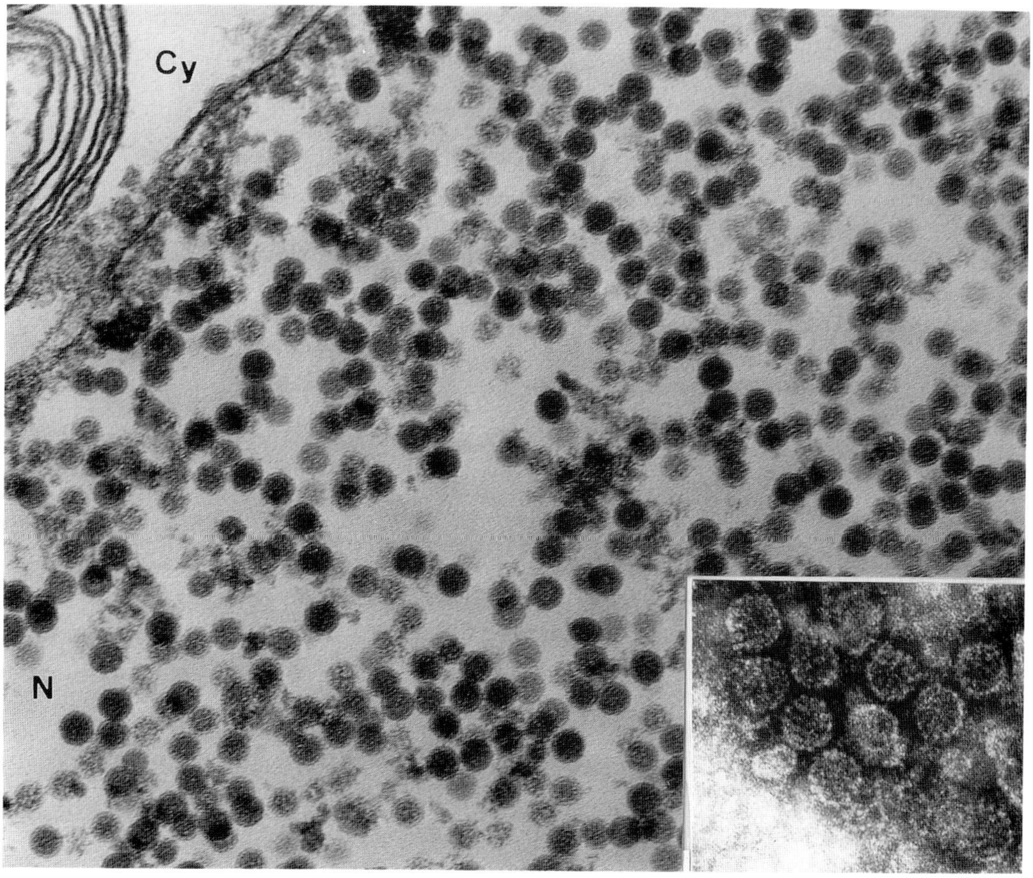

Figure 23.4 Electron micrograph of human polyomavirus–infected HDF cell, isolated from a urine sample that was inoculated into HDF cell culture. *N* = nucleus, *Cy* = cytoplasm (100,800×). Inset: Same virions stained with PTA (169,000×). (Original specimen supplied by M. Hirsch, Harvard Medical School)

or tissue in situ hybridization techniques, but commercial kits are not yet available.

ELECTRON MICROSCOPIC EXAMINATION

An example of a polyomavirus isolated from the urine of an immunosuppressed patient grown in HDF cell culture is illustrated in figure 23.4. Virions are usually scattered in the nucleus of the infected cells. Negatively stained virions of the same sample are shown in the inset. Although SV_{40} is not a human virus, it is often present in primary rhesus monkey kidney cell cultures as a source of endogenous virus contamination. Therefore, it is important to recognize its presence. For comparison, an example of SV_{40} in

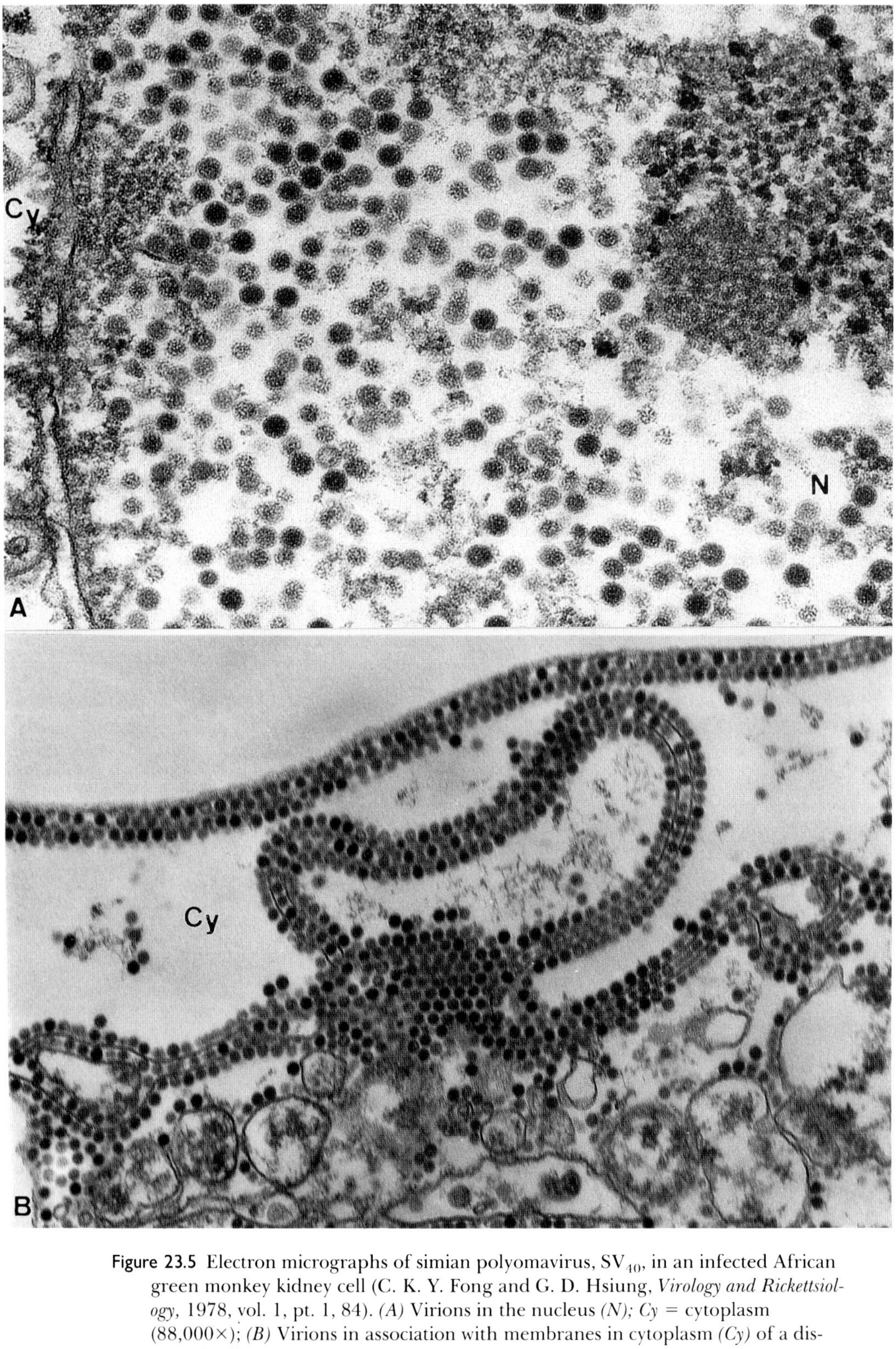

Figure 23.5 Electron micrographs of simian polyomavirus, SV_{40}, in an infected African green monkey kidney cell (C. K. Y. Fong and G. D. Hsiung, *Virology and Rickettsiology,* 1978, vol. 1, pt. 1, 84). *(A)* Virions in the nucleus *(N)*; *Cy* = cytoplasm (88,000×); *(B)* Virions in association with membranes in cytoplasm *(Cy)* of a disrupted cell (48,000×)

monkey kidney cells is illustrated in figure 23.5. Although EM is not routinely used in the diagnosis of polyomaviruses, it is an important tool in diagnosis confirmation of this group of viruses.

Serodiagnosis

Antibodies to human polyomaviruses, BK and JC, can be measured by hemagglutination-inhibition or ELISA assay methods, but these tests are not routinely used for diagnostic purposes.

SUPPLEMENTARY READING

Amortegui, A. J., and Meyer, M. P. In-situ hybridization for the diagnosis and typing of human papillomavirus. *Clin. Biochem.* 23:301–6 (1990).

Arthur, R. R. Polyomaviruses. In Lennette, E., ed., *Laboratory Diagnosis of Viral Infections,* pp. 613–25. New York: Marcel Dekker, 1992.

Crum, C. P.; Barber, S.; and Roche, J. K. Pathobiology of papillomavirus-related cervical diseases: Prospects for immunodiagnosis. *Clin. Microbiol. Rev.* 4:270–85 (1991).

David, F.; Levy, R.; and Lucotte, G. Detection and typing of human papillomavirus DNA from cervical biopsies by the slot-blot hybridization method. *Mol. Cell. Probes* 4:53–61 (1990).

Eckhart, W. Polyomaviridae and their replication. In Fields, B. N., and Knipe, D. M., eds., *Fields Virology,* 2d ed., pp. 1593–1607. New York: Raven, 1990.

Gardner, S. D.; Field, A. M.; Coleman, D. V.; and Huline, B. New human papovavirus (BK) isolated from urine after renal transplantation. *Lancet* 1:1253–57 (1971).

Howley, P. M. Papillomavirinae and their replication. In Fields, B. N., and Knipe, D. M., eds., *Fields Virology,* 2d ed., pp. 1625–50. New York: Raven, 1990.

———. Role of the human papillomavirus in human cancer. *Cancer Res.* 51 (18 supp.): 5019S–22S (1991).

Jawetz, E.; Melnick, J. L.; Adelberg, E. A.; Brooks, G. F.; Butel, J. S.; and Nicolas Ornston, L., eds. Tumor viruses and oncogenes. In *Medical Microbiology,* 19th ed., pp. 549–82. Norwalk: Appleton and Lange, 1991.

McCance, D. J. Papillomaviruses. In Balows, A.; Hausler, W. J., Jr.; Herrmann, K. L.; Isenberg, H. D.; and Shadomy, H. J., eds. *Manual of Clinical Microbiology,* 5th ed., pp. 998–1004. Washington, D.C.: American Society for Microbiology, 1991.

Marshall, W. F.; Telenti, A.; Proper, J.; Aksamit, A. J.; and Smith, T. F. Rapid detection of polyomavirus BK by a shell vial cell culture assay. *J. Clin. Microbiol.* 28:1613–15 (1990).

Melnick, J. L. Papovavirus group. *Science* 135:1128–30 (1962).

Padgett, B. L.; Walker, D. L.; ZuRhein, G. M.; Eckroade, R. J.; and Dessel, B. H. Cultivation of papova-like virus from human brain with progressive multifocal leukoencephalopathy. *Lancet* 1:1257–60 (1971).

Ranow, R. F. Nucleic acid hybridization as a diagnostic tool for the detection of human papillomaviruses. *Adv. Exp. Med. Biol.* 263:89–109 (1990).

Reid, R. Human papillomaviruses. *Obstetrics and Gynecology Clinics of North America.* Philadelphia: W. B. Saunders, 1987.

Reid, R.; Greenberg, M. D.; Lorincz, A.; Jenson, A. B.; Laverty, C. R.; Husain, M.; Daoud, Y.; Zado, B.; White, T.; Cantor, D.; and Goldrath, M. Should cervical cytology testing be augmented by cervicography or human papillomavirus deoxyribonucleic acid detection? *Am. J. Obstet. Gynec.* 164:1461–71 (1991).

Sawchuk, W. S. Ancillary diagnostic tests for detection of human papillomavirus infection. *Dermat. Clinics* 9:277–86 (1991).

Shah, K. V. Papillomaviruses. In Lennette, E., ed., *Laboratory Diagnosis of Viral Infection,* 2d ed., pp. 591–612. New York: Marcel Dekker, 1992.

———. Polyomaviruses. In Fields, B. N., and Knipe, D. M., eds., *Fields Virology,* 2d ed., pp. 591–612. New York: Raven, 1990.

zur Hausen, H. Papillomaviruses as carcinoma viruses. *Adv. in Viral Oncol.* 8:1–26 (1989).

24

PARVOVIRIDAE

M. L. LANDRY

Parvoviridae is a family of small DNA viruses classified into three genera: *Parvovirus, Dependovirus,* and *Densovirus,* according to their mode of replication and host range (table 24.1).

Parvoviruses are nonenveloped, icosahedral, single-stranded DNA viruses, 20–25 nm in diameter. Replication occurs in the nucleus of infected cells, and productive infection of autonomous parvoviruses is restricted to rapidly dividing cells. Virions contain a single molecule of either plus- or minus-strand DNA encapsidated in each virion.

HUMAN PARVOVIRUS B19

In 1975, Y. E. Cossart and colleagues reported the first evidence that parvoviruses infect humans. A parvovirus-positive serum gave a false-positive result by hepatitis B surface antigen (HB_sAg) testing, was examined by EM, and spherical particles typical of parvovirus were discovered (figure 24.1). Since this serum was number 19 in panel B of the control sera for hepatitis B surface antigen testing, it became known as B19. Subsequent studies have found parvovirus B19 to be the etiologic agent of erythema infectiosum or fifth disease of childhood, associated with transient aplastic crisis in patients with hemoglobinopathies, a flulike illness with arthralgia in adults, and nonimmune hydrops fetalis in pregnancy. Furthermore, patients with acquired or congenital immunodeficiencies may suffer chronic infections with parvovirus B19, resulting in a prolonged pure red cell aplasia.

Parvovirus B19 appears to be transmitted by respiratory secretions, re-

ISBN 0-300-05845-4.

Table 24.1 Parvoviridae: Mode of Replication and Host Range

Genus	Mode of Replication	Representative Virus	Host
Parvovirus	Autonomous	B19 RA-1 Minute virus of mice (MVM)[a]	Humans Humans Mice
Dependovirus	Require helper virus for replication	Adeno-associated virus (AAV)[b]	Humans Birds
Densovirus	Autonomous	Arthropod parvovirus	Insects

a. Other parvoviruses of different animal species, including feline parvovirus (FPV), canine parvovirus (CPV), etc., are associated with a variety of clinical diseases of veterinary importance.

b. AAV also occur in bovine, canine, and other animal species.

sulting in viremia and viral shedding from the respiratory tract. In volunteer studies, IgM antibodies to parvovirus B19 appear in the serum during the second week after inoculation, followed within a few days by IgG antibodies. The period of viremia may be asymptomatic or may be associated with malaise, fever, and headache. Rash, arthritis, and arthralgia begin coincident with the appearance of antibody and the cessation of viremia and virus shedding. Parvovirus B19 has been found to replicate in erythroid progenitor cells in the bone marrow, resulting in areticulocytosis from the period of peak viremia until several days after the appearance of antibody response. In normal hosts, the transient decrease in hemoglobin is not clinically significant, but in patients with hemoglobinopathies, transfusion support is often required during this period.

Since isolation of parvovirus B19 requires specialized cell cultures that are not routinely available and clinical symptoms generally become manifest coincident with the development of IgM antibodies, diagnosis in the clinical laboratory has relied upon detection of IgM and IgG antibodies. Methods for the detection of viral antigen and nucleic acid have also been developed but are not routinely available.

Specimen Source

Virus is present in serum during the period of viremia and in bone marrow aspirates during the period of areticulocytosis. Highest titers are observed during aplastic crisis in patients with hemoglobinopathies or in persistent infection in immunocompromised hosts. Virus can also be isolated from cord blood, amniotic fluid, fetal ascites and tissues, and urine and throat washes.

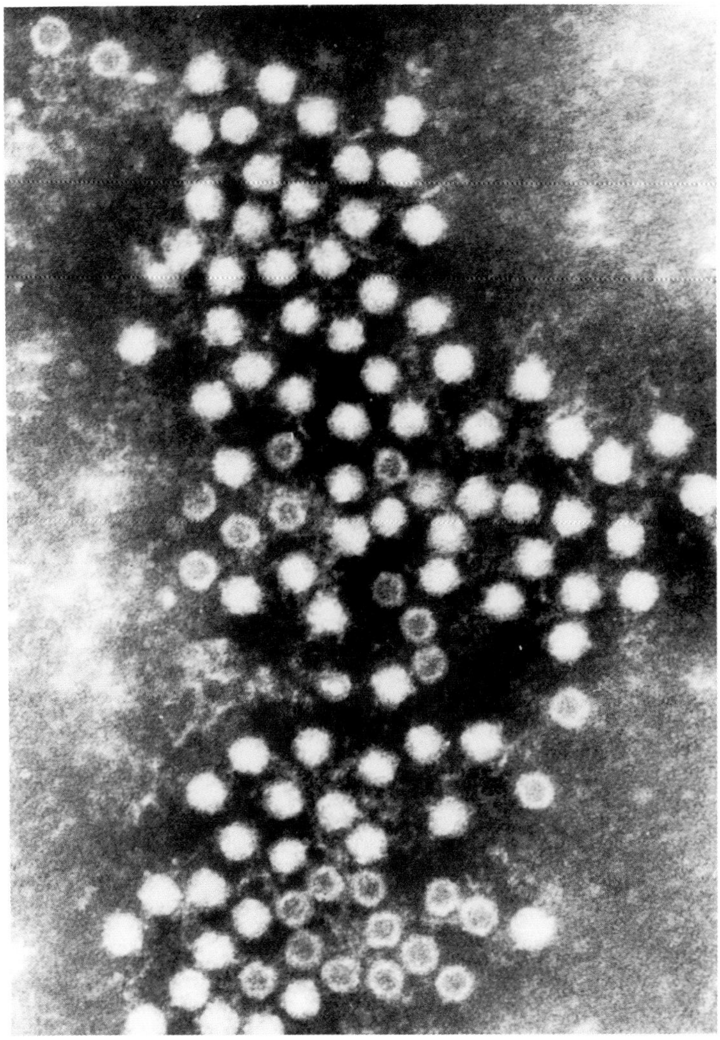

Figure 24.1 An immunoelectron micrograph showing aggregates of parvovirus B19 in serum. (Reproduced with permission, B. J. Cohen, *Parvoviruses and Human Disease*, CRC Press, 1988, 76)

Virus Isolation and Propagation

Isolation of parvovirus B19 in cell culture is not routinely available. It has been successfully propagated in suspension cultures of erythropoietin-stimulated human erythroid bone marrow cells from patients with hemolytic anemia, hemopoietic cells derived from human fetal liver, and human cord blood. Propagation in an erythropoietin-dependent megakaryocytic leukemia cell line has also been accomplished. To date, use of these systems in a clinical laboratory is not practical.

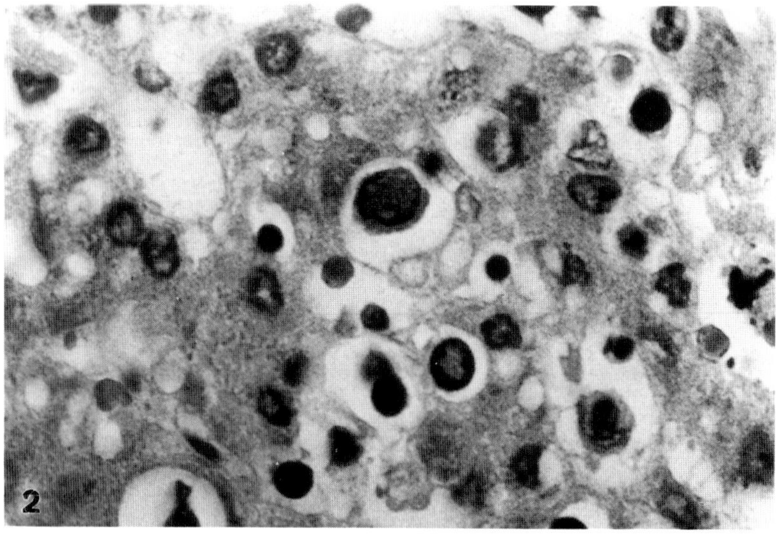

Figure 24.2 A fetal liver section of H&E-stained preparation showing eosinophilic intranuclear inclusions in erythroid cells within hepatic sinusoids (250×). (Reproduced with permission, R. A. Franciosi and P. Tattersall, *Human Path.* 19:489 [1988])

Detection of Parvovirus B19

LIGHT AND ELECTRON MICROSCOPY

Parvovirus may be visualized in various body fluids using EM and the pseudoreplica technique but may be difficult to distinguish morphologically from enteroviruses. Immune electron microscopy, using specific antiserum or monoclonal antibody to B19, facilitates detection of the virus by aggregating virus particles (see figure 24.1). Eosinophilic intranuclear inclusions in erythroid cells within hepatic sinusoids have been noted in fetal liver tissue (figure 24.2).

VIRAL ANTIGEN DETECTION

Viral antigen has been detected in serum using a variety of methods, such as counterimmunoelectrophoresis (CIE), RIA, and ELISA in an antigen-capture assay. These methods have been largely supplanted by both antibody determination and nucleic acid detection techniques.

VIRAL NUCLEIC ACID DETECTION

Detection of viral nucleic acid is useful in cases where antibody response is impaired, such as immunodeficiency states, when a tissue diagnosis is desired as in hydrops fetalis or in studies of pathogenesis. Parvovirus B19 DNA has been detected by hybridization with CDNA probes, riboprobes, and oligonucleotide probes. Radiolabeled probes have been replaced by biotin- or digoxigenin-labeled nonradioactive probes. Parvovirus B19 diagnosis has

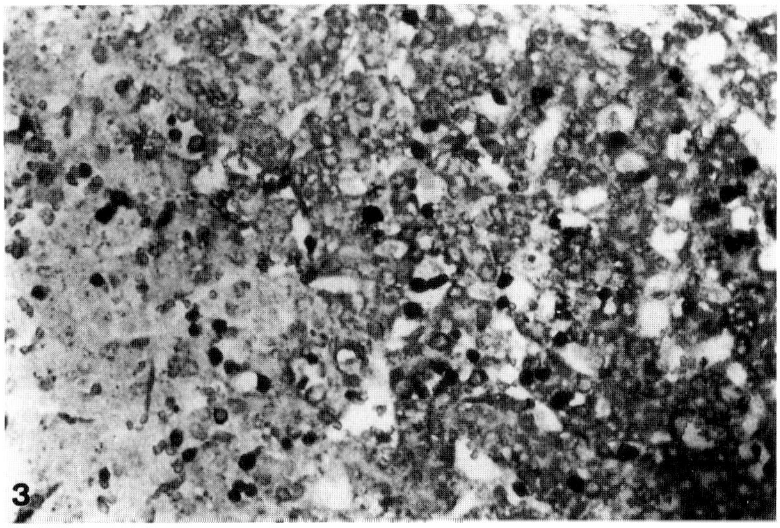

Figure 24.3 A fetal liver section showing nuclei of erythroid cells hybridized with parvovirus B19 DNA biotinylated probe (160×). (Reproduced with permission, R. A. Franciosi and P. Tattersall, *Human Path.* 19:489 [1988])

been accomplished by in situ hybridization of infected tissues (figure 24.3) and by spot hybridization using serum, throat swabs, fetal ascitic fluid, and tissues. As techniques are simplified and specific nonradioactive probes become commercially available, more widespread clinical use can be expected.

The polymerase chain reaction has also been applied to parvovirus diagnosis using a variety of specimen types, and as expected, is significantly more sensitive than spot hybridization.

Serodiagnosis

In the immunocompetent host, primary parvovirus B19 infection is suspected with the development of a characteristic rash, arthralgias, aplastic crisis, or hydrops fetalis. These symptoms generally appear at approximately the same time parvovirus specific–IgM antibody is detectable in serum, followed within a few days by the appearance of IgG. Thus, antibody determination methods provide a rapid and sensitive means of diagnosis. Both IgM and IgG detection are used to diagnose acute infections, whereas IgG alone is used to determine immune status.

An antibody capture ELISA for the detection of parvovirus B19–specific IgM and IgG has been developed at the Centers for Disease Control and provides the most sensitive and specific results to date. By this method, microtiter wells are coated with antihuman IgM or IgG, after which the human serum to be tested is added. Parvovirus B19 antigen is then added. After overnight incubation, the anti–parvovirus B19 monoclonal antibody is added, followed by the anti-mouse biotinylated antibody, then streptavidin-peroxidase and a substrate. Since cell culture systems are not yet readily available for production of parvovirus B19 antigens, the antigen source for the ELISA was originally human serum. Recombinant antigen produced in a baculovirus system is now available and has replaced human serum as the ELISA antigen source. Work is in progress to develop commercial kits using antibody capture in a simplified format that will make parvovirus B19 serodiagnosis more widely available.

Other Parvoviruses Infecting Humans

Parvovirus-like particles have been visualized in stools of patients, however, the exact nature of these viruses and their pathogenicity are unclear. A small parvovirus-like agent, designated RA-1, has also been recovered from the synovial tissue of a patient with rheumatoid arthritis. RA-1 is distinct from B19, and antibodies to RA-1 have not been clearly associated with any clinical syndrome in humans.

Adeno-associated viruses (AAV) require coinfection with a helper virus, such as adeno- or herpesvirus, for their own replication. AAV have been widely studied by molecular biologists because of their unusual replication and small genome, but little is known about their epidemiology and pathogenesis. Although AAV have not been conclusively associated with human disease, antibody surveys indicate that human infections are common. AAV have been isolated from humans in outbreaks of adenovirus infection, but there have been no unique features of disease attributable to AAV.

SUPPLEMENTARY READING

Anderson, L. J. Human parvoviruses. *J. Inf. Dis.* 161:603–8 (1990).

Brown, K. E.; Mori, J.; Cohen, B. J.; and Field, A. M. In vitro propagation of parvovirus B19 in primary fetal liver culture. *J. Gen. Virol.* 72:741–845 (1991).

Brown, T.; Anand, A.; Ritchie, L. D.; Clewley, J. P.; and Reid, T. M. Intrauterine parvovirus infection associated with hydrops fetalis. *Lancet* 2:1022–34 (1984).

Clewley, J. P. Detection of human parvovirus using a molecularly cloned probe. *J. Med. Virol.* 15:173–81 (1985).

———. Polymerase chain reaction assay of parvovirus B19 DNA in clinical specimens. *J. Clin. Microbiol.* 27:2647–51 (1989).

Cohen, B. J. Laboratory tests for the diagnosis of infection with B19 virus. In Pattison, J. R., ed., *Parvoviruses and Human Disease*. Boca Raton: CRC Press, 1988.

Cossart, Y. E.; Field, A. M.; Cant, B.; and Widdows, D. Parvovirus-like particles in human sera. *Lancet* 1:72–73 (1975).

Cotmore, S. F., and Tattersall, P. Characterization and molecular cloning of a human parvovirus genome. *Science* 226:1161–65 (1984).

Franciosi, R. A., and Tattersall, P. Fetal infection with human parvovirus B19. *Human Path.* 19:489–91 (1988).

Kock, W. C., and Adler, S. P. Detection of human parvovirus B19 DNA by using the polymerase chain reaction. *J. Clin. Microbiol.* 38:65–69 (1990).

Naides, S. J. Parvoviruses. In Specter, S., and Lancz, G. J., eds., *Clinical Virology Manual,* 2d ed., pp. 547–69. New York: Elsevier, 1992.

Ozawa, K.; Kurtzman, G.; and Young, N. Replication of the B19 parvovirus in human bone marrow cell cultures. *Science* 233:883–86 (1986).

Pattison, J. R. *Parvoviruses and Human Disease*. Boca Raton: CRC Press, 1988.

———. Parvoviruses. In Fields, B. N., and Knipe, D. M., eds., *Fields Virology,* 2d ed., pp. 1765–83. New York: Raven, 1990.

Paver, W. K.; Caul, E. O.; Ashley, C. R.; and Clarke, S. K. A small virus in human faeces. *Lancet* 1:237–39 (1973).

Shimomura, S.; Komatsu, N.; Frickhofen, N.; Anderson, S.; Kajigaya, S.; and Young, N. S. First continuous propagation of B19 parvovirus in a cell line. *Blood* 79:18–24 (1992).

Simpson, R. W.; McGinty, L.; Simon, L.; Smith, C. A.; Godzeski, C. W.; and Boyd, R. J. Association of parvoviruses with rheumatoid arthritis of humans. *Science* 223:1425–28 (1984).

Tyrrell, D. A. J. Human parvovirus, fifth disease and marrow aplasia. *Arch. Dis. Childhood* 59:197–98 (1984).

Zerbini, M.; Musiani, M.; Venturoli, S.; Gallinella, G.; Gibellini, D.; Gentilomi, G.; and LaPlaca, M. Rapid screening for B19 parvovirus DNA in clinical specimens with a digoxigenin-labeled DNA hybridization probe. *J. Clin. Microbiol.* 28:2496–99 (1990).

25

POXVIRIDAE

G. D. HSIUNG

POXVIRUSES

Three serologically indistinguishable poxviruses, such as vaccinia, variola major (smallpox), and variola minor (alastrim), are associated with different forms of human disease. Variola has been eradicated from the world since 9 December 1979. In fact, the WHO Committee on orthopoxvirus infections has agreed that all variola virus stocks should be destroyed by 31 December 1993. Vaccinia virus is not, in general, highly pathogenic for humans, although it may produce a severe generalized infection following vaccination and, in rare instances, ocular infection. Clinically, infection of humans by monkeypox virus is similar to that of smallpox: most patients are not severely ill, although deaths have been reported in association with exposure to the virus. Cowpox infection in humans, when it occurs, is a mild disease.

Morphologically, all poxviruses are similar in shape and structure: they are brick-shaped and large in size, 230 × 300 nm. Virions of this group can be easily recognized in crust extracts or in vesicle fluids by means of electron microscopy. Similarly, large numbers of poxvirus virions are often observed in biopsy tissues obtained from molluscum contagiosum lesions (figure 25.1). The latter is an unclassified poxvirus. Vaccinia virus can be propagated in a variety of cell cultures, including RhMK, Hep-2, Vero, and HDF. A large number of typical poxvirus virions are seen in an infected cell (figure 25.2).

Poxvirus induces eosinophilic cytoplasmic inclusions or Guarnieri bodies in infected cells. Following light microscopic examination of stained smears obtained from skin or mucous membrane lesions, aggregates of virions or

ISBN 0-300-05845-4.

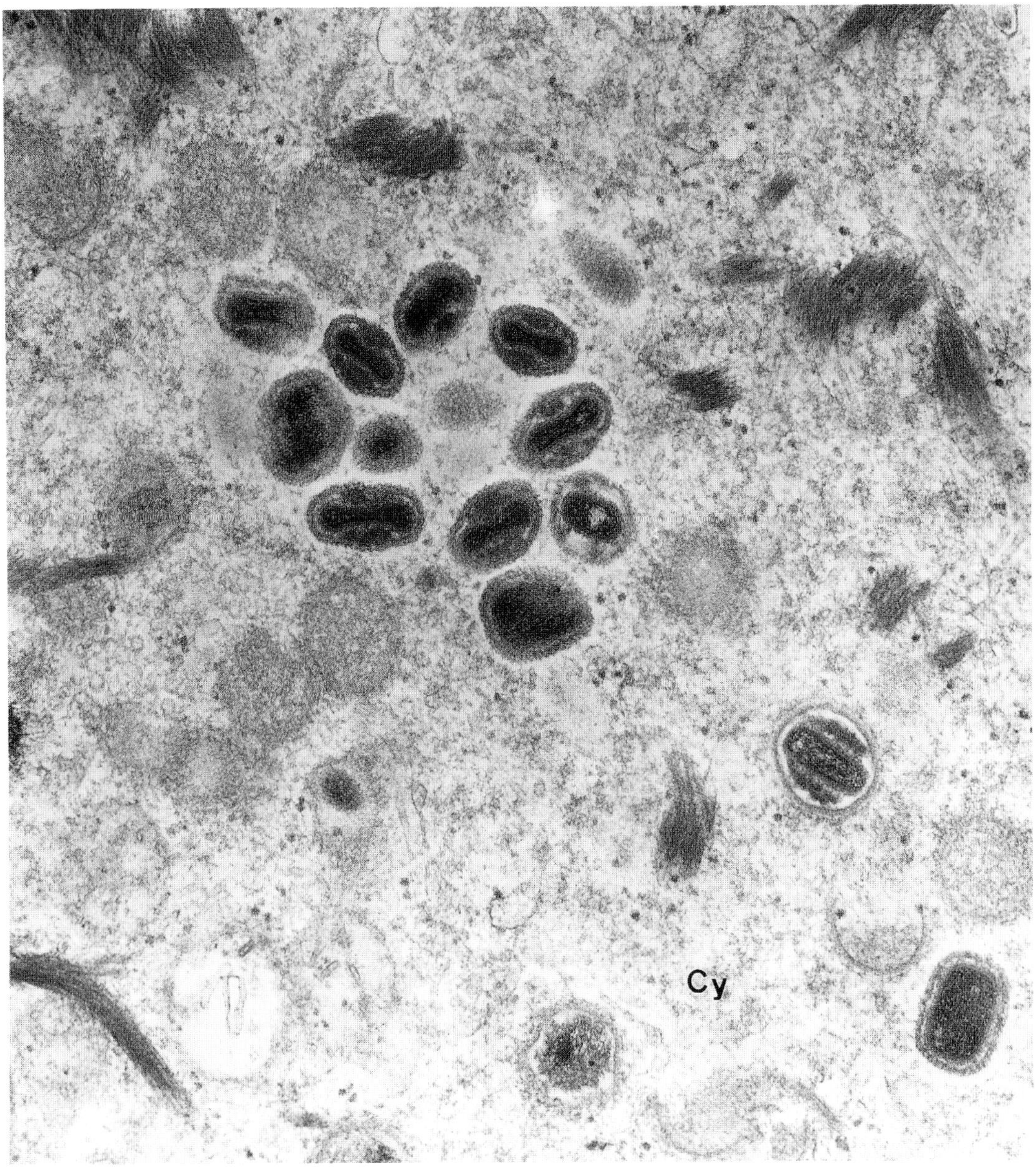

Figure 25.1 Electron micrograph of poxvirus virions in a molluscum contagiosum tissue section (48,000×)

elementary bodies can be seen. This method was first used by Paschen in 1906.

Poxviruses are stable even at room temperature for several months, and refrigeration is not necessary when specimens for virus isolation are shipped by mail. Hep-2, Vero, and human diploid fibroblasts (HDF) cell cultures

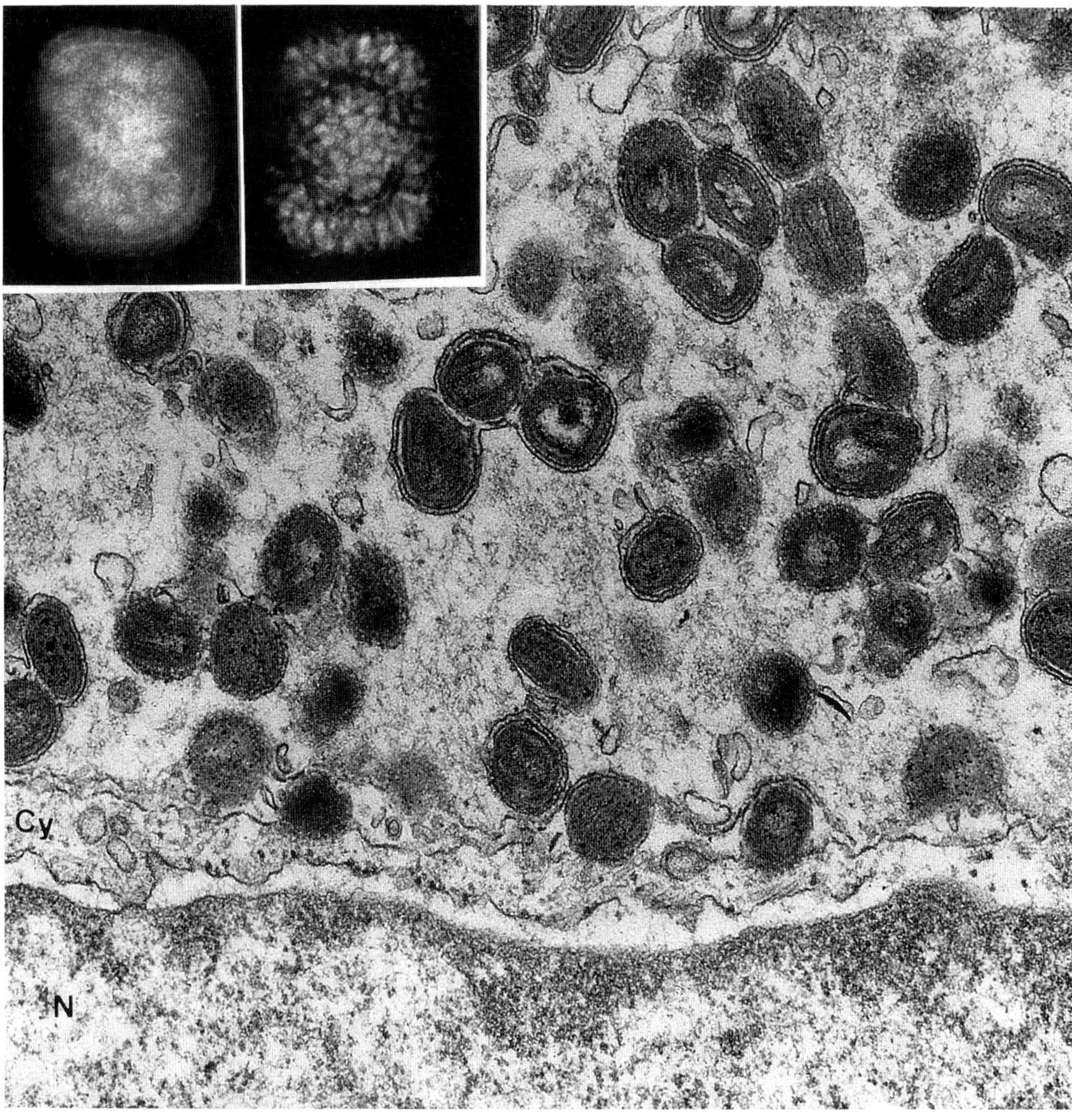

Figure 25.2 Electron micrographs of vaccinia virus in an infected RhMK cell showing many poxvirus virions in the cytoplasmic inclusion (48,000×). *N* = nucleus, *Cy* = cytoplasm. Insets: Vaccinia virus virions stained with PTA showing different appearance of virus structure (100,800×). (Modified from C. K. Y. Fong and G. D. Hsiung, *Virology and Rickettsiology*, 1978, vol. 1, pt. 1, 99)

have been found useful for the propagation of vaccinia and monkeypox viruses. Since smallpox caused by variola virus has been eradicated, the following description is concerned mostly with vaccinia virus, which is currently used as a vector for carrying genomic material from several viruses in the preparation of vaccines by recombinant DNA techniques.

VACCINIA VIRUS

Specimen Source and Inoculation

Scrapings from skin lesions, vesicle fluids, and crusts are good sources for vaccinia virus isolation. These can be inoculated directly into monkey kidney, Hep-2, and a variety of other cells. The CAM of 12- to 14-day-old embryonated chicken eggs are highly sensitive to vaccinia virus infection.

Vaccinia Virus Isolation, Propagation, and Assay

CYTOPATHIC EFFECT IN CELL CULTURES WITH FLUID MEDIUM

Vaccinia virus produces CPE in a variety of cell cultures of both primary and passaged cell lines within 18–24 hours of infection. Cytopathic effect induced by vaccinia virus in Hep-2 (figure 25.3) and RhMK cells (figure 25.4) is characterized by multinucleated giant cells. Extensive distortion of the monolayers is typical of vaccinia virus infection. Eosinophilic cytoplasmic inclusions are easily distinguishable in the infected RhMK cells fixed and stained with H&E (figure 25.5). These inclusions are Feulgen-positive, indicating the presence of DNA.

POCK FORMATION ON CAM OF EMBRYONATED EGGS

Inoculation of the CAM of embryonated chicken eggs is the most reliable laboratory test for the isolation of the poxviruses. Embryos 12 to 14 days old are most suitable for propagation; after incubation at 36°C or 38.5°C, vaccinia virus pocks appear in two to three days. Pocks on CAM produced by vaccinia virus are large and ulcerated.

ANIMAL PATHOGENICITY

Vaccinia virus readily produces lesions on the skin of rabbits and kills infant mice after intracerebral inoculation (table 25.1).

Identification of Poxviruses

An early presumptive diagnosis of poxvirus infection can be made most rapidly by observing virus particles in negatively stained preparations of vesicle fluid under EM examination. Virus aggregates—that is, elementary

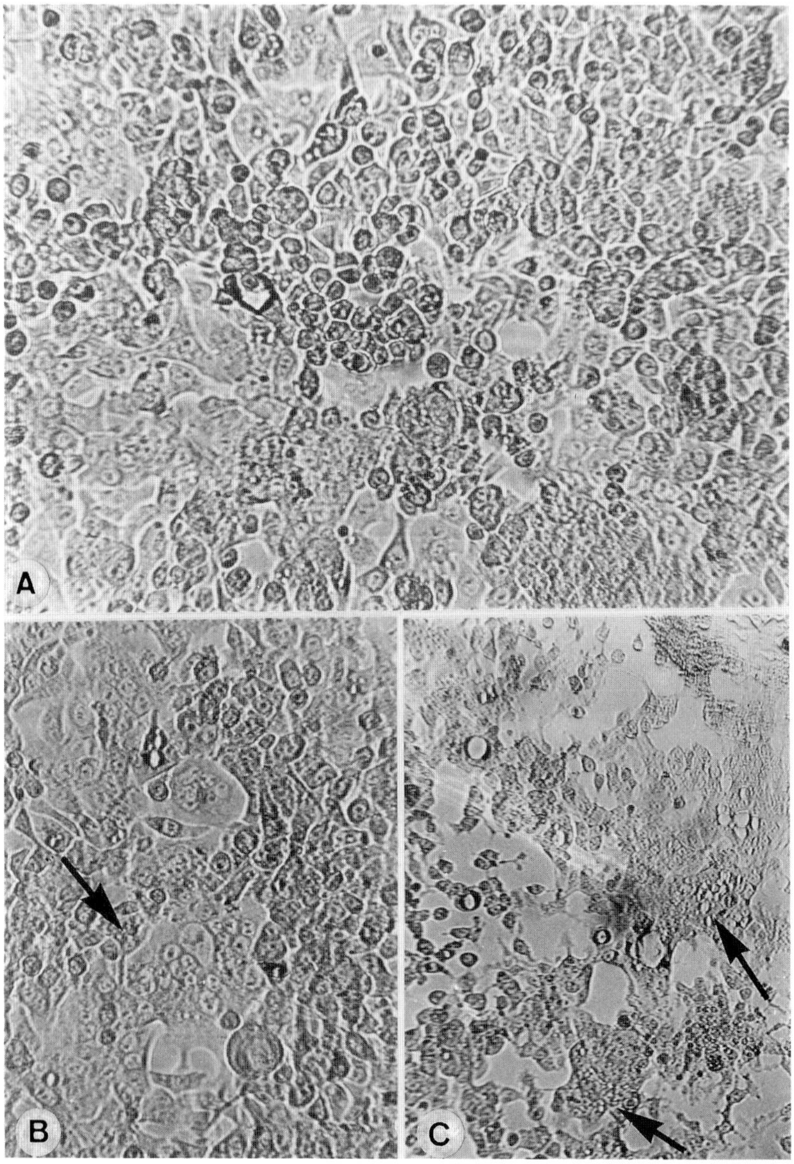

Figure 25.3 Cytopathic effect induced by vaccinia virus in Hep-2 cell cultures (100×): *(A)* Uninfected Hep-2 culture; *(B)* Early CPE. Note area of multinucleated cells (arrow); *(C)* Late CPE. Note area of multinucleated cells (arrows)

bodies—in a Giemsa-stained smear of scrapings from the base of skin lesions can also be easily recognized. Smears from skin lesions or vesicle fluid can be stained by the IF technique with immune serum to vaccinia virus. Better results are obtained with cultured cells that have been infected with the isolate for 16 to 24 hours, followed by the IF staining technique. Poxvirus

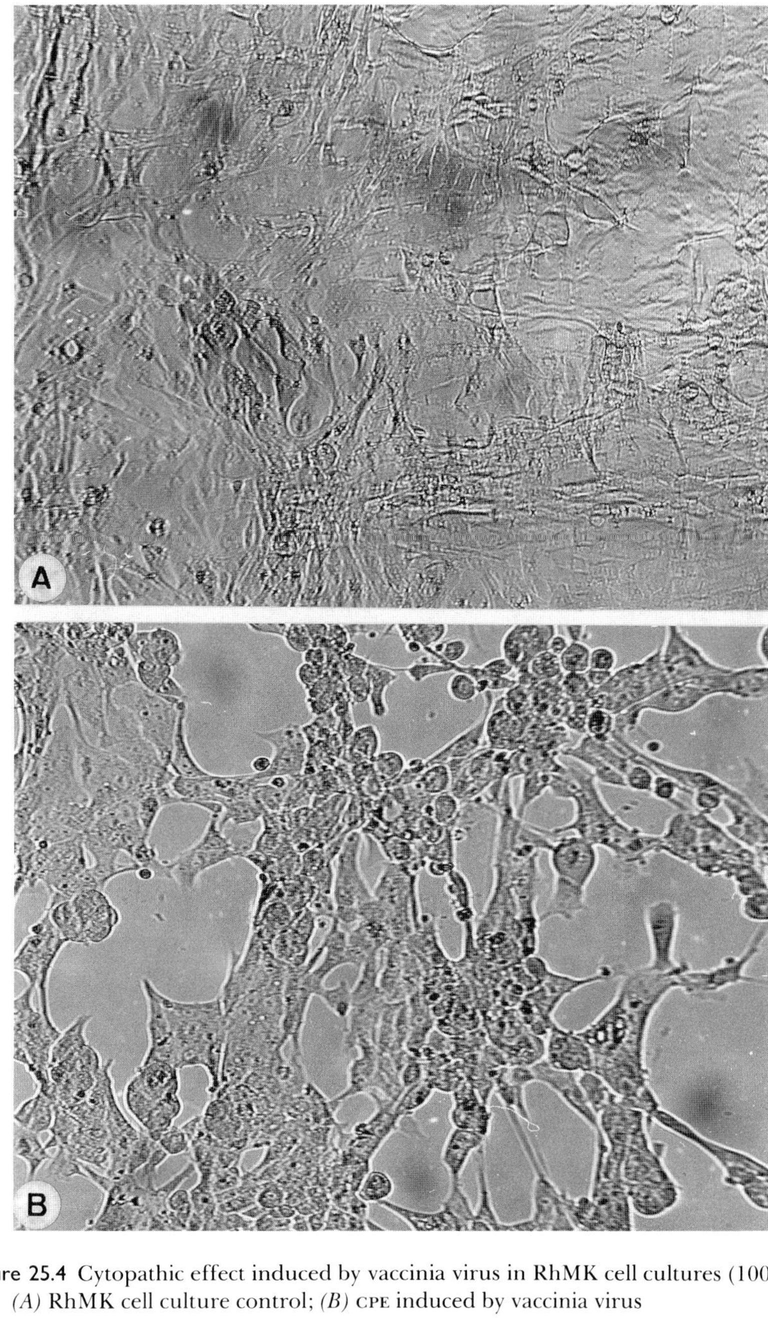

Figure 25.4 Cytopathic effect induced by vaccinia virus in RhMK cell cultures (100×): *(A)* RhMK cell culture control; *(B)* CPE induced by vaccinia virus

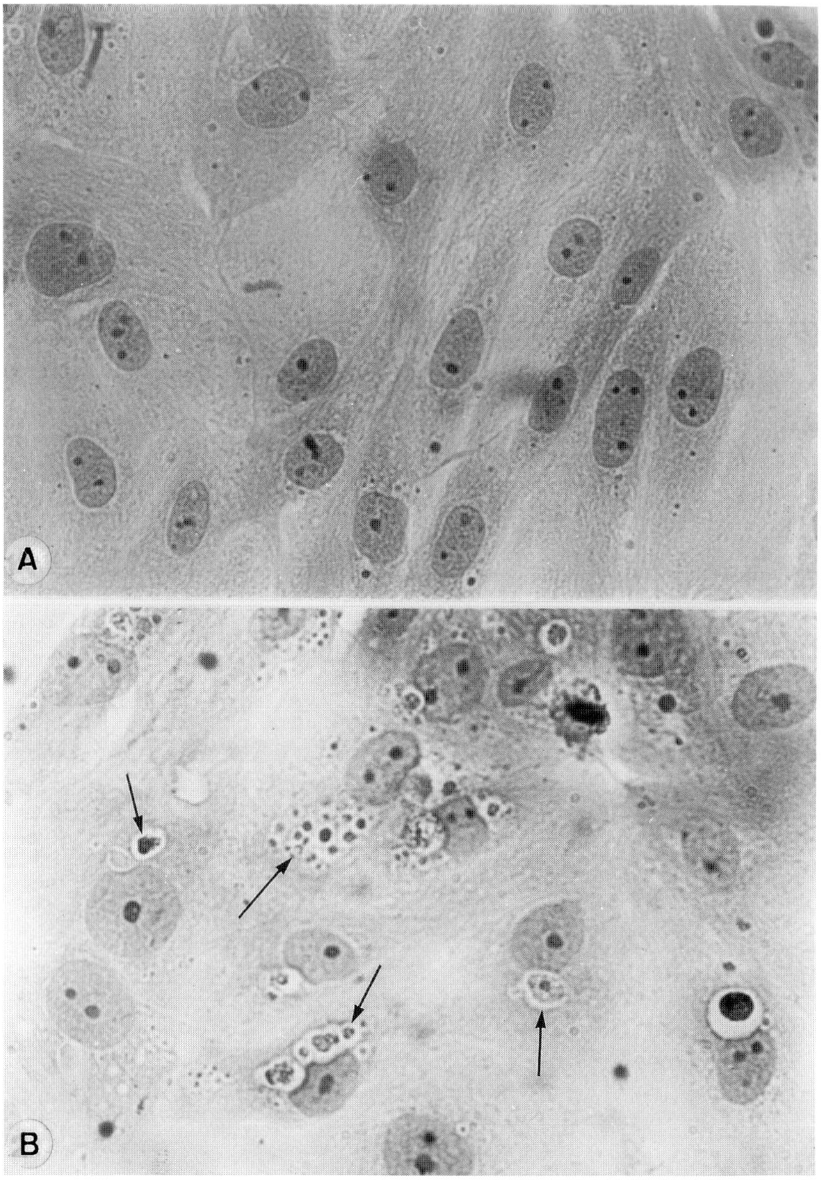

Figure 25.5 Eosinophilic intracytoplasmic inclusions induced by vaccinia virus in RhMK cell cultures (H&E, 400×): *(A)* Uninfected RhMK culture; *(B)* Eosinophilic intracytoplasmic inclusions (arrows) in RhMK cells

Table 25.1 Differential Characteristics of Herpes Simplex, Varicella-Zoster, Vaccinia, and Variola Viruses[a]

				Cell Culture				Embryonated Eggs	Animals
				CPE					
Virus Family	Virus	Direct Examination of Smears or Scrapings	Reaction −20°C Temperature	MK	HDF	Hep-2	Inclusion Bodies	Pocks on Chorioallantoic Membrane	Inoculation on Mice or Rabbits
Herpesviridae	Herpes simplex	Multinucleated giant cells with intranuclear inclusions	Labile	+/−	+	+	Intranuclear inclusions (Cowdry type A)	Pinpoint for type 1; large, clear for type 2	Death
	Varicella-zoster	Multinucleated giant cells with intranuclear inclusions	Labile	−	+[b]	−	Intranuclear inclusions (Cowdry type A)	None	No effect
Poxviridae	Variola[a]	Elementary bodies (virus particles)	Stable	+/−	+/−	+/−	Intracyto-plasmic inclusions (Guarnieri bodies)	Small, discrete, circular	No effect
	Vaccinia	Elementary bodies (virus particles)	Stable	+	+	+	Intracyto-plasmic inclusions (Guarnieri bodies)	Large, diffused, hemorrhagic, necrotic	Death or lesions

\+ virus-induced change; − no change

a. Since variola virus is eradicated worldwide, its inclusion here is only for comparison.

b. Tissue cells from human embryonic lung or skin are preferable, and infectious virus material is best passaged serially from infected cells.

strains can be differentiated by pock morphology on CAM; vaccinia virus pocks are diffused and necrotic (see table 25.1).

The final identification of poxvirus isolates may be made by immunofluorescence or CF test. A specific type of variola, alastrim, or vaccinia virus, however, cannot be identified because of the serologic relations within the poxvirus group.

Differentiation of Herpes Simplex, Varicella-Zoster, Vaccinia, and Variola Viruses (table 25.1)

Since herpes simplex virus and vaccinia virus occasionally cause severe generalized vesicular eruptions in individuals with atopic eczema and, rarely, varicella-zoster virus may do the same, isolation of the virus in cell culture for purposes of identification is important. When facilities are available, electron microscopic examination of negatively stained preparations of lesion fluid can permit the diagnosis within hours since herpesvirus and poxvirus virions are easily distinguished from each other (figure 25.2 inset and compare with figure 22.1 inset). Indirect immunofluorescent staining technique using specific antiserum to each virus type is the most useful and rapid method for distinguishing HSV, VZV, and vaccinia virus infection in cells taken from lesions. Under these circumstances, other useful biologic properties that can be compared include the type of pock produced on the CAM of chick embryos, pathogenicity in mice and rabbits, and the type of inclusion produced in infected cells, each of which is specifically characteristic of each virus (table 25.1).

Serodiagnosis

Serologic methods are not suitable for the quick and accurate diagnosis of poxvirus infections. When no specimen is available for virologic studies, however, a 4-fold rise in antibody titer can be measured by IFA, ELISA, RIA, and the neutralization test. Neutralizing antibody titers may persist for years. None of these serologic methods, however, will differentiate infection with vaccinia from other poxviruses, because of their close antigenic relatedness.

SUPPLEMENTARY READING

Behbehani, A. M. The smallpox story: Historical perspective. *Am. Soc. Microbiol. News* 57:571–76 (1991).

———. The small pox story: Life and death of an old disease. *Microbiol. Rev.* 47:455–502 (1982).

Dowdle, W. R. Exotic viral diseases. *Yale J. Biol. Med.* 53:109–15 (1980).

Esposito, J. J., and Nakano, J. H. Poxvirus infections in humans. In Balows, A., ed. in chief, *Manual of Clinical Microbiology,* 5th ed., pp. 858–67. Washington, D.C.: American Society for Microbiology, 1991.

Henderson, D. A. Principles and lessons from the smallpox eradication programme. *Bull. WHO* 65:535–46 (1987).

Hruby, D. E. Vaccinia virus vectors: New strategies for producing recombinant vaccines. *Clin. Microbiol. Rev.* 3:153–70 (1990).

Mahy, B. W.; Esposito, J. J.; and Venter, J. C. Sequencing the smallpox virus genome: Prelude to destruction of a virus species. *Am. Soc. Microbiol. News* 56:577–80 (1991).

Moss, B. Vaccinia virus: A tool for research and vaccine development. *Science* 252:1661–67 (1991).

Williams. M. G.; Almeida, J. D.; and Howatson, A. F. Electron microscope studies on viral skin lesions: A simple and rapid method of identifying virus particles. *Arch. Derm.* 86:290–97 (1962).

PART FIVE

RNA AND DNA MIXED VIRUS GROUPS

26

HEPATITIS VIRUSES

G. STACK AND J. R. BOVE

NOMENCLATURE AND CLASSIFICATION

Hepatitis viruses are a heterogeneous group of genetically unrelated viruses. They are classified together because of a shared hepatotropism and because their primary clinical effect is hepatitis. While other viruses, such as some coxsackieviruses, herpesviruses, and some arboviruses can infect the liver, they are not classified as hepatitis viruses, because their primary effects involve other target organs. Five categories of hepatitis viruses have been identified, designated alphabetically from A to E. Each is a member of a different virus family (table 26.1). The dates of their discovery as well as the techniques used in their initial characterization are given in table 26.1. While these five viruses account for the majority of known cases of viral hepatitis, the possibility remains that other hepatitis viruses may yet be identified.

PROPERTIES OF HEPATITIS VIRUSES

Hepatitis A Virus

The hepatitis A virus (HAV) is classified as an enterovirus in the Picornaviridae family (tables 26.1, 26.2). Its virion is 27 nm in diameter, nonenveloped (figure 26.1), icosahedral in symmetry, and composed of multiple copies of four nucleocapsid subunits, the polypeptides VP-1 to VP-4. It was originally classified as enterovirus 72 with one serotype. The RNA genome is linear, single-stranded, and 7,478 nucleotides long. It contains one large open reading frame that encodes for a single large polypeptide, which is cleaved post-translationally into 11 proteins, the four structural nucleocapsid subunits and seven nonstructural proteins.

ISBN 0-300-05845-4.

Table 26.1 Nomenclature and Classification of Hepatitis Viruses

Nomenclature	Abbrev.	Virus Classification: Family and Genus	Date First Reported	Method of Recognition	Reference
Hepatitis A virus	HAV	Picornaviridae *enterovirus*	1973	Immune EM	Feinstone et al., 1973
Hepatitis B virus	HBV	Hepadnaviridae *hepadnavirus*	1961–70	Serology, EM	Allison and Blumberg, 1961
Hepatitis C virus	HCV	Flaviviridae	1989	Molecular cloning	Choo et al., 1989
Hepatitis D virus	HDV	Satellite, viroid-like	1977	Immunofluorescence	Rizzetto et al., 1977
Hepatitis E virus	HEV	Caliciviridae	1990	Molecular cloning	Reyes et al., 1990

Table 26.2 Properties of Hepatitis Viruses

Hepatitis Virus	Virus Morphology	Genome Structure	Incubation Period	Chronicity	Mode of Transmission
HAV	27 nm, icosahedral, nonenveloped	Linear, single-stranded RNA	2–6 wks	No	Fecal-oral
HBV	42 nm, spherical, enveloped	Circular, double-stranded DNA	6 wks–6 mos	Yes 5–10%	Parenteral
HCV	30–60 nm, enveloped	Linear, single-stranded RNA	2–12 wks	Yes 50%	Parenteral
HDV	35–37 nm, enveloped (hybrid w/HB_sAg coat)	Circular, single-stranded RNA	3–7 wks (coinfection w/HBV or superinfection of HBV carriers necessary)	Yes	Parenteral
HEV	27–34 nm, nonenveloped	Linear, single-stranded RNA	2–9 wks	No	Fecal-oral

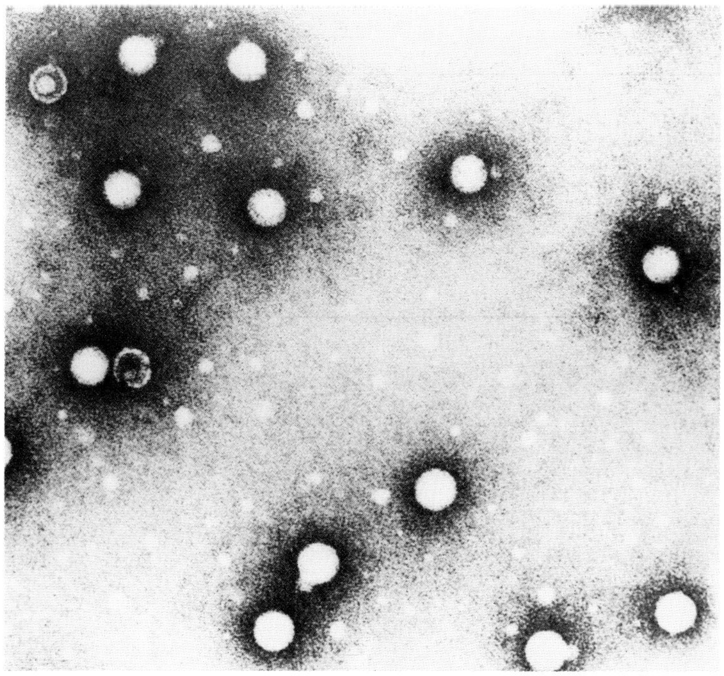

Figure 26.1 Electron micrograph of hepatitis A virions in human stool (150,000×). (Reproduced with permission, R. Dulbecco and H. S. Ginsberg, ed., *Virology*, Hagerstown: Harper and Row, 1980, 1219)

HAV is transmitted predominantly by the fecal-oral route. Contamination of food and water with fecally shed viruses and poor hygiene, especially in endemic areas, result in HAV spread. The incubation period is typically about one month. The clinical features of hepatitis A are usually mild and often subclinical, especially in children. The risk of more severe infections increases in adults. Overall, about 45% of the urban U.S. population has serologic evidence of prior HAV infection. The prevalence rate increases with age and is highest in lower socioeconomic groups. Patients with symptomatic and icteric cases can usually expect full clinical recovery one to two months after the onset of the illness. HAV infections do not lead to chronic liver disease, and there is no carrier state.

Hepatitis B Virus

The hepatitis B virus (HBV), unlike the other known hepatitis viruses, has a DNA genome (table 26.2). It is classified as hepadnavirus type I in the Hepadnaviridae family (table 26.1). HBV virion is 42 nm in diameter, spherical, and has a lipid-envelope (figure 26.2). The intact virion is known historically as the *Dane particle*. The major surface proteins of the HBV envelope are a

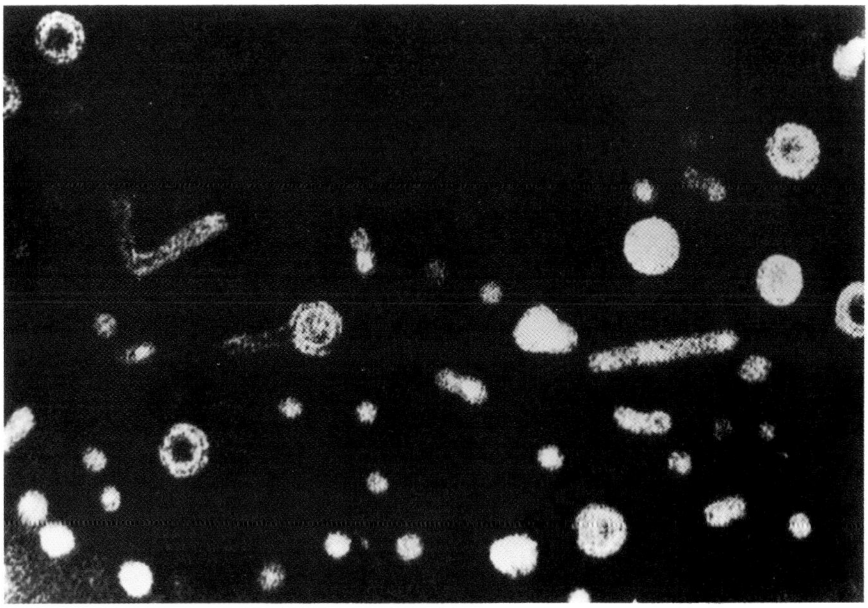

Figure 26.2 Electron micrograph of hepatitis B virions in human serum along with 22-nm spherical particles and filamentous forms (PTA staining; 220,000×)

24,000 Dalton protein and its 28,000 Dalton glycosylated product. Epitopes on the major surface protein are known as hepatitis B surface antigen (HB_sAg). Heterogeneity in the sequence of this surface protein has created subtype-specific epitopes. For example, while all known surface antigens contain the *a* epitope, HB_sAg will contain either d or y and either w or r subdeterminants. Within the envelope is a 27-nm icosahedral nucleocapsid core that contains a 3,200 nucleotide-long, circular, predominantly double-stranded DNA genome. In addition, the nucleocapsid core contains a 21,000 Dalton protein with epitope(s) known as the hepatitis B core antigen (HB_cAg). The hepatitis B e antigen (HB_eAg) appears also to be present in the nucleocapsid core. During acute infection, subviral HB_sAg particles can be detected in the blood-stream of affected individuals. These are 22-nm- diameter, spherical or filamentous forms that are typically present in great excess compared to the mature intact virion (figure 26.2). These structures are composed of excess HB_sAg and do not contain DNA or HB_cAg.

HBV is transmitted primarily by the blood. Most other body fluids also can potentially transmit the virus, although probably at lower efficiencies. Percutaneous and parenteral transmissions are most common, while oral transmission appears inefficient. Exchanges of blood and body fluids during personal and intimate contact, perinatal-maternal neonatal spread, and unsafe needle practices in parenteral drug use are examples of transmission routes. The incubation period for hepatitis B after viral exposure is commonly two to three months. The viremic phase is more prolonged than in hepatitis A and is typically present prior to symptomatic illness and then declines during clinical recovery. An estimated 5%–10% of patients remain viremic and become chronic carriers for years or for life. These patients are at risk for chronic persistent and chronic active hepatitis, cirrhosis, or hepatocellular carcinoma. A rare but serious complication of HBV infection is acute fulminant hepatitis or fulminant hepatitis in the setting of chronic infection. Fulminant hepatitis results in massive hepatic necrosis and has a high fatality rate. Fulminant hepatitis is often associated with HDV coinfection or superinfection.

Hepatitis C Virus

Hepatitis C virus (HCV) has a single-stranded RNA genome with properties similar to those of the Flaviviridae family (tables 26.1, 26.2). Biochemical studies and filtration data indicate that the virion is lipid-enveloped and about 30–60 nm in diameter. The RNA genome is about 10 kb long and appears to consist of a single, large, open reading frame. Viral proteins apparently are produced by cleavage of a single large 3,011 amino acid precursor (figure 26.3). Structural proteins, including a nucleocapsid core protein and envelope glycoprotein(s), presumably are cleaved from the amino terminus of the precursor. Four, or possibly five, nonstructural proteins, including those involved in virion replication, appear to be cleaved from the remainder of the precursor polypeptide.

HCV is known to be transmitted parenterally. Other possible routes of transmission are not well documented. Individuals at increased risk of HCV infection are those with exposure to HCV-infected blood or blood products, such as intravenous users, transfusion recipients (particularly prior to anti-HCV screening of donated blood), hemodialysis patients, and hemophiliacs who receive factor concentrates that have not been processed for viral removal or inactivation. Community-acquired or sporadic cases of hepatitis C in patients with no identifiable risk factors account for about 40% of cases. Needle sharing in intravenous drug use represents the most commonly identified risk factor in the United States, where it accounts for over 40% of studied HCV infections. Data regarding sexual transmission of HCV are conflicting. HCV appears to be transmitted inefficiently by sexual contact. Maternal-neonatal transmission also appears to be low. The typical incubation period of HCV, 2–12 weeks, is intermediate between that of HAV and HBV. Infections are usually clinically mild. Many patients are asymptomatic and only about 25% are icteric. The most common symptoms are fatigue and anorexia. Hepatitis C infection has a high rate of progression to chronicity. A chronic carrier state appears common since about 90% of apparently healthy, asymptomatic blood donors with antibody to HCV remain capable of transmitting HCV infection to transfusion recipients. In addition, about

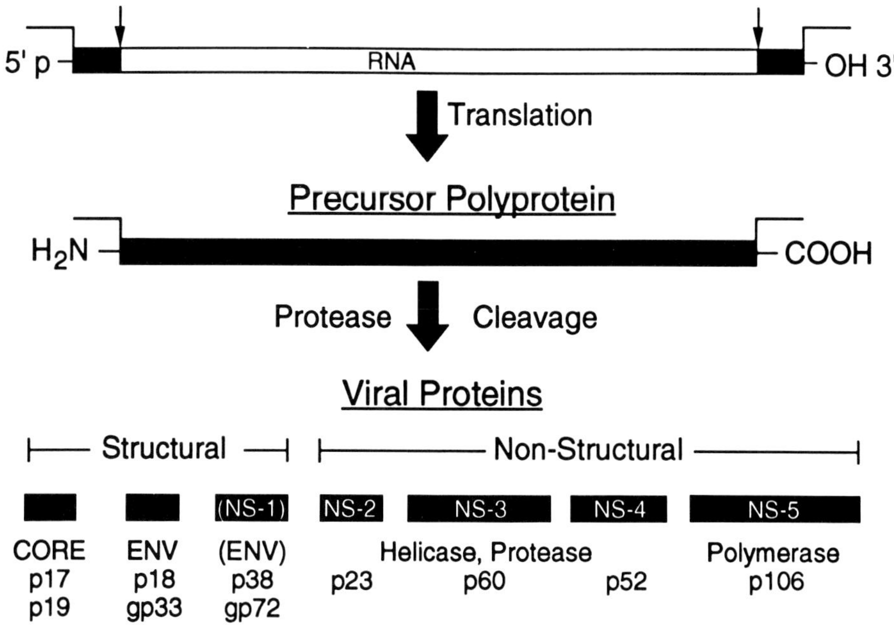

Figure 26.3 Proposed scheme for hepatitis C virus RNA genome and processing of a hepatitis C virus precursor polyprotein. The structural region encodes nucleocapsid (CORE) and envelope (ENV) sequences. The postulated functions encoded by nonstructural (NS) regions are indicated. (Modified from L. R. Overby and M. Houghton, Hepatitis Viruses, in E. H. Lennette, ed., *Laboratory Diagnosis of Viral Infections,* 2d ed. New York: Marcel Dekker, 1991, 412)

50%–60% of acute infections, even if clinically mild, lead to chronic hepatitis, as evidenced by elevated alanine aminotransferase (ALT) levels for greater than one year. On liver biopsy, about 40% of patients with chronic ALT elevations have chronic active hepatitis, and about 20% have cirrhosis. The percentage that progress to hepatocellular carcinoma is unknown. The prevalence of HCV antibodies, however, is high in patients with hepatocellular carcinoma. The course of chronic active hepatitis and cirrhosis in hepatitis C appears more indolent than in alcoholic cirrhosis. Although overt portal hypertension appears uncommon, liver-related mortality is possible.

Hepatitis D (Delta) Virus

The hepatitis D or delta virus (HDV) virion has a defective RNA genome, is 35–37 nm in diameter, and is enveloped (figure 26.4). It is defective in the sense that HDV requires a helper function from HBV for infectivity. In fact, the HDV virion is a hybrid structure of HDV and HBV. The lipid envelope contains HB_sAg, while the nucleocapsid core consists of the HDV RNA genome and hepatitis D antigen (HDAg). The RNA genome, which encodes HDAg, is single-

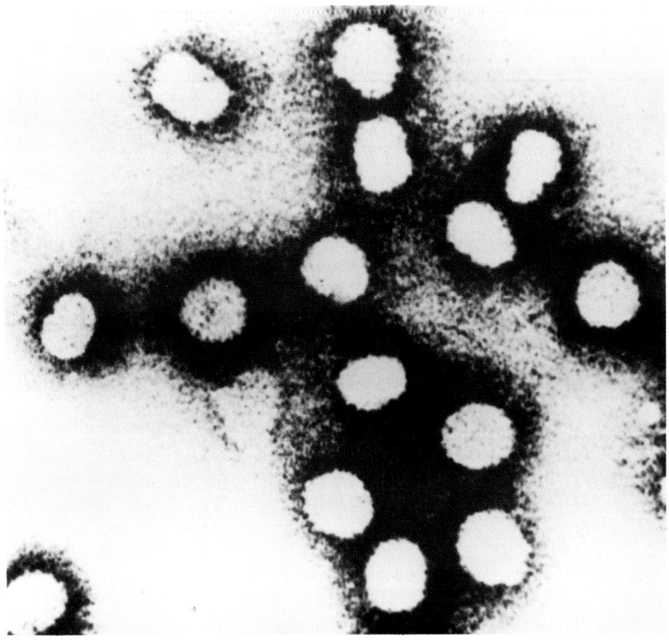

Figure 26.4 Electron micrograph of hepatitis D virions (300,000×). (Reproduced with permission, L. R. Overby and M. Houghton, Hepatitis Viruses, in E. H. Lennette, ed., *Laboratory Diagnosis of Viral Infections,* 2d ed. New York: Marcel Dekker, 1991, 414)

stranded, circular, and about 1.7 kb long (table 26.2). HDV has some features in common with plant satellite viruses or viroids.

HDV, a blood-borne virus, is spread by percutaneous or parenteral exposure to virus-contaminated blood. It is endemic among hepatitis B–infected patients in the Mediterranean region where it was first described. HDV infection occurs only in individuals with acute or chronic hepatitis B infection. Without HBV, HDV replication and infectivity are deficient. HDV superinfection in chronic hepatitis B has been associated with the development of fulminant hepatitis and increased mortality.

Hepatitis E Virus

The hepatitis E virus (HEV) virion is 27–34 nm in diameter, nonenveloped, and contains RNA (table 26.2, figure 26.5). The RNA genome is linear, single-stranded, and 7.6 kb long. HEV resembles caliciviruses. It is spread enterically like HAV and has a clinical course similar to that of hepatitis A. High mortality, however, has been reported with HEV infection during pregnancy. Epidemics of hepatitis E have been reported in India, Asia, Central America, and Africa owing to the consumption of fecally contaminated water. Chronic infections with HEV have not been described.

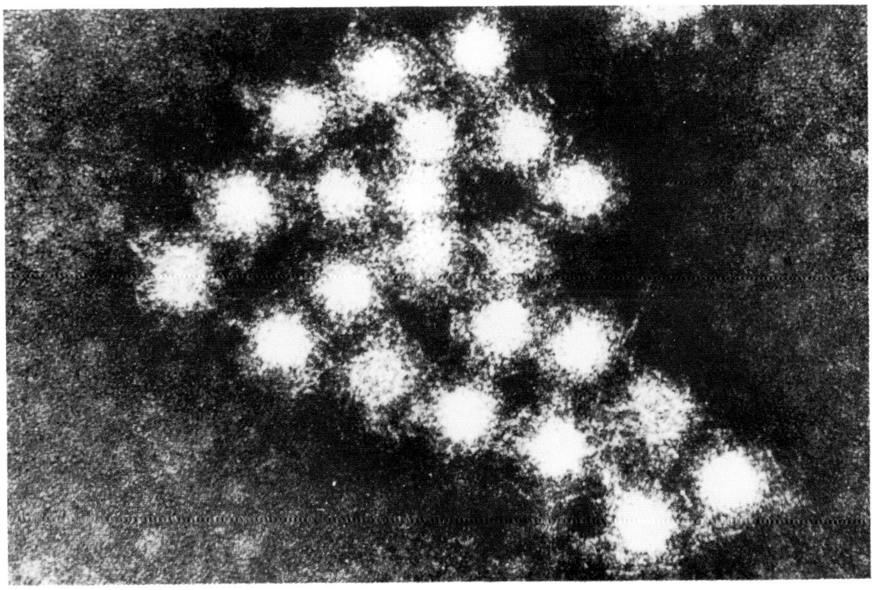

Figure 26.5 Immune electron micrograph of aggregated virions of hepatitis E virus in human stool (222,566×). (Reproduced with permission, L. R. Overby and M. Houghton, Hepatitis Viruses, in E. H. Lennette, ed., *Laboratory Diagnosis of Viral Infections*, 2d ed. New York: Marcel Dekker, 1991, 415)

LABORATORY DIAGNOSIS OF HEPATITIS VIRUS INFECTION

Hepatitis viruses cannot be grown readily in cell cultures and have been difficult to study by classical virological techniques. Electron microscopy has been used to visualize the virions of HAV and HEV in stool, as well as HBV and HDV in serum. Still, EM is largely used as a tool in hepatitis research and is not easily applied to large-scale diagnostic testing. As a result, the ability to diagnose clinical infection with these viruses relies on various immunologic assays (IA) rather than on direct viral detection. Almost all serological tests for hepatitis viruses are designed to detect the presence or absence of antibodies to various viral proteins (table 26.3). Thus, they are done on serum or plasma, depend upon the ability of the infected person to mount measurable antibody response, and can be false-negative in infected patients who are immunosuppressed, such as occurs in HIV infection. As is usual with antibody responses, IgM is formed in early infection, followed by IgG. The ability to differentiate between these two forms of antibody is helpful when attempting to differentiate between acute and chronic infection. The most commonly used commercial test is the enzyme immunoassay method. Radioimmunoassay, hemagglutination, and latex agglutination have also been used for some markers, but less frequently. In clinical practice, immunoassay results are reported as either negative or positive. Titration values are not

Table 26.3 Laboratory Diagnostic Tests for Hepatitis Virus Infections

Hepatitis Virus	Diagnositic Marker	Method[a,b]	Significance of Positive Result
HAV	HAV virion or RNA	EM or NA	Etiologic agent of "infectious" hepatitis
	Total anti-HAV	IA	Detectable at onset of symptoms, lifetime persistence
	IgM anti-HAV	IA	Indicates recent infection with hepatitis A; detectable for 4–6 months after infection
HBV	HBV virion or DNA	EM or NA	Etiologic agent of "serum" hepatitis; originally known as Dane particle
	HB_sAg	IA or IHC	Soluble antigen; indicates presence of chronic HBV infection; originally known as Australia antigen
	HB_eAg	IA	Soluble antigen; correlates with HBV replication, high-titer HBV in serum, and infectivity of serum
	HB_cAg	IHC	Detectable in hepatic tissue only; not a serum antigen
	Anti-HB_s	IA	Indicates past infection with and immunity to HBV, passive antibody from HBIG, or immune response to hepatitis B vaccine
	Anti-HB_e	IA	Prescence in serum is a sign of recovery in acute HBV infection; in chronic HB_sAg carrier, it indicates lower titer of HBV
	Total anti-HB_c	IA	Indicates prior infection with HBV at some undefined time
	IgM anti-HB_c	IA	Indicates recent infection with HBV; detectable for 4–6 months after infection
HCV	HCV RNA	NA[c]	Etiologic agent of parenterally transmitted non-A, non-B hepatitis
	Anti-HCV	IA	Indicates present or past infection with HCV; probably infectious state
HDV	HDV virion or RNA	EM, NA	Etiologic agent of delta hepatitis

(*continued*)

Table 26.3 (Continued)

Hepatitis Virus	Diagnositic Marker	Method[a,b]	Significance of Positive Result
	HDAg	—	No commercial test available; detectable in early acute delta infection
	Anti-HDV	IA	Indicates present or past infection with delta virus
HEV	HEV virion or RNA	EM, NA	Etiologic agent of enterically transmitted non-A, non–B hepatitis

Adapted from MMWR 39:1–26, 1990.

a. EM electron microscopy
IA immunoassay
IHC hepatic immunohistochemistry
NA nucleic acid detection

b. EM and NA are predominantly used for research purposes. Positive results in NA assays suggest, but do not directly measure, the presence of intact virions.

c. No confirmed detection of HCV by EM reported as of press time.

reported, and unlike other viral antibody tests, one does not look for evidence of rising or falling titers. The detection of viral nucleic acids in clinical specimens is now possible owing to the availability of molecular probes resulting from the cloning and sequencing of the hepatitis viruses. These molecular techniques are predominantly research tools at this time but may become increasingly useful in the clinical laboratory as techniques are streamlined and automated. Some details regarding testing for specific viral infections are presented below.

Serodiagnosis for HAV

Commercial EIA's for both IgM and total (i.e., IgG plus IgM) anti-HAV are available for clinical use. The IgM class anti-HAV is the earliest serologic marker and appears about two weeks after exposure, often after the patient has become asymptomatic. The IgM anti-HAV is detectable for about two months and is gradually replaced by IgG antibodies, which persist for the life of the patient. Thus, the diagnosis of hepatitis A in the acute or early recovery phase is made by determination of IgM anti-HAV. The presence of predominantly IgG anti-HAV indicates past infection with persisting immunity.

HBV Antigen Detection and Serodiagnosis for HBV and HDV

There are serologic tests for two HBV antigens, surface (HB_sAg) and e (HB_eAg). The core antigen (HB_cAg) can be found in hepatic tissue by immunohistochemistry, but not in serum. The HB_sAg can be detected free in plasma or serum as well as in hepatic tissue. A positive test for HB_sAg indicates either present or chronic infection and is generally taken as evidence that the patient is infectious, either acutely or as a chronic carrier. Circulating HB_eAg is a marker of active viral replication and definitely indicates infectivity.

Antibody tests for a more extensive array of hepatitis B viral markers are available. The pattern of antibody responses to three viral antigens, surface (anti-HB_s), core (anti-HB_c), and e (anti-HB_e), is useful in clinical diagnosis and staging (table 26.4).

Antibodies to the HB_sAg are formed in response to either infection or immunization with hepatitis B vaccine. These antibodies may persist for years, although in an occasional patient the antibodies disappear in time. The antibody response to hepatitis B infection is always marked by the formation of antibodies to the core protein (anti-HB_c). In acute infection the antibodies are IgM. These are replaced by IgG antibodies in chronic infection or in recovered hepatitis B. The IgG antibodies persist for life. Because hepatitis B vaccine does not contain core proteins, no anti-HB_c is seen in

Table 26.4 Interpretation of Test Results for Hepatitis B Virus Infection

	Serologic Markers			
Interpretation[a]	HB_sAg	Anti-HB_s	Total Anti-HB_c[b]	IgM Anti-HB_c
Uninfected	–	–	–	–
Vaccination	–	+	–	–
Early acute infection[c]	+	–	–	–
Acute infection[c,f]	+	–	+	+
Acute infection in core "window" period	–	–	+	+
Resolving acute infection[f]	–	+	+	+
Chronic infection[d,g]	+	–	+	–
Chronic infection with heterotypic anti-HB_s[d,e,g]	+	+	+	–
Prior infection (or rarely chronic infection with low-level HB_sAg)[h]	–	+	+	–

a. Other interpretations not listed include possible false positive and negative results.
b. Both IgM and IgG are measured in total anti-HB_c.
c. HB_eAg will be positive early in infection.
d. HB_eAg positivity indicates high infectivity; absence of HB_eAg with positive anti-HB_e indicates low infectivity.
e. Anti-HB_s may be directed against a HB_sAg subtype determinant different from that of the detected HB_sAg, e.g., anti-HB_s specific for d determinant with HB_sAg of subtype ay.
f. The presence of anti-HDV would indicate coinfection with HDV.
g. The presence of anti-HDV would indicate coinfection or superinfection with HDV.
h. Anti-HB_s may disappear with time in some cases.

response to vaccination. Thus, the finding of anti-HB_c always means hepatitis B infection, either past or present. Antibodies to the e antigen (anti-HB_e) are generally considered to be a sign of recovered hepatitis B. They are not seen in chronic active hepatitis where the patient is considered to be infectious.

Antibodies to hepatitis D (anti-HDV) are seen in patients who are infected with hepatitis B and are coinfected or superinfected with hepatitis D. They are a useful tool in epidemic settings and in patients such as intravenous drug users who are at increased risk of delta coinfection.

Serodiagnosis for HCV

EIA's for the detection of antibodies to HCV have been developed using recombinant-derived antigens. Current tests, known as second generation tests, are both more sensitive and more specific than the earlier first-generation tests. The antigenic configuration of these newer tests includes several hepatitis C antigens, all recombinant-derived. Included are proteins expressed by both structural and nonstructural genes. Different manufacturers use different terms (5-1-1, c-100c, c33c, c22, HC-34, and HC-31) for these various proteins. "Positive" results with the EIA tests can be evaluated using supplemental tests that employ dot blot or similar methods and combinations of isolated single proteins. The combination of second-generation EIA testing and supplemental testing is both highly specific and sensitive for hepatitis C.

Detection of HEV in Stool

Diagnostic immunoassays for HEV infection are not commercially available at this time. Researchers have used cloned viral sequences as hybridization probes to detect viral nucleic acid in stool. In addition, immunoelectron microscopy has been used to detect HEV in human stool.

SUPPLEMENTARY READING

Allison, A. C., and Blumberg, B. S. An isoprecipitation reaction distinguishing human serum protein types. *Lancet* 1:634 (1961). (This is the first report of the "Australia antigen," later identified as HB_sAg. Definitive connection of this antigen to hepatitis and discovery of the virion itself [Dane particle] came later.)

Alter, H. J. Descartes before the horse: I clone, therefore I am—The hepatitis C virus in current perspective. *Ann. Intern. Med.* 115:644–49 (1991).

Alter, M. J.; Hadler, S. C.; Judson, F. N.; Mares, A.; Alexander, W. J.; Hu, P. Y.; Miller, J. K.; Moyer, L. A.; Fields, H. A.; Bradley, D. W.; and Margolis, H. S. Risk factors for acute non-A, non-B hepatitis in the United States and association with hepatitis C virus infection. *JAMA* 264:2231–35 (1990).

Bonino, F.; Hoyer, B.; Ford, E.; Shih, J. W.-K.; Purcell, R. H.; and Gerin, J. L. The δ (delta) agent; HB_sAg particles with d antigen and RNA in the serum of an HBV carrier. *Hepatology* 1:127–31 (1981).

Bonino, F.; Negro, F.; Brunetto, M. R.; and Verme, G. Hepatitis delta virus infection. *Prog. Liver Dis.* 9:485–96 (1990).

Bradley, D. W. Enterically-transmitted non-A, non-B hepatitis. *Br. Med. Bull.* 46:442–61 (1990).

Choo, Q.-L.; Kuo, G.; Weiner, A. J.; Overby, L. R.; Bradley, D. W.; and Houghton, M. Isolation of a CDNA clone derived from a blood-borne non-A, non-B viral hepatitis genome. *Science* 244:359–62 (1989).

Choo, Q.-L.; Richman, J. H.; Han, J. H.; Berger, K.; Lee, C.; Dong, C.; Gallegos, C.; Coit, D.; Medina-Selby, A.; Barr, P. J.; Weiner, A. J.; Bradley, D. W.; Kud, G.; and Houghton, M. Genetic organization and diversity of the hepatitis C virus. *Proc. Natl. Acad. Sci.* (USA) 88:2451–55 (1991).

Cohen, J. I.; Ticehurst, J. R.; Purcell, R. H.; Buckler-White, A.; and Baroudy, B. M. Complete nucleotide sequence of wild-type hepatitis A virus: Comparison with different strains of hepatitis A virus and other picornaviruses. *J. Virol.* 61:50–59 (1987).

Feinstone, S. M.; Kapikian, A. Z.; and Purcell, R. H. Hepatitis A: Detection by immune electron microscopy of a virus-like antigen associated with acute illness. *Science* 182:1026–28 (1973).

Gust, I. D., and Feinstone, S. M. Hepatitis A. *Prog. Liver Dis.* 9:371–78 (1990).

Hoofnagle, J. H., and DiBisceglie, A. M. Serologic diagnosis of acute and chronic viral hepatitis. *Semin. Liver Dis.* 11:73–83 (1991).

Kaneko, S.; Miller, R. H.; DiBisceglie, A. M.; Feinstone, S. M.; Hoofnagle, J. H.; and Purcell, R. H. Detection of hepatitis B virus DNA in serum by polymerase chain reaction: Application for clinical diagnosis. *Gastroenterology* 99:799–804 (1990).

Kools, A. M. Hepatitis A, B, C, D, and E. Update on testing and treatment. *Postgrad. Med.* 91:109–14 (1992).

Miller, R. H., and Purcell, R. H. Hepatitis C virus shares amino acid sequence similarity with pestiviruses and flaviviruses as well as members of two plant virus supergroups. *Proc. Natl. Acad. Sci.* (USA) 87:2057–61 (1990).

Overby, L. R., and Houghton, M. Hepatitis viruses. In Lennette, E. H., ed., *Laboratory Diagnosis of Viral Infections*, 2d ed., pp. 403–41. New York: Marcel Dekker, 1991.

Reyes, G. R., and Baroudy, B. M. Molecular biology of non-A, non-B hepatitis agents: Hepatitis C and hepatitis E viruses. *Adv. Virus Res.* 40:57–102 (1991).

Reyes, G. R.; Purdy, M. A.; Kim, J. P.; Luk, K-C.; Young, L. M.; Fry, K. E.; and Bradley, D. W. Isolation of a CDNA from the virus responsible for enterically transmitted non-A, non-B hepatitis. *Science* 247:1335–39 (1990).

Rizzetto, M.; Canese, M. G.; Arico, S.; Crivelli, O.; Trepo, C.; Bonino, F.; and Verme, G. Immunofluorescence detection of a new antigen-antibody system (δ/anti-δ) associated with hepatitis B virus in liver and serum of HB_sAg carriers. *Gut* 18:997–1003 (1977).

Takemura, F.; Ishii, T.; Fujii, N.; and Uchida, T. Complete nucleotide sequence of hepatitis B virus. *Nucleic Acids Res.* 18:4587 (1990).

Tam, A. W.; Smith, M. M.; Guerra, M. E.; Huang, C. C.; Bradley, D. W.; Fry, K. E.; and Reyes, G. R. Hepatitis E virus (HEV): Molecular cloning and sequencing of the full-length viral genome. *Virology* 185:120–31 (1991).

Wang, K.-S.; Choo, Q.-L.; Weiner, A. J.; Ou, J.-H.; Najarian, R. C.; Thayer, R. M.; Mullenbach, G. T.; Denniston, K. J.; Gerin J. L.; and Houghton, M. Structure, sequence and expression of the hepatitis delta (δ) viral genome. *Nature* 323:508–14 (1987).

27

ARBOVIRUSES: ALPHAVIRUSES, FLAVIVIRUSES, AND BUNYAVIRUSES

N. KARABATSOS

TAXONOMIC STATUS AND NOMENCLATURE

Arthropod-borne viruses (arboviruses) constitute a group of taxonomically heterogeneous viruses, but only three families will be discussed in this chapter: Togaviridae, Flaviviridae, and Bunyaviridae. Members of the genera Alphavirus and Flavivirus were formerly classified within the family Togaviridae, which originally included arbovirus serologic groups A and B, respectively (table 27.1). Currently, alphaviruses constitute a genus in the family Togaviridae, which includes 28 members, and flaviviruses represent a genus in the family Flaviviridae, which consists of 68 members. Four genera comprise viruses of the family Bunyaviridae. These include the bunyaviruses, nairoviruses, phleboviruses, and hantaviruses. The *Bunyavirus* genus consists of viruses belonging to 18 different antigenic groups in addition to several antigenically unassigned viruses.

There are 12 alphaviruses known to produce important diseases in humans found in the United States or in Central and South America. These include eastern and western equine encephalomyelitis, Venezuelan equine encephalomyelitis, Mayaro, and to a lesser extent, Everglades viruses. Disease-inducing flaviviruses active in the Americas or the Caribbean include St. Louis encephalitis and the four dengue serotypes—yellow fever, Powassan, Rocio, and Rio Bravo viruses. In the *Bunyavirus* genus, those reported to cause human illness are LaCrosse, Jamestown Canyon, and snowshoe hare viruses, all members of the California antigenic group, which are active in the United States and in Canada.

Table 27.1 Registered Arboviruses as of June 1992

Nucleic Acid Type	Virus Family and Genus	Particle Size (nm)	Antigenic Group(s)	Total Number of Members	Commonly Known Virus Members
RNA	Togaviridae *Alphavirus*	60–70	A	28	Eastern equine encephalitis, Western equine encephalitis, Venezuelan equine encephalitis, Sindbis
	Possible member		Ungrouped	1	
	Flaviviridae *Flavivirus*	40–50	B	68	Dengue 1–4, Japanese encephalitis, St. Louis encephalitis, tick-borne encephalitis, West Nile yellow fever, Powassan
	Possible member		Ungrouped	1	
	Bunyaviridae				
	Bunyavirus	80–100	Bunyamwera supergroup; 18 groups + 3 antigenically unassigned viruses	138	Bunyamwera California encephalitis LaCrosse Oropouche
	Nairovirus		6 groups	24	Crimean-Congo hemorrhagic fever Nairobi sheep disease
	Phlebovirus		Phlebotomus fever group	38	Naples sandfly fever Sicilian sandfly fever Rift Valley fever
			Uukuniemi group	6	Uukuniemi
	Hantavirus		Hantaan group	5	Hantaan, Seoul, Puumula

"Bunyavirus-like" (unassigned, probable, or possible members)		9 groups +	21	Bhanja, Lssyk-Kul
		22 ungrouped viruses	22	
Reoviridae				
Orbivirus	65–80	15 groups +	62	Bluetongue, Kemerovo, African horse sickness
		7 ungrouped viruses	7	
Coltivirus	80	Colorado tick fever group	2	Colorado tick fever, Eyach
Unassigned or possible members		Ungrouped	6	
Rhabdoviridae				
Vesiculovirus	170 × 70	Vesicular stomatitis group	18	Vesicular stomatitis, Indiana; Vesicular stomatitis, New Jersey
Lyssavirus	180 × 75	Rabies virus group + 2 other groups	16	Lagos bat, Kotonkan, bovine ephemeral fever
Unassigned or possible members		6 groups +	19	
		17 ungrouped viruses	17	
Arenaviridae				
Arenavirus	110–130	Tacaribe group	12	Lassa, Junin, Machupo
Filoviridae				
Filovirus	790–970 × 80	Marburg group	2	Marburg; Ebola
Nodaviridae				
Nodavirus	29–30	Nodamura virus group	1	
Coronaviridae				
Coronavirus	60–220	Ungrouped	1	

(*continued*)

Table 27.1 *(Continued)*

Nucleic Acid Type	Virus Family and Genus	Particle Size (nm)	Antigenic Group(s)	Total Number of Members	Commonly Known Virus Members
	Paramyxoviridae				
	Paramyxovirus	150	Ungrouped	1	
	Unassigned or possible member		Ungrouped	1	
	Orthomyxoviridae	90–110	Influenza D viruses; Thogoto virus group	1	
			Ungrouped	1	
DNA	Herpesviridae	100–200	Ungrouped	1	
	Possible members		Ungrouped	1	
	Poxviridae				
	Possible members	200–450 × 140–260	Ungrouped	2	
	African swine fever*	125–300	Ungrouped	1	African swine fever
	Unclassified		3 groups +	4	
			8 ungrouped viruses	8	
				536	

*Removed from family Iridoviridae; a new official family name has not been designated.

By definition, an arbovirus is a virus that can be biologically transmitted by an arthropod vector. After the vector feeds on an infected vertebrate and virus multiplication occurs in the vector (the extrinsic incubation period), that vector is capable of transmitting the infection, by bite, to another vertebrate. This definition excludes those viruses that can be transmitted by arthropods but do not multiply in the vector. To date there are 535 registered arboviruses, most of which are known or suspected to be arthropod-borne (table 27.1), but conclusive evidence for virus replication in a vector is available for relatively few.

MORPHOLOGY AND BIOCHEMICAL PROPERTIES

Alphaviruses and flaviviruses share several physicochemical properties. Flaviviruses are small (40–50 nm in diameter), and alphaviruses are medium (60–70 nm) in size. Their virions each possess a lipid envelope and are sensitive to lipid solvents and detergents and contain single-stranded RNA. Replication of alphaviruses and flaviviruses occurs in the cytoplasm of infected cells; maturation and envelopment take place by budding at the plasma membrane (figure 27.1) or through intracytoplasmic membranes, most commonly the endoplasmic reticulum (figures 27.2 and 27.3).

Virions of bunyaviruses are larger and pleomorphic (80–100 nm diameter), with an envelope on which glycoprotein surface projections are located, and contain three segments of single-stranded RNA. They are also sensitive to lipid solvents and detergents. These viruses replicate in the cytoplasm of infected cells, and they mature and acquire an envelope by budding into smooth-surface vesicles in the Golgi region, although some plasma membrane budding has also been observed (figure 27.4).

SPECIMEN SOURCE AND INOCULATION

Serum from patients during the acute phase of febrile illnesses is the most productive source for recovering certain arboviruses; it can be used for virus isolation either undilute or diluted 1:10 and 1:100 in a protein-containing diluent. It is important to inoculate unknown specimens into mice at two, or preferably more, dilutions (undiluted to 10^{-3}) because some arboviruses, if inoculated at high concentrations, may induce potent inhibitors that restrict or completely mask symptoms of infection. In addition, a high concentration of defective particles in the specimen may induce autointerference, while the presence of antibody in tissue specimens may disguise symptoms of the virus or produce a prozone effect at low dilutions. A number of the encephalitis-

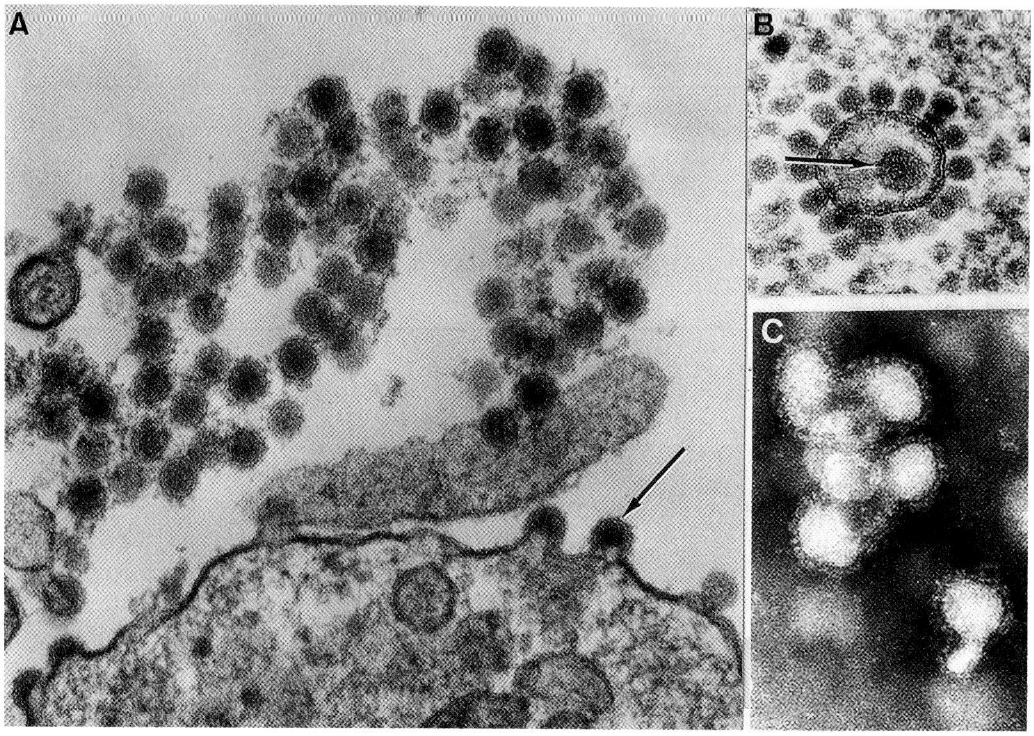

Figure 27.1 Electron micrographs of an alphavirus (Sindbis virus) in an infected Vero cell: *(A)* Thin section of Sindbis virus–infected Vero cell showing extracellular virions and nucleocapsids in the process of budding at the cell membrane (100,800×); *(B)* Nucleocapsids and enveloped virion (arrow) in the cytoplasm of the cell (100,800×); *(C)* Sindbis virus virions stained with PTA (168,000×)

causing arboviruses, however, can be isolated from CSF collected during the acute phase and rarely from blood.

When arbovirus infection is a suspected cause of death, virus can generally be recovered from brain tissue taken at autopsy. Processed specimens should be inoculated into newborn mice or susceptible primary and serially propagated cell cultures with a minimum of delay, since these viruses are labile when temperatures and pH vary.

Although inoculation of newborn mice remains as the method of choice for isolation of most important arboviruses, other isolation systems have gained prominence and have become important adjuncts. These include the use of serially propagated mammalian or mosquito cell cultures and intrathoracic mosquito inoculation, which is especially useful for the isolation of dengue viruses.

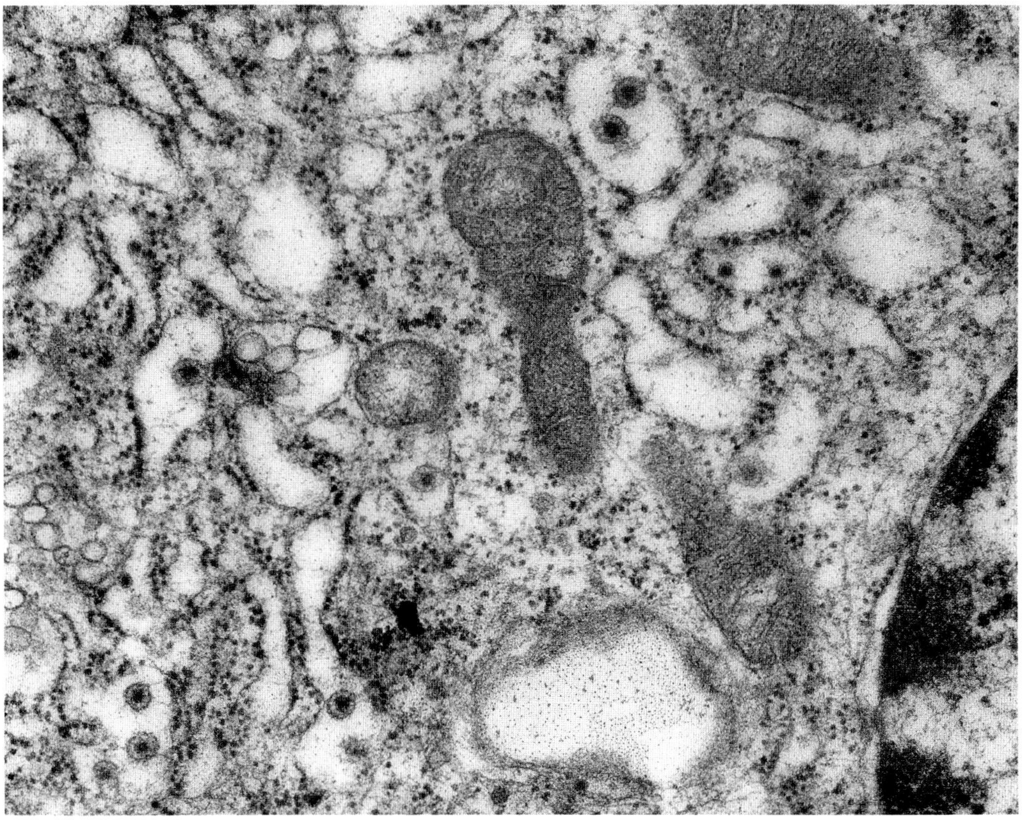

Figure 27.2 Electron micrograph of a flavivirus (Japanese encephalitis virus) in an infected BHK-21 cell showing virions in the dilated endoplasmic reticulum (76,500×). (Courtesy J. D. Wright)

VIRUS ISOLATION, PROPAGATION, AND ASSAY

Mouse Pathogenicity

Until recently, the newborn mouse was considered a universal host system for isolation of arboviruses. Newborn mice, one to four days old, are inoculated by the intracerebral (IC) route (0.01–0.02 ml) or in combination with the intraperitoneal route (0.03–0.05 ml). Except for a few alphaviruses that have an extremely short incubation period, illness or hindlimb paralysis usually occurs two to five days after inoculation. Some arboviruses have a prolonged incubation period in suckling mice, and an observation period of 14–21 days may be necessary. This is particularly true for the dengue viruses. Mice showing signs of illness should be sacrificed immediately, brain tissue

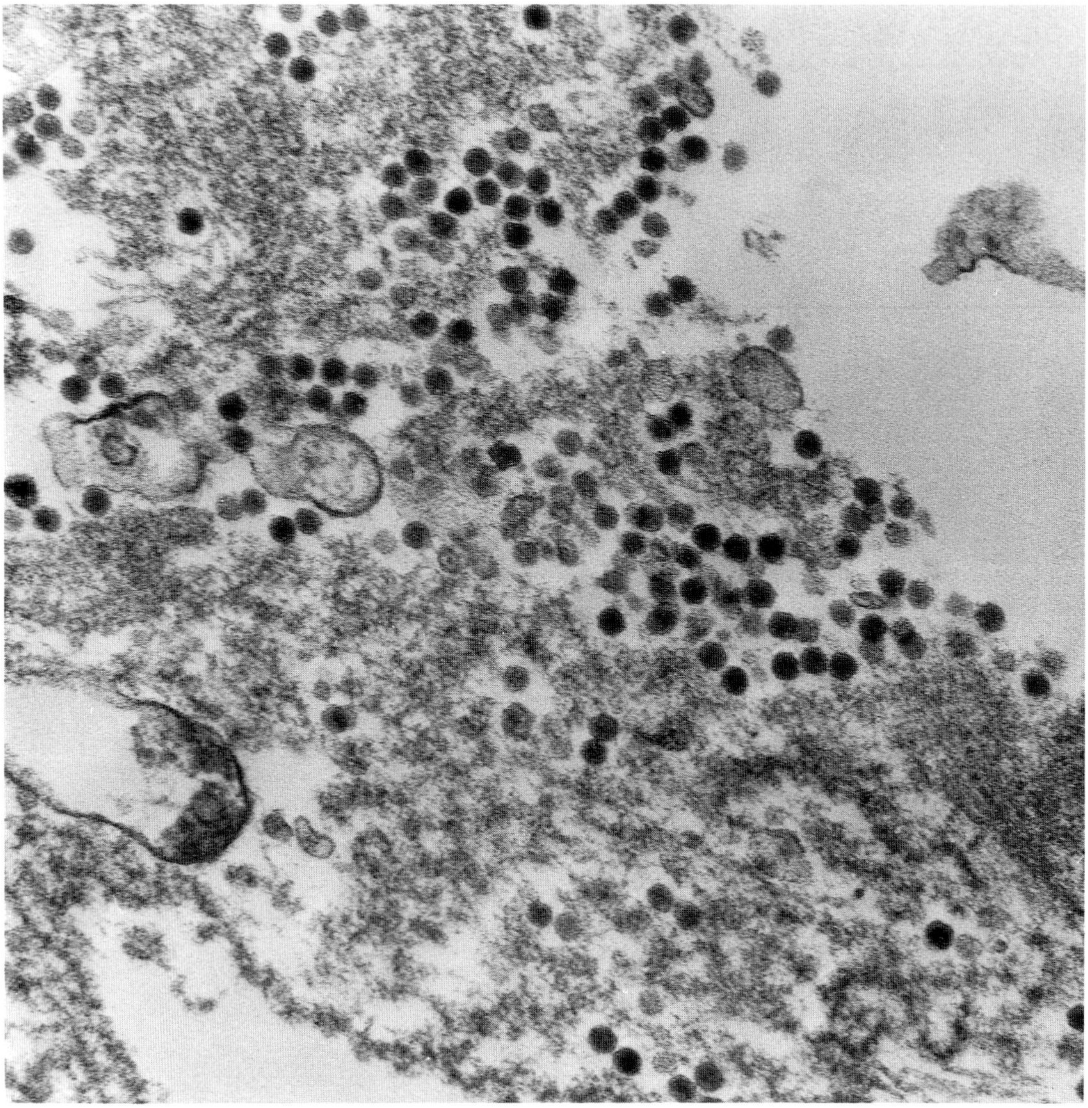

Figure 27.3 Electron micrograph of a flavivirus (St. Louis encephalitis virus) in an infected dolphin cell line showing extracellular virions and budding into cytoplasmic vacuoles. (99,000×). (Courtesy D. Medina)

harvested, and a 10% suspension made with buffered diluent containing 0.75% bovine albumin at a pH between 7.2 and 8.0. These brain suspensions are passaged by the IC route to other suckling mice to confirm the isolation and for virus identification by serologic means. Certain bunyaviruses are found in higher titers in the liver rather than the brain of sick mice.

In areas where arboviruses are endemic or in the region of an outbreak, mosquitoes are often trapped and tested for virus isolation. Pools of up to

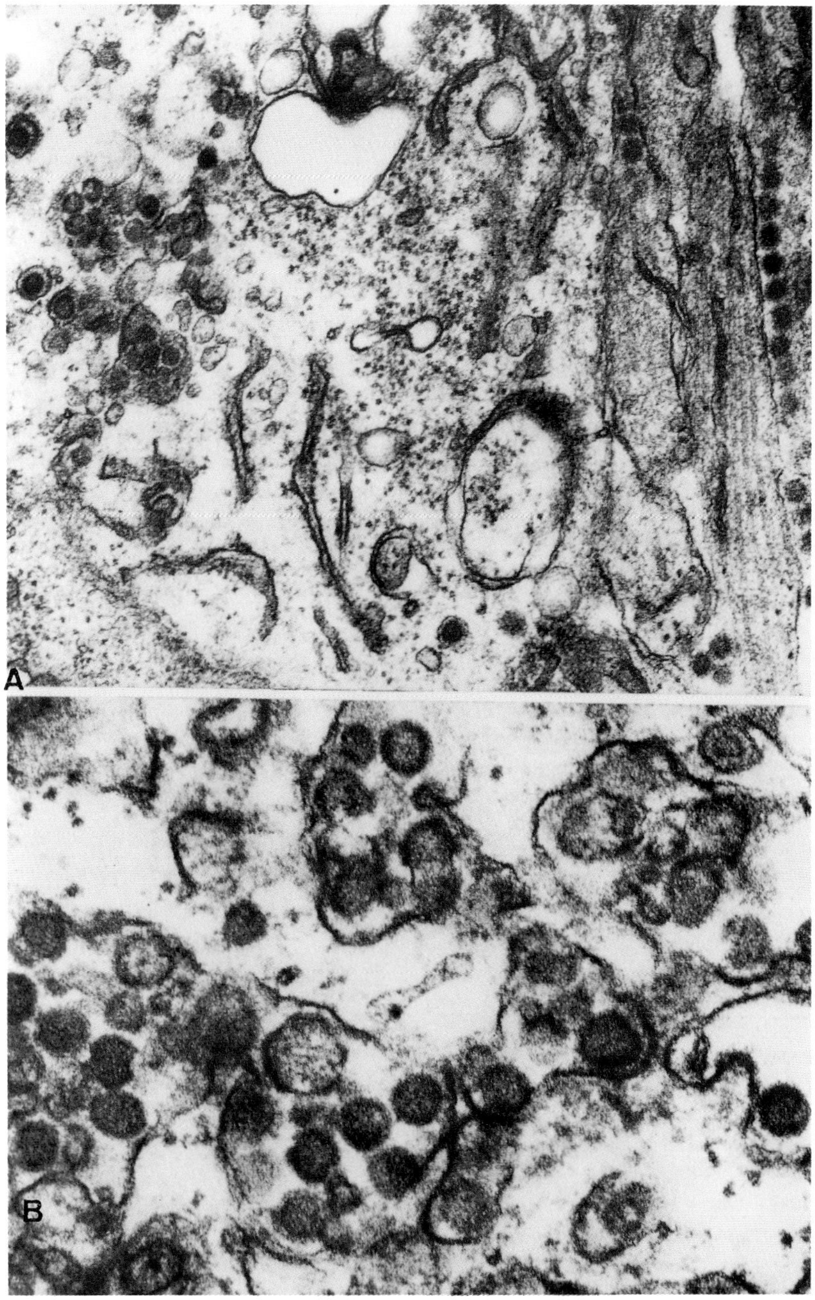

Figure 27.4 Section of brain tissue from a suckling mouse, inoculated with Bunyamwera virus (reproduced with permission, F. A. Murphy et al., *Intervirology* 1:297–316 [1973]). *(A)* Virions are contained within Golgi vesicles (left) of a neuron and in extracellular spaces between neuronal processes (right). Extracellular virions appear fuzzy (35,000×); *(B)* Bunyamwera virus, at higher magnification, within Golgi cisternae. The moderately dense center, membranous envelope, and rather smooth surface outline are typical of intracellular virions (89,000×).

25–100 mosquitoes are ground in 2 ml of BA-1 diluent (Medium-199 with 1% bovine albumin) and clarified by centrifugation at 12,100 × *g* for 30 minutes. Extracts of mosquitoes can be inoculated into mice or cell cultures as described above.

The sentinel mouse procedure (exposure of newborn mice to mosquitoes in the field) is useful for isolating some arboviruses, especially in enzootic areas of the New World tropics. It is useful, however, only in areas where virus-infected mosquitoes that feed on rodents are found. After exposure for several days, the mice are brought back to the laboratory and observed for signs of illness. Sick mice are processed as above.

Cytopathic Effect in Cell Cultures

Most of the arboviruses tested replicate in continuous cell lines. Presently, primary cell cultures are no longer routinely used. Cell lines frequently used are Vero, BHK-21, and CER (hamster cells). Insect cell cultures are increasingly being used with success. For example, a mosquito cell line, C6/36 (derived by cloning of parental *Aedes albopictus* cell cultures), or intrathoracically inoculated mosquitoes are even more sensitive for isolation of dengue viruses than is the continuous rhesus monkey kidney cell line, LLC-MK_2. The latter, however, is preferred to newborn mice. In addition, mosquito cell cultures have been found to be more sensitive than newborn mice for isolation of yellow fever virus.

Several flaviviruses have been reported to produce a distinctive CPE characterized by syncytium formation in continuous lines of mosquito cell cultures. These include Japanese encephalitis virus, West Nile virus, and dengue virus types 1–4. Other arboviruses that have been reported to produce CPE in invertebrate cell cultures include St. Louis encephalitis and yellow fever viruses. CPE in invertebrate cells, however, is an inconsistent property.

Plaque Formation in Cell Cultures

Currently, plaque formation in plastic plates is used almost exclusively for virus assay and neutralization tests. Continuous cell lines most frequently employed for plaque formation include Vero and the E6 clone of Vero, BHK-21, LLC-MK_2, CER, PS (cloned porcine stable cells), and clones of CV-1 cells.

PREPARATION OF MONOCLONAL ANTIBODIES

An extremely important advance took place in the identification of arboviruses: the development of hybridoma cell lines that secrete monospecific antibodies to the various virus strains. Hybridoma cell lines are prepared by fusing splenic lymphocytes from immunized or hyperimmunized mice with mouse plasmacytoma cells (NS-1). Clones of fused cells are isolated, and the antibodies secreted by these clones are characterized. (Refer to Roehrig et al., 1980, for a detailed protocol.) Among the alphaviruses and flaviviruses and certain of the bunyaviruses, this important technology has been applied to analyzing and characterizing the antigenic structure of surface proteins of some of the important pathogens in these groups. In addition, type-specific monoclonal dengue virus antibodies have been isolated and used to rapidly type and identify dengue isolates.

IDENTIFICATION OF ISOLATES

Most alphaviruses and flaviviruses possess common group antigens as measured by one or more of the following: hemagglutination-inhibition, complement-fixation, and immunofluorescence assay techniques. They can therefore be grouped by technique using grouping immune reagents or group reactive monoclonal antibodies. Each of the alphaviruses, flaviviruses, and bunyaviruses is commonly identified by neutralization tests either in cell cultures (plaque-reduction neutralization) or in mice. Immunologic cross-reactions are more evident in IF and HI, and are less so in CF tests. Most cross-reactions, though not all, are minimized by the neutralization test.

Mouse Protection Neutralization Test

Neutralization tests in mice are used to identify certain arboviruses, particularly when an isolate is not known to induce CPE or form plaques readily in a cell culture system. The procedures are similar to those described below for the plaque-reduction neutralization test.

Plaque-Reduction Neutralization Test

Effective neutralization is often dependent upon a heat-labile complement-like factor. Thus, fresh, normal human serum is added to the heat-inactivated test human or animal serum to enhance the neutralization test. Virus dilutions are made in phosphate-buffered saline containing 0.75% bovine albumin. Detailed procedures for the plaque-reduction neutralization test are given in chapter 6.

Immunofluorescent Staining Reaction

The IF test is currently used as a rapid and accurate method for the identification of dengue viruses or to classify viruses into given serogroups with the use of polyvalent grouping antisera. The impression smears made from virus-infected head squashes of mosquitoes are stained with fluorescein-conjugated anti-mouse antibody following reaction with dengue type-specific monoclonal antibody.

Hemagglutination Test

Prior to identification of an arbovirus isolate by the HI test, an HA antigen is prepared (see below). Because HA activity by arboviruses is pH-dependent, the optimal pH must be determined each time the test is performed.

PREPARATION OF HA ANTIGENS FROM INFECTED MOUSE BRAIN TISSUE

1. Weigh infected mouse brain tissue and homogenize for 1 minute in a chilled electric blender (Sorvall Omni-mixer*) with 4 volumes of cold 8.5% sucrose in distilled water.
2. To 20 volumes of chilled acetone being vigorously agitated, add 1 volume of homogenate by rapidly expelling from a chilled syringe with a large needle submerged below the surface of the acetone.
3. Shake the mixture vigorously and allow tissue to settle for 10–30 minutes in a refrigerator or in an ice bath.
4. Aspirate the supernatant and discard. Add chilled acetone to the pink precipitate. The quantity of acetone added should be equal to the original volume of acetone used.
5. Shake vigorously and allow residue to settle for 60 minutes at 4°C.
6. Remove and discard the supernatant fluid. Add a small quantity of fresh acetone to the sediment in order to transfer extracted residue.
7. Transfer sediment to thick-walled centrifuge bottles immersed in an ice bath and dry by attaching to a vacuum pump for 1 hour.
8. Rehydrate the dried sediment with borate saline, pH 9.0, in a volume equal to 0.4 of the volume of the original homogenate. Let stand overnight in the refrigerator.
9. Centrifuge the rehydrated preparation at 12,100 × *g* for 30 minutes.

The supernatant fluid representing the finished hemagglutinating antigen can be stored in the frozen state or lyophilized. Since sucrose exerts a protective effect, all sucrose-acetone-extracted antigens contain residual infective virus. If serology is to be performed with an inactivated antigen, proceed from step 7 above, as follows:

8. Rehydrate the dried sediment with borate saline, pH 9.0, containing 0.1 M Tris buffer in a volume equal to 0.4 of the volume of the original homogenate. Let stand overnight in the refrigerator.

*The use of trade names is for identification only and does not constitute endorsement by the Public Health Service or by the U.S. Department of Health and Human Services.

9. Centrifuge the rehydrated preparation at 12,100 × *g* for 30 minutes.
10. Prepare a 10% solution of beta-propiolactone (BPL) in cold PBS. Add sufficient BPL to antigen to give a final concentration of 0.3% for the inactivation of alphaviruses or 0.1% for all other arbovirus antigen preparations. Stir antigen mixture for 72 hours at 4°C.
11. Test for loss of infectivity by inoculation of newborn mice.

TITRATION OF THE HA ANTIGEN AT DIFFERENT PH VALUES IN MICROTITER PLATES

1. Prepare serial 2-fold dilutions of HA antigen in borate-buffered saline, pH 9.0, containing 0.4% bovine serum albumin (formula below), using 0.05 ml volumes. To test for HA activity at different pH values, prepare several identical sets of antigen dilutions in plastic panels.
2. Prepare a working suspension of goose red blood cells by diluting the 8% stock suspension (prepared in dextrose gelatin Veronal [DGV] diluent; formula below) to 1:24 in the various pH adjusting diluents (pH 5.8–7.0 in increments of 0.1 pH units).
3. Add 0.05 ml of the working RBC suspension to each of the antigen dilutions in the plastic panels and cover plates. Also include a set of RBC in diluents of different pH values as controls. Incubate at 22°C or at room temperature for 60 minutes or until RBC settle.
4. The highest antigen dilution showing complete agglutination is taken as the endpoint. The optimal pH is that yielding the highest HA titer and clearest HA pattern.
5. Calculate the dilution at the optimal pH that gives 8–16 units per 0.05 ml or 4–8 units per 0.025 ml.

Currently, HA antigens for a number of the arboviruses may be prepared in cell cultures. Harvested infected cell culture fluids are concentrated by polyethylene-glycol precipitation and extracted with cold acetone (Della Porta and Westaway, 1972; Grimstad et al., 1987). Recently, a protocol has been devised for the preparation of HA antigen for the four serotypes of dengue virus in C6/36 mosquito cell lines. This protocol has been so successful that it has become the laboratory's routine method for preparing HA antigens for the four serotypes of dengue virus. The procedure is identical to that described for the preparation of Jamestown Canyon HA antigen (see Grimstad el al., 1987), except that C6/36 mosquito cell cultures are used for the dengue viruses.

Hemagglutination-Inhibition Test

Determine the pH at which the antigen is optimally reactive and the approximate titer of the preparation (as described above). The antigen is then tested against antisera to recognized viruses. It is essential to include homologous titrations with known antigen and the same known antisera for comparison.

1. Prepare serial 2-fold dilutions of antiserum, which has been pretreated with kaolin or acetone to remove inhibitors (see below), in borate-buffered saline containing 0.4% bovine albumin, pH 9.0.
2. Dilute the HA antigen with the same diluent as above to contain 8 units per 0.025 ml. The dilution necessary to give this number of units is calculated from the titer at that pH showing the highest HA activity.
3. Add 0.025 ml of the virus antigen containing 8 HA units to 0.025 ml of each dilution of antiserum. Incubate at room temperature for 2 hours, keeping the plate covered during this period.
4. Dilute an 8% suspension of male domestic goose erythrocytes to 1:24 in the adjusting diluent that yields the optimal pH for HA activity. Add 0.05 ml of the diluted RBC suspension to each virus-antiserum mixture.
5. Prepare three controls: antiserum with diluent and RBC (antiserum control); diluent plus RBC (RBC control); and a titration of the diluted HA preparation in order to verify the exact number of units employed (antigen control).
6. Incubate test and homologous titrations, as well as all controls, at room temperature for 30–60 minutes. The endpoint titers are read as the highest dilutions of antisera completely inhibiting the agglutination of 8 HA units of antigen.

REMOVAL OF NONSPECIFIC INHIBITORS IN SERA BY ACETONE TREATMENT

1. Dilute sera to 1:10 with PBS.
2. Chill diluted serum in an ice bath and add 12 volumes of cold acetone.
3. Extract for 5 minutes with periodic shaking.
4. Centrifuge for 5 minutes at 11,060 × *g* and remove the supernatant by pouring off or by aspiration.
5. Reextract the precipitate as before.
6. Again, remove the acetone and dry the sediment for 1 hour with a vacuum pump or allow to air dry overnight at room temperature.
7. Reconstitute the dried material in borate saline, pH 9.0, to give the equivalent of a 1:10 serum dilution.
8. To remove naturally occurring agglutinins for goose erythrocytes, chill treated sera in an ice bath and add 0.1 ml of packed, washed goose RBC per 5 ml of serum at the 1:10 dilution.
9. Allow adsorption to take place for 20 minutes with occasional shaking.
10. Centrifuge for 10 minutes at 400 × *g* in the cold. Remove supernatant fluid and test by the HI reaction.

Direct Detection of Viral Antigen or Genomic Substances

Methods designed to detect viral antigen or genomic sequences directly in specimens are becoming more prominent in the laboratory diagnosis of arboviral infections. They have shown advantages in enhancing the specificity and rapidity of diagnosis.

DETECTION OF VIRAL ANTIGEN

ELISA has been applied for the direct detection of viral antigens by capturing antigen onto a solid phase with a monoclonal or polyclonal antibody of defined specificity. Captured antigen is detected with an enzyme-conjugated or biotin-tagged antibody. This technique has been used to detect antigen for several arboviruses directly in the blood of patients. The sensitivity of antigen capture ELISA for yellow fever virus has been reported as similar to that of viral isolation in cell cultures, mosquitoes, or mice, and results are obtained three to seven days earlier. This procedure also has been applied for the direct detection of arboviral antigens in mosquitoes, ticks, and sandflies.

The in situ ELISA represents a combination of methods that uses the sensitivity of cell cultures for viral isolation with the rapidity of ELISA to identify isolates amplified by cell culture. Viruses that are at subdetectable levels in clinical specimens are amplified by growth in cell culture in 96-well plates. After fixation, the resulting cell-associated antigens are detected by ELISA in the same cell culture wells originally inoculated with the clinical specimen.

Certain arboviruses may be directly detected in infected cells or tissues by IF or other immunohistochemical procedures. Liver biopsy specimens may be tested for suspected yellow fever virus by IF or immunohistochemical staining. The same methods may be applied for the detection of arboviruses that cause encephalitis in brain tissues from patients with central nervous system disease.

Furthermore, IF techniques have been used to detect Colorado tick fever virus in infected erythrocytes from peripheral blood and Ross River virus in monocytes of synovial fluid from inflamed joints.

DETECTION OF VIRAL GENOME

In situ hybridization has been applied for the detection of genomic sequences of several arboviruses in mosquitoes, "biting midges," and yellow fever virus genome in formalin-fixed, paraffin-embedded liver tissue from yellow fever patients. For yellow fever virus detection, in situ hybridization is more advantageous than immunohistochemical detection of viral antigens because of the greater stability of viral RNA in specimens.

A polymerase-chain reaction for the detection of arboviral genomic sequences has been developed for yellow fever, dengue, Japanese encephalitis,

St. Louis encephalitis viruses, and several hantaviruses. Since most arboviruses have positive-sense, single-stranded RNA genomes, the arboviral RNA in the clinical sample must first be isolated and transcribed to CDNA before amplification by PCR. Primers are used to initiate the PCR process and to amplify certain targeted genomic sequences. In general, these genomic sequences are conserved and do not vary greatly. Amplified DNA products are identified by examining the mobility of electrophoretic bands after restriction enzyme digestion and by dot hybridization with virus-specific probes. The PCR assay has been reported to detect yellow fever genomic sequences in experimentally infected suckling mouse brains at a level of approximately 30 plaque-forming units (T. M. Brown, G. J. Chang, C. B. Cropp, K. E. Robbins, and T. F. Tsai, pers. comm., 1992). The application of this assay to the testing of clinical specimens, however, is still preliminary.

Serodiagnosis

The HI, neutralization, and CF tests are used for the presumptive diagnosis of arbovirus infection in recently ill patients from whom no virus was isolated during the acute phase. The CF antibody titer rises more slowly than the HI and the neutralizing antibodies but decreases rapidly, usually within a few months.

ELISA for Antiarboviral IgM and IgG Determination

Although several state health laboratories continue to use the HI test as their primary serodiagnostic tool for arbovirus infections, some do not. Some have replaced the HI test with the enzyme-linked immunosorbent assay (ELISA) in an IgM antibody capture format. The IgM capture ELISA has considerable advantage: the test detects viral specific IgM antibodies produced within a few days of a primary infection. Thus, not only is the test indicative of a recent infection, it also eliminates the need for a convalescent-phase specimen in some instances. This test also provides the optimum method for IgM detection. It is very sensitive, and false-positive reactions owing to rheumatoid factor are minimized. IgM antibody in serum or cerebrospinal fluid (CSF) is captured onto the solid phase with an antispecies antibody. If viral-specific IgM antibody has been captured, it will bind to the arbovirus antigen added (sucrose-acetone-extracted mouse brain preparation). The bound arbovirus antigen is then detected by adding specific antiviral antibody as a detector, usually composed of group-reactive monoclonal antibodies.

Viral-specific IgG can be detected by ELISA, either by the antibody capture method or by an indirect method that uses an antihuman immunoglobulin conjugate. Generally, purified or partially purified antigens are required, unless one uses the antibody capture technique. In some instances, the IgG ELISA has been used instead of the HI test. It is as sensitive as—if not more sensitive—than the HI test, but its specificity is less than that of the HI test.

Immunofluorescence Assays for Antiarboviral IgM and IgG

IF is a simple, inexpensive test that has been shown to be as specific as, but more sensitive than, the HI test for the serologic diagnosis of certain flavivirus infections. Furthermore, IF has been recommended as the procedure of choice for the serologic diagnosis of Lassa fever virus infections. It also has been adopted for the determination of IgM antibody but was shown to be less sensitive than when it is applied for the determination of IgG antibody. Although it was found to be less time-consuming and more sensitive than the CF test, it is somewhat subjective, labor-intensive, and does not readily lend itself to automation.

Acute and convalescent paired sera are used for antibody determinations. A 4-fold or greater rise in antibody titer to a given arbovirus is presumptive evidence that the virus used in the test or a related virus was responsible for the illness. Since infection with some arboviruses induces antibody responses that persist for many years, perhaps for life, antibody determinations in a given population can be used for determining the prevalence of infections.

REAGENTS

Borate-Buffered Saline, pH 9.0

The final solution consists of 0.05 M borate–0.12 M NaCl. Prepared by diluting with distilled water 80 ml 1.5 M NaCl, 100 ml of 0.05 M H_3BO_3, and 23 ml 1.0 M NaOH to 1 liter.

pH-Adjusting Diluents for Arbovirus HA Test

The following tabulation gives the percentage combinations of phosphate buffers. Each combination (A + B), when mixed with an equal volume of borate saline pH 9.0 (above), gives the final pH indicated in the tabulation.

Solution A (ml) 0.15 M NaCl– 0.2 M Na_2HPO_4	Solution B (ml) 0.15 M NaCl– 0.2 M NaH_2PO_4	Final pH
3.0	97.0	5.75
12.5	87.5	6.0
22.0	78.0	6.2
32.0	68.0	6.4

(continued)

Solution A (ml) 0.15 M NaCl– 0.2 M Na_2HPO_4	Solution B (ml) 0.15 M NaCl– 0.2 M NaH_2PO_4	Final pH
45.0	55.0	6.6
55.0	45.0	6.8
64.0	36.0	7.0
72.0	28.0	7.2
79.0	21.0	7.4

Dextrose Gelatin Veronal (DGV)	
Veronal (barbital)	0.58 g
Gelatin	0.60 g
Sodium Veronal (sodium barbital)	0.38 g
$CaCl_2$ (anhydrous)	0.02 g
$MgSO_4 \cdot 7H_2O$	0.12 g
NaCl	8.5 g
Dextrose	10.0 g
Distilled water	1,000.0 ml

The Veronal and gelatin are dissolved in water by heating. Other reagents are added and the mixture is autoclaved at 10 lb for 10 minutes.

SUPPLEMENTARY READING

American Committee on Arthropod-Borne Viruses. Arbovirus names. *Am. J. Trop. Med. Hyg.* 18:731–34 (1969).

Bishop, D. H. L., and Shope, R. E. Bunyaviridae. In Fraenkel-Conrat, H., and Wagner, R. A., eds., *Comprehensive Virology*, vol. 14, pp. 1–156. New York: Plenum, 1979.

Burke, D. S.; Nisalak, A.; and Ussery, M. A. Antibody capture immunoassay detection of Japanese encephalitis virus immunoglobulin M and G antibodies in cerebrospinal fluid. *J. Clin. Microbiol.* 16:1034–42 (1982).

Calisher, C. H., and Beaty, B. J. Arboviruses. In Lennette, E. H., ed., *Laboratory Diagnosis of Viral Infections*, 2d ed. pp. 243–79, rev. and expanded. New York: Marcel Dekker, 1990.

Calisher, C. H.; Berardi, V. P.; Muth, D. J.; and Buff, E. P. Specificity of immunoglobulin M and G antibody responses in humans infected with eastern and western equine encephalitis viruses: Application to rapid serodiagnosis. *J. Clin. Microbiol.* 23:369–72 (1986).

Casey, H. L. Adaptation of LBCF method to microtechnique, in standardized diagnostic complement fixation method and adaptation to micro test. Public Health Monograph 74, Public Health Service Publication 1228, pt. 2. Washington, D.C.: U.S. Government Printing Office, 1965.

Chambers, T. J.; Hahn, C. S.; Galler, R.; and Rice, C. M. Flavivirus genome, organization, expression and replication. *Ann. Rev. Microbiol.* 44:649–88 (1990).

Chappel, W. A.; Halonen, P. E.; Toole, R. F.; Calisher, C. H.; and Chester, L. Preparation of LaCrosse virus hemagglutinating antigen in BHK-21 suspensions cell cultures. *Appl. Microbiol.* 18:433–37 (1969).

Chappel, W. A.; Sasso, D. R.; Toole, R. F.; and Monath, T. Labile serum factor and its effects on arbovirus neutralization. *Appl. Microbiol.* 21:79–83 (1971).

Clarke, D. H., and Casals, J. Techniques for hemagglutination and hemagglutination-inhibition with arthropod-borne viruses. *Am. J. Trop. Med. Hyg.* 7:561–73 (1958).

Della-Porta, A. J., and Westaway, E. G. Rapid preparation of hemagglutinins of togaviruses from infected cell culture fluids. *Appl. Microbiol.* 23:158–60 (1972).

Francki, R. I. B.; Faquet, C. M.; Knudson, D. L.; and Brown, F., eds. Classification and nomenclature of viruses. Fifth Report of the International Committee on Taxonomy of Viruses. Archives of Virology, supp. 2. New York: Springer-Verlag (1990).

Gentry, M. K.; Henchel, E. A.; McGowan, J. M.; Brandt, W. E.; and Dalrymple, J. M. Identification of distinct antigenic determinants on dengue-2 virus using monoclonal antibodies. *Am. J. Trop. Med. Hyg.* 31:548–55 (1982).

Grimstad, P. R.; Artsob, H.; Karabatsos, N.; and Calisher, C. H. Production and use of a hemagglutinin for detecting antibody to Jamestown Canyon virus. *J. Clin. Microbiol.* 25:1557–59 (1987).

Heinz, F. X., and Roehrig, J. T. Flaviviruses. In van Regenmortel, M. H. V., and Neurath, A. R., eds., *Immunochemistry of Viruses*. Vol. 2, *The Basis for Serodiagnosis and Vaccines,* pp. 289–305. Amsterdam: Elsevier, 1990.

Henchel, E. A.; McGowan, J. M.; Seguin, M. C.; Gentry, M. K.; and Brandt, W. E. Rapid identification of dengue virus isolates by using monoclonal antibodies in an indirect immunofluorescence assay. *Am. J. Trop. Med. Hyg.* 32:164–69 (1983).

Howe, D. K.; Vodkin, M. H.; Novak, R. J.; Mitchell, C. J.; and McLaughlin, G. L. Detection of St. Louis encephalitis virus in mosquitoes by use of the polymerase chain reaction. *J. Am. Mosquito Control Assoc.* 8:333–35 (1992).

Howe, D. K.; Vodkin, M. H.; Novak, R. J.; Shope, R. E.; and McLaughlin, G. L. Use of the polymerase chain reaction for the sensitive detection of St. Louis encephalitis viral RNA. *J. Virol. Meth.* 36:101–10 (1992).

Karabatsos, N., ed. *International Catalogue of Arboviruses including Certain Other Viruses of Vertebrates*. 3d ed. San Antonio: American Society for Tropical Medicine and Hygiene, 1985.

Karabatsos, N., and Mathews, J. H. Serological reactions of fractionated hamster immunoglobulins with California group viruses. *Am. J. Trop. Med. Hyg.* 29:1420–27 (1980).

Lanciotti, R. S.; Calisher, C. H.; Gubler, D. J.; Chang, G. J.; and Vorndam, A. V. Rapid detection and typing of dengue viruses from clinical samples by using reverse transcriptase polymerase chain reaction. *J. Clin. Microbiol.* 30:545–51 (1992).

Lindsey, H. S.; Calisher, C. H.; and Mathews, J. H. Serum dilution neutralization test for California group virus identification and serology. *J. Clin. Microbiol.* 4:503–10 (1976).

Monath, T. P. Flaviviruses. In Fields, B. N., and Knipe, D. M., eds., *Fields Virology,* pp. 763–814. New York: Raven, 1990.

———. Pathobiology of the flaviviruses. In Schlesinger, S., and Schlesinger, M. J., eds., *The Togaviridae and Flaviviridae,* pp. 375–440. New York: Plenum, 1986.

———, ed. *The Arboviruses: Epidemiology and Ecology,* vols. 1–5. Boca Raton: CRC Press, 1989.

———, ed. *St. Louis Encephalitis*. Washington, D.C.: American Public Health Association, 1980.

Monath, T. P.; Ballinger, M. E.; and Miller, B. R. Detection of yellow fever viral RNA by nucleic acid hybridization and viral antigen by immunocytochemistry in fixed human liver. *Am. J. Trop. Med. Hyg.* 40:663–68 (1989).

Monath, T. P.; Cropp, C. B.; Muth, D. J.; and Calisher, C. H. Indirect fluorescent antibody test for the diagnosis of yellow fever. *Trans. Roy. Soc. Trop. Med. Hyg.* 75:282–86 (1981).

Monath, T. P., and Nystrom, R. R. Detection of yellow fever virus in serum by enzyme immunoassay. *Am. J. Trop. Med. Hyg.* 33:151–57 (1984).

Monath, T. P.; Nystrom, R. R.; Bailey, R. E.; Calisher, C. H.; and Muth, D. J. Immunoglobulin M antibody capture enzyme-linked immunosorbent assay for diagnosis of St. Louis encephalitis. *J. Clin. Microbiol.* 20:784–90 (1984).

Morita, K.; Tanaka, M.; and Igarashi, A. Rapid identification of dengue virus serotypes by using polymerase chain reaction. *J. Clin. Microbiol.* 29:2107–10 (1991).

Olson, J. G.; Scott, T. W.; Lorenz, L. H.; and Hubbard, J. L. Enzyme immunoassay for detection of antibodies against eastern equine encephalomyelitis virus in sentinel chickens. *J. Clin. Microbiol.* 29:1457–61 (1991).

Peters, C. J., and Dalrymple, J. M. Alphaviruses. In Fields, B. N., and Knipe, D. M., eds., *Fields Virology,* pp. 713–61. New York: Raven, 1990.

Roehrig, J. T. Togaviruses. In van Regenmortel, M. H. V., and Neurath, A. R., eds., *Immunochemistry of Viruses.* Vol. 2, *The Basis for Serodiagnosis and Vaccines,* pp. 217–78. Amsterdam: Elsevier, 1990.

———. The use of monoclonal antibodies in studies of the structural proteins of togaviruses and flaviviruses. In Schlesinger, S., and Schlesinger, M. J., eds., *The Togaviridae and Flaviridae,* pp. 251–78. New York: Plenum, 1986.

Roehrig, J. T.; Corser, J. A.; and Schesinger, M. J. Isolation and characterization of hybridoma cell lines producing monoclonal antibodies against the structural proteins of Sindbis virus. *Virology* 101:41–49 (1980).

Saluzzo, J. F.; Monath, T. P.; Cornet, M.; Deubel, V.; and Digoutte, J. P. Comparison de différentes techniques pour la détection du virus de la fièvre jaune dans les prélèvements humains et les lots de moustiques: Intérêt d'une méthode rapide de diagnostic par ELISA. *Ann. Inst. Pasteur Virol.* 136:115 (1985).

Theiler, M., and Downs, W. G. *The Arthropod-Borne Viruses of Vertebrates: An Account of the Rockefeller Foundation Virus Program,* pp. 1951–70. New Haven: Yale University Press, 1973.

Thompson, W. H., and Calisher, C. H., eds. *California Serogroup Viruses.* New York: Alan R. Liss, 1983.

Tsai, T. F. Abroviruses. In Rose, N. R.; DeMacario E. C.; Fahey, J. L.; Friedman, H.; and Penn, E. G., eds., *Manual of Clinical Laboratory Immunology,* 4th ed., pp. 606–18. Washington, D.C.: American Society for Microbiology, 1992.

———. Arboviral infections in the United States. *Infectious Disease Clinics of North America,* vol. 5, pp. 73–102 (1991).

Tsai, T. F.; Bolin, R. A.; Montoya, M.; Bailey, R. E.; Francy, D. B.; Jozan, M.; and Roehrig, J. T. Detection of St. Louis encephalitis virus antigen in mosquitoes by capture enzyme immunoassay. *J. Clin. Microbiol.* 25:370–76 (1987).

Xiao, S.-Y.; Chu, Y.-K.; Knauert, F. K.; Lofts, R.; Dalrymple, J. M.; and LeDuc, J. W. Comparison of hantavirus isolates using a genus-reactive primer pair polymerase chain reaction. *J. Gen. Virol.* 73:567–73 (1992).

PART SIX

CELL CULTURE

28

CELL CULTURE PREPARATION

V. F. CHAN AND G. D. HSIUNG

The technique of culturing animal cells in monolayers on glass or plastic has provided the virologist with an important and versatile tool for isolating viruses of medical and public health importance. Cell culture systems have been the mainstay in virus detection. The more the number and variety of cell culture systems used, the greater the chance of isolating viral pathogens and potentially novel agents. Cost effectiveness, however, limits the number and kind of cell culture systems maintained by individual virology laboratories. It is therefore important to choose the most sensitive cell culture system(s) to provide the most rapid diagnosis with the least amount of technical labor.

CELL CULTURES COMMONLY USED IN VIROLOGY LABORATORIES

For diagnostic virology laboratories, the question to address is which of the many cell culture systems would be the most practical to maintain for routine diagnostic work. Factors that help in decision making include individual preferences based on experience, the availability of animal tissues for primary cell culture (cells obtained directly from an animal, embryo, or placenta, and so on), and capacity to maintain different cell lines, each having its own growth requirements. In addition, diploid cell lines may be subcultured up to 50 passages only, whereas heteroploid continuous cell lines may be subcultured indefinitely.

In general, in a diagnostic laboratory, a satisfactory combination of cell

ISBN 0-300-05845-4.

Table 28.1 Cell Cultures Available for Virus Isolation

Cell Culture	Common Name	Tissue Origin	Useful for Isolation of:
Primary	HEK	Human embryo kidney	Adenovirus, HSV, entero- (polio-, cox-, echo-)
	MK	Monkey kidney[a]	Entero-, rhino-, reo-, influenza, parainfluenza, mumps, measles
	RK	Rabbit kidney	HSV
	GPE	Guinea pig embryo	HSV, certain coxsackie A
	CE	Chicken embryo	HSV-2, certain arboviruses
Passaged diploid, cell strains	HDF, WI-38, MRC 5	Human diploid fibroblast (embryonic lung or newborn foreskin)	CMV, VZV, HSV, entero- (polio-, cox-, echo-), rhino-, RSV, mumps
Passaged cell lines from human cancer cells	H9[b]	Human T-lymphocytes	HIV
	Hep2	Human larynx carcinoma	RSV, adenovirus, HSV, pox-, polio-, coxsackie B, certain rhinoviruses
	A549	Human lung carcinoma	RSV, adenovirus, HSV
	HeLa	Human cervix carcinoma	Rhinovirus, adenovirus, RSV, HSV
	RD	Human rhabdomyosarcoma	Certain echo-, coxsackie A, poliovirus
Passaged cell lines derived from kidney cell cultures	BGMK	Buffalo green monkey kidney	Polio-, coxsackie B
	Vero	African green monkey kidney	Rubella, arbovirus, RSV, measles, echovirus, HSV
	RK-13	Rabbit kidney	Rubella
	BHK-21	Baby hamster kidney	Rubella, vaccinia, arbovirus

a. rhesus, cynomulgus, African green, etc.
b. H9 = a human T-lymphocyte cell line

cultures to be maintained consists of a primary monkey kidney cell culture, a human diploid cell strain (cells maintain diploid markers derived from human fetal tissue), and/or a human cell line (HeLa or Hep-2) having heteroploid markers derived from human cancer tissue. If available, primary human kidney or human amnion may also be used. Table 28.1 lists those cell culture systems useful in the isolation of viruses associated with human diseases. Most, if not all, of these cell cultures are available commercially.

GENERAL CRITERIA FOR RECOGNITION OF VIRAL REPLICATION IN CELL CULTURE

Many viruses induce visible cytopathic effect in cell cultures. Cytopathic effect is caused by the death of cells by viral infection and is identified either by the cells rounding up and detaching from the plastic or glass surface, or by cells fusing and forming syncytia. Distinctive CPE induced by certain viruses in selective cell culture systems can provide a presumptive identification of the virus isolate. In some instances, CPE is not evident, but if the cells are overlaid with a solid or semisolid medium, the virus will cause plaque formation—a zone of cell death caused by a sequential viral infection of the surrounding cells. Other techniques used to detect virus when no gross CPE is visible include: interference (used for rubella virus, see chapter 17); hemadsorption (used for influenza and parainfluenza viruses, see chapter 15); hemagglutination (used for influenza and arboviruses, see chapters 9, 15, and 27); cytochemical staining (used for herpesviruses, measles, and rabies, see chapters 11 and 19); immunoassays, such as immunofluorescent and immunoperoxidase staining (see chapter 7); enzyme immunoassay or radioimmunoassay, see chapter 8); and most recently, nucleic acid hybridization procedures (see chapter 12).

CPE that is not virus induced (false-positive) could be due to the toxic effects of clinical specimens—notably blood, urine, stool, or bacteria toxin. Although true virus-induced CPE can be carried through serial passages, toxic effects are not. Other conditions that might be confused with virus isolation from clinical specimen are the presence of endogenous viral contaminants—particularly the simian viruses in primary monkey kidney cell cultures, avian viruses in chick embryo cell cultures, guinea pig viruses in GPE cell cultures, and mycoplasma in passaged cell lines. For further information, see the next chapter (chapter 29).

PREPARATION OF PRIMARY KIDNEY CELL CULTURE

Observance of strict aseptic techniques cannot be overemphasized. Use a clean laboratory gown, wash hands thoroughly, work under a biologic safety hood, and deliver solutions and media with a mechanical pipette-aid. Large volumes of media can be measured using sterile graduated cylinders.

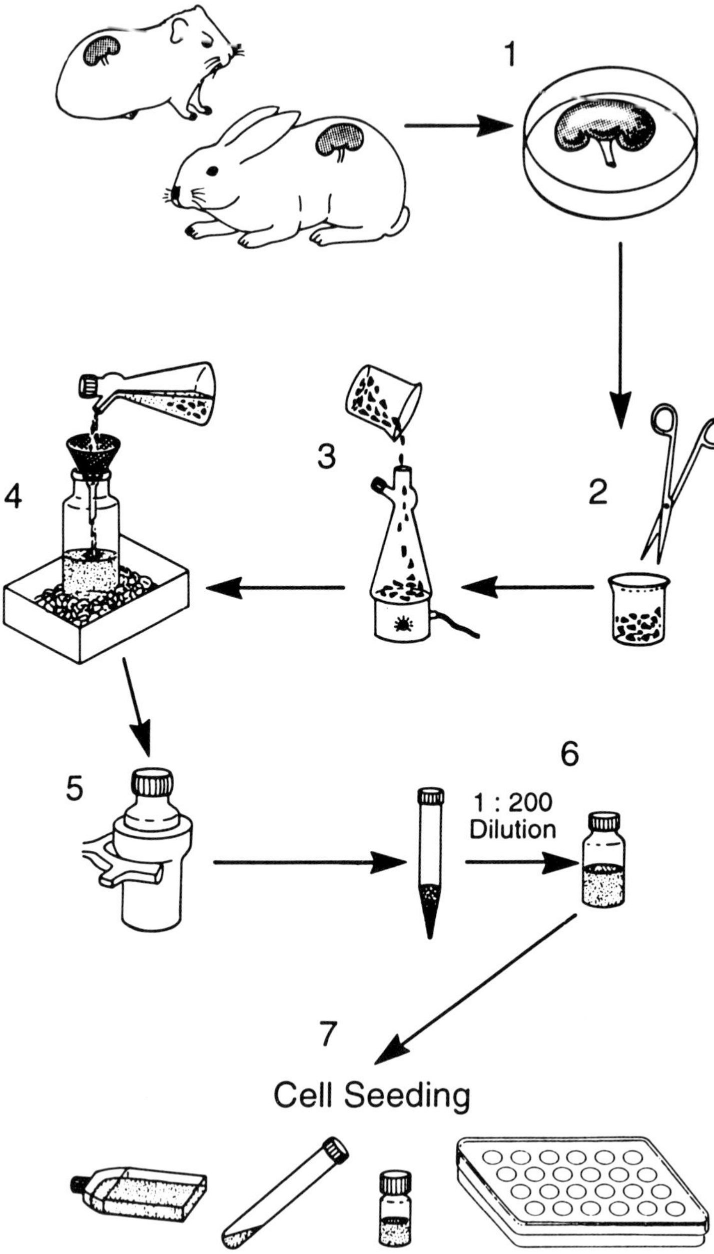

Figure 28.1 Diagram for preparation of primary kidney cell culture showing equipment and procedures (steps 1–7; see text)

Basic techniques in preparing a primary cell culture are briefly described as follows:

Supplies and Equipment

All glassware and plasticware including instruments and magnetic stirring bars must be sterile.

Scissors and forceps
Trypsinizing flask
Magnetic stirrer base
Magnetic stirring bars, Teflon or glass-coated
Sterile gauze squares, 4 × 4 inch
Funnels
Plastic petri dishes of various sizes
Plastic centrifuge tubes 15, 50, 250 ml
Graduated cylinders 50, 250, 500, and 1,000 ml
Pipettes 1, 2, 5, 10 ml
Containers, depending upon need: Dram vial with coverslips, culture tubes, plastic plates (96-, 24-, 12-, 6-, 4-wells, etc.), tissue culture flasks (25 cm^2 [T25], 75 cm^2 [T75], and 150 cm^2 [T150])

Media and Solutions

1. Trypsin (0.25%) in PBS + penicillin (200 u/ml) and streptomycin (200 μ/ml), and other antibiotics, as preferred (see table 30.1).
2. Growth medium, prewarmed to 37°C, usually Eagle's minimum essential medium (MEM) in Hanks' BSS + 10% fetal bovine serum (EHS; for formula, see chapter 30.)
3. Maintenance medium: MEM in Earle's BSS + 2% fetal bovine serum (EES; for formula, see chapter 30).
4. Hanks' BSS.
5. Trypan blue solution for cell counts (see chapter 30).

Procedures for Primary Kidney Cell Culture Preparation (figure 28.1; steps 1–7)

Step 1: Shave the animal (monkey, rabbit, or guinea pig) over the abdomen, just below the rib site. Apply 70% alcohol to shaved skin. Anesthetize and sacrifice the animal; make an incision over abdomen; remove the kidneys aseptically (one side at a time) and put them into a sterile petri dish containing 10 ml of Hanks' BSS.

Step 2: Decapsulate kidneys with sterile scissors and forceps. Remove connective tissues and blood clots. Cut kidneys in half and excise the medulla (whitish portion at the pelvis). Wash twice with Hanks' BSS. Cut cortex into small pieces (about 1 mm^3).

Step 3: Transfer cut tissue to a trypsinizing flask. Wash twice with Hanks' BSS to remove residual blood. Add approximately 25 ml prewarmed trypsin solution to the trypsinizing flask and stir with magnetic stirring bar for 5 minutes. Discard first batch of dispersed cortical kidney cells. Add approximately 25–30 ml fresh prewarmed trypsin solution and stir for 15 minutes.

Step 4: Filter the cell suspension through two or three layers of sterile gauze into a 250 ml centrifuge tube containing approximately 10 ml of growth medium. Keep cell suspension in an ice bath or refrigerator. Repeat steps 3 and 4 until tissues are exhausted. Alternatively, trypsinize overnight at 4°C with 150 ml of trypsin (for two kidneys) at a rate just short of bubbling. The following morning, filter the cell suspension into a 250 ml centrifuge tube and proceed as follows:

Step 5: Centrifuge the cell suspension at 750 rpm for 15 minutes at 4°C. Discard supernatant and resuspend packed cells in growth medium immediately, using approximately 10–15 ml for each centrifuge tube of packed cells. Shake well or resuspend cells by alternate suction and expulsion through a pipette attached to a mechanical pipet-aid.

Step 6: Transfer cell suspension into a 15-ml graduated centrifuge tube and recentrifuge at 750 rpm for 10 minutes; discard supernatant; resuspend packed cells with growth medium at 1:200 dilution for rabbit kidney cells, 1:250 dilution for guinea pig kidney cells, and 1:300 for monkey kidney cells.

Step 7: Seed cell suspension by means of an automatic pipette as follows:

Volume	*Container*
1.0 ml	Dram-shell vial or culture tube
	Plastic plates
0.2 ml/well	96-well
1.0 ml/well	24- or 4-well (16 mm diameter)
2.0 ml/well	12-well
5.0 ml/well	6-well
	Plastic flasks
2–5 ml	T25
10–20 ml	T75
25–50 ml	T150

Note: For more accurate cell counts in the suspension, perform an actual cell count as follows:

1. Add 0.1 ml trypan blue solution (see chapter 30) to 0.9 ml of the cell suspension at step 6 above. Mix well and then fill the chamber of a hemacytometer.
2. Count unstained (viable) cells (stained cells are dead) in 10 squares of the two chambers (5 large squares from each chamber). Mean number of cells in 10 squares $\times 10^4$ equals total number of cells/ml of the cell suspension.
3. Adjust cell concentration to 3×10^5 cells/ml with growth medium (EHS).

Maintenance of Primary Monkey Kidney Cell Cultures

If the plating efficiency of the seeded cells is good, a confluent monolayer should form 48–72 hours after seeding. Check by examining under a light microscope. Visually, the pH of the medium changes from pH 7.2 (orange-red) to an acidic pH (yellow) with active cell growth. Change the growth medium to the maintenance medium (EES), composed of MEM-Earle's BSS and supplemented with 2% fetal bovine serum plus the antibiotics (as in the growth medium). Monolayers are best when used during the logarithmic phase (fastest rate) of growth—about three to four days after the seeding time, when the young cells are most sensitive to virus infection.

PREPARATION OF PRIMARY EMBRYO CELL CULTURE

This method can be used for the preparation of avian and mammalian embryo cell cultures. The concentration of trypsin solution, however, should be adjusted according to the animal species used. For example, guinea pig embryo cells are extremely sensitive to trypsin digestion, therefore a 0.005% trypsin is used and the cells should be removed from the trypsin solution as soon as possible. Otherwise, the cells will disintegrate. For avian embryos, a 0.25% trypsin solution is satisfactory since avian embryo cells are quite resistant to trypsin digestion.

Procedures for Embryo Cell Preparation

Step 1: Select chicken embryos 9 to 10 days old, duck embryos 12 to 14 days old, or guinea pig embryos 30–35 days' gestation.

Step 2: Clean egg shells with 70% ethanol. Open shells over the air sac, remove embryos, decapitate, discard head and feet, and place torsos in a sterile container. Wash torsos in cold Hanks' BSS, then mince with scissors. For the guinea pig embryos, shave the belly of the pregnant guinea pig mother, wash with 70% alcohol, and anesthetize with fluorothane. Aseptically remove the embryos from the bicornate uterus and place them in sterile Hanks' BSS. Process as described above.

Step 3: Transfer tissue fragments to a trypsinizing flask and wash three times with cold Hanks' BSS to remove blood as thoroughly as possible. Wash once with prewarmed 0.005% (for GPE) or 0.25% trypsin (for CE). Add 50–100 ml prewarmed appropriate trypsin solution to the tissue fragments in a trypsinizing flask and stir with a magnetic stirring bar for 45 minutes at 37°C (only 5 minutes for GPE cells).

Step 4: Filter cell suspension through six layers of sterile gauze and collect filtrate in a 250-ml centrifuge tube.

Step 5: Centrifuge at 1,000 rpm for 15 minutes. Discard trypsin solution, wash cells twice with cold Hanks' BSS, and recentrifuge at 1,000 rpm for 15 minutes. Discard supernatant. Add 30–40 ml cold Hanks' BSS or growth medium (EHS) to the packed cells and break the clumps by gentle pipetting.

Step 6: Refilter cell suspension and collect in a graduated centrifuge tube. Centrifuge at 1,000 rpm for 15 minutes and discard the supernatant. Resuspend the packed cells in cold EHS containing 10% fetal bovine serum at a 1:10 dilution (1 ml packed cells per 9 ml EHS). The 10% cell suspension can be kept as stock at 4°C for 3–4 days.

Step 7: For seeding, dilute to 1:40 with cold growth medium containing 5%–10% fetal bovine serum (final concentration, 1:400). Seed 25 ml per T75 flask or 1 ml per dram-shell vial or culture tube. Incubate cultures at 35°C for 48–72 hours before use.

SERIAL PASSAGED CELL LINES

The subculturing method that follows can be used for many serially passaged cell lines, such as Hep-2, HeLa, or A549, as well as for secondary (subculture of primary) cell cultures. To store cells, both primary and passaged cell lines can be preserved by the method described below, but the recovery percentage of primary cells is usually not as good as that of passaged cell lines.

Subculture of Hep-2 Cells

1. Discard supernatant fluid from 5-day-old Hep-2 cell culture flask (T75).
2. Add 3.0 ml trypsin-versene (EDTA) (0.25–0.02%) solution (for formula, see chapter 30), prewarmed at 37°C, to each T75 cell culture flask.

3. Incubate at 37°C for 30–60 seconds.
4. Shake flask to dislodge cells, immediately add 5–10 ml growth medium (EHS + 10% FBS).
5. Transfer cell suspension into a 15-ml centrifuge tube; centrifuge at 1,000 rpm for 10 minutes. Discard supernatant.
6. Add 5–10 ml of growth medium and break the cell clumps by gentle pipetting.
7. Prepare a 1:4 dilution of the suspension (for diploid cells, make a 1:2 dilution), but if the monolayer cells are to be used in 1–2 days, dilute the cell suspension only 1:2.
8. Seed flasks with 10 ml, tubes with 1 ml, and incubate at 37°C.
9. At 3- to 4-day intervals, replace fluid with Eagle's maintenance medium with 2% fetal bovine serum (EES, chapter 30).

Preservation of Cells (Manually)

1. To 0.5 ml packed Hep-2 cells, add cooled medium consisting of 8.5 ml of EHS without antibiotics, plus 1 ml of filtered sterile dimethylsulfoxide (DMSO). Final cell concentration should at least be around 1.0–1.5 $\times 10^6$ cells per ml. Distribute into 1-ml freezer vials, 1 ml per vial.
2. Pack the tubes in an insulated box.
3. Cool cells in box slowly in 3 steps:
   a. 4°C for 2 hours
   b. −20°C for 2 hours or until fluid is frozen
   c. −70°C for long-term storage either in a regular −70°C freezer or, preferably, in liquid nitrogen at −196°C. For the latter, sealed freezer vials are necessary (commercially available).
4. Steps 2 and 3 can be simplified if a Nitrofreezer is available. Place vials in the holder (cane) that sits in the liquid N_2 vapor phase for 24 hours. Cell will slowly freeze automatically. After 24 hours, move the vial holder to the liquid N_2 portion.
5. To revive cells from a frozen state:
   a. Defrost cells in suspension rapidly in a 37°C water bath.
   b. Wash outside of vial with 70% ethanol.
   c. Remove DMSO medium from cell suspension by centrifugation at 1,000 rpm for 10 minutes.
   d. Add slowly Eagle's growth medium with 10% fetal bovine serum (EHS) to the original 5% cell suspension so that final cell concentration is 0.2%.
   e. Seed 10–20 ml of this suspension per T75 flask and incubate at 37°C.
   f. Replace medium with fresh EHS 24 hours after incubation at 37°C to remove any additional DMSO that may still be present in the cell suspension.

29

ENDOGENOUS VIRAL AND MYCOPLASMAL CONTAMINANTS IN CELL CULTURES

G. D. HSIUNG AND V. F. CHAN

From time to time, primary cell cultures or passaged cell lines become contaminated with adventitious viruses or mycoplasma. Recognizing their presence is important in diagnostic virology because the interpretation of results may be affected by these contaminants. Endogenous viruses may be manifest as visible CPE that mimics the changes induced by viruses in specimens (false-positive), or they may chronically remain and silently infect the cells, thus inhibiting the replication of a pathogen in specimens (false-negative). Furthermore, endogenous animal viruses, especially the simian B virus, may be hazardous and pose a real threat to handlers of these cell cultures.

This chapter provides a brief discussion of certain characteristics of the more common endogenous viral contaminants and mycoplasmal organisms found in animal cell cultures. It is designed to provide information for the recognition of the presence of some of these viral and mycoplasmal agents.

Uninoculated primary cell cultures made from simian and nonsimian tissues frequently contain viral endogenous agents. Because of these viruses, it is desirable to use germ-free animals or eggs from pathogen-free chicken flocks for cell culture. Endogenous simian viruses that contaminate primary cell cultures derived from monkeys appear to be the most numerous and constitute a variety of virus types within each virus group. Thus, the following description of simian viruses can be used as a guide for recognizing endogenous viruses present in cell cultures derived from other animal species.

ISBN 0-300-05845-4.

SIMIAN VIRUSES

The widespread use of cell cultures prepared from monkey kidneys in the 1960s and 1970s resulted in numerous reports of endogenous simian viruses recoverable from uninoculated cultures. These viruses, including SV_{40}, SV_5, and foamy viruses, are relatively common, and their presence can create problems in the diagnosis of human infection. The simian viruses representing those that have been reported in each virus family are listed in table 29.1.

In the following sections, the properties of representative simian viruses from each major virus family are presented as a guide for the recognition of these agents in primary monkey cell cultures. Once an endogenous virus is recognized, the cultures are considered unsuitable for the isolation of a virus from clinical specimens and should be disposed of properly.

Simian DNA Viruses

SIMIAN PAPOVAVIRUS

SV_{40}, or the vacuolating virus of monkeys, is one of the most common latent viruses of rhesus monkeys. Cytoplasmic vacuoles characteristic of SV_{40} can be recognized readily when African green monkey, patas monkey, or baboon kidney cell cultures are infected with this virus type (figure 29.1, *B1)*. Difficulties have been encountered with SV_{40}-infected rhesus monkey cells since CPE is not apparent in them. With IF staining, however, or in H&E-stained preparations, basophilic intranuclear inclusions can easily be recognized in all infected monkey cells (figure 29.1, *B2)*. Because of its ability to induce tumors in experimental animals, SV_{40} has attracted the interest of many investigators in cancer research for decades.

SIMIAN ADENOVIRUSES

These viruses are most commonly recovered from rhesus or cynomolgus monkey kidney cultures or from monkey stools. They produce CPE in rhesus monkey kidney cultures that is characterized by the rounding and clumping of the cells (figure 29.1, *C1)*. Basophilic intranuclear inclusions are easily identifiable in preparations stained with H&E (figure 29.1, *C2)*.

SIMIAN HERPESVIRUS

The B virus of monkeys is a highly dangerous pathogen for humans, although it is of relatively low pathogenicity for rhesus monkeys. It can be isolated from the throat, the central nervous system, and the kidney tissues of rhesus monkeys even though no disease is observed in the monkeys. The virus induces a large, rounded cluster of degenerated cells in tissue culture (figure 29.1, *D1)* and large, clear circular plaques under agar overlay. Monkey cytomegalovirus is another member of the simian herpesvirus group.

Table 29.1 Major Families of Simian Viruses

Nucleic Acid Type	Size: Range	Size: nm	Ether[a] Sensitivity	Virus Family	Representative Simian Viruses[b]
DNA	Small	<50	Res.	Papovaviridae	SV_{40}
	Medium	50–100	Res.	Adenoviridae	SV_{1}, SV_{11}, SV_{15}, SV_{17}, SV_{20}, SV_{23}, SV_{25}, SV_{27}, SV_{30-34}, SV_{36-39}, SA_{7}, SA_{17}, SA_{18}
	Large	>100	Sens.	Herpesviridae	B, M, SA_{8}, SA_{6}, CMV, SA_{15}
			Res.	Poxviridae	Yaba, Monkeypox
RNA	Small	<50	Res.	Picornaviridae	SV_{2}, SV_{6}, SV_{16}, SV_{18}, SV_{26}, SV_{29}, SV_{35}, SV_{42-49}, SA_{4}
	Medium	50–100	Res.	Reoviridae	SV_{12}, SV_{59}, SA_{3}, SA_{16}, SA_{11} (rotavirus)
	Large	>100	Sens.	Paramyxoviridae	SV_{5}, SV_{41}, Measles, SA_{1}
				Retroviridae	Foamy virus types 1–7

Modified from G. D. Hsiung, *Ann. N.Y. Acad. Sci.* 162:485 (1969).

a. Res. resistant, Sens. sensitive

b. SV simian virus, SA simian agent

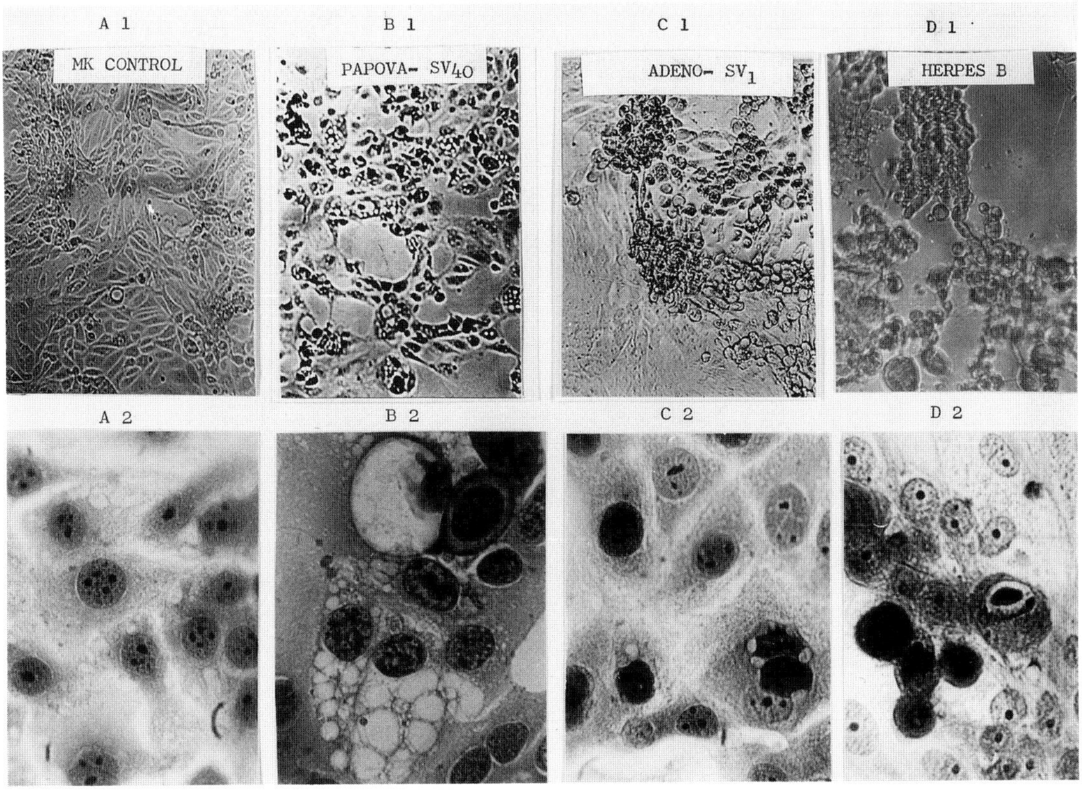

Figure 29.1 Simian DNA virus-induced cellular changes. Upper row: Simian virus–induced CPE in monkey kidney cell cultures (100×). Lower row: H&E-stained preparations of monkey kidney cells (400×) (G. D. Hsiung, *Bact. Rev.* 32:192 [1968]). *(A1* and *A2)*. Uninfected RhMK cell cultures; *(B1)* SV_{40} papovavirus–induced CPE in MK cells; *(B2)* SV_{40} papovavirus-induced intranuclear inclusions; *(C1)* SV_1 adenovirus–induced CPE in RhMK cells; *(C2)* SV_1 adenovirus–induced intranuclear inclusions; *(D1)* Herpes B virus-induced CPE in RhMK cells; *(D2)* Monkey cytomegalovirus–induced intranuclear inclusions in rhesus monkey cells

Typical type A intranuclear inclusions are easily recognizable (figure 29.1, *D2)*. Although isolation of monkey CMV does not occur often, high titers of neutralizing antibody to African green monkey CMV is frequently detected in sera collected from green monkeys.

Simian RNA Viruses

SIMIAN ENTEROVIRUSES

In culture, the simian enteroviruses induce CPE characterized by small pleomorphic cells (figure 29.2, *A)* similar to those induced by human enteroviruses in monkey cell cultures.

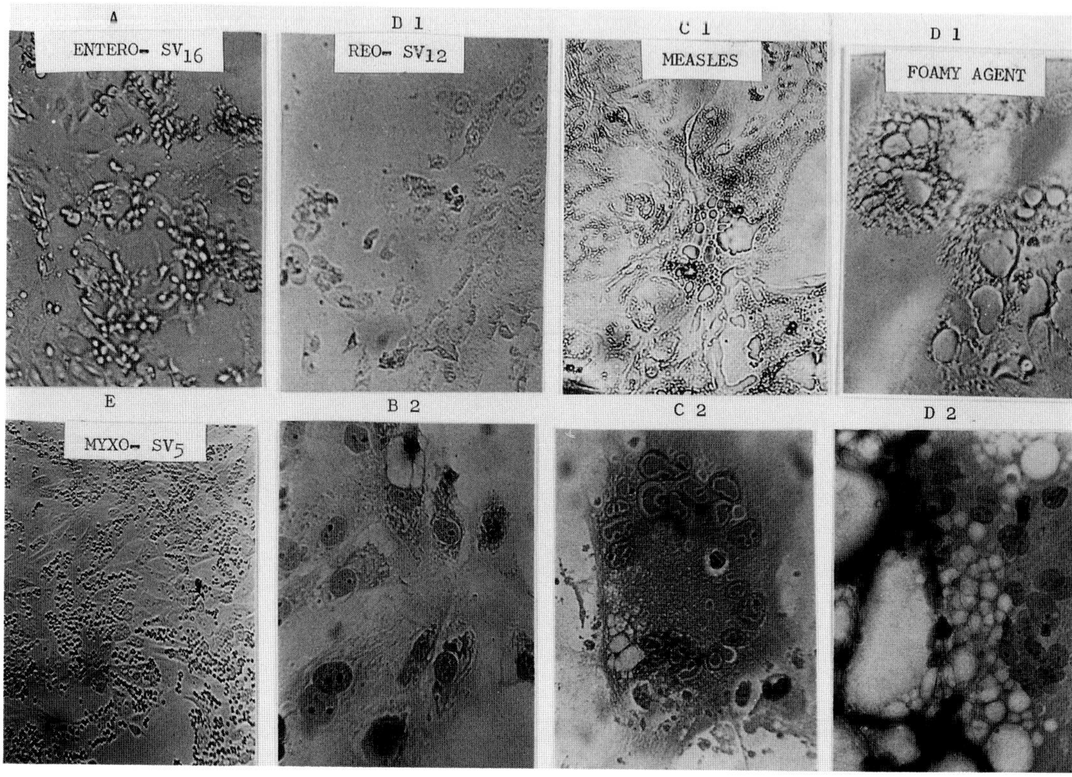

Figure 29.2. Simian RNA virus-induced cellular changes. Upper row: Simian virus-induced CPE in monkey kidney cell cultures (100×). Lower row: H&E-stained preparations of infected monkey kidney cells (400×), except *E*, 100×) (G. D. Hsiung, *Bact. Rev.* 32:193 [1968]). *(A)* Monkey enterovirus SV_{16}–induced CPE in RhMK cells; *(B1* and *B2)* Monkey reovirus SV_{12}–infected RhMK cells; note intracytoplasmic inclusions in *B2; (C1* and *C2)* Monkey measles virus–infected AGMK cells; note intranuclear inclusions in the nuclei of a syncytial cell in *C2; (D1* and *D2)* Foamy virus–infected RhMK cells; note absence of intranuclear inclusions in *D2; (E)* Monkey parainfluenza virus SV_5–infected RhMK cells; note hemadsorption of guinea pig erythrocytes

SIMIAN REOVIRUSES

Reovirus-infected cells are very granular (figure 29.2, *B1).* Distinct intracytoplasmic, eosinophilic inclusions are apparent in stained preparations of infected rhesus cultures (figure 29.2, *B2).* SA-11, a simian rotavirus, has been used to produce antiserum that is used in the ELISA assay for the detection of human rotavirus. The latter does not replicate well in cell cultures.

SIMIAN PARAINFLUENZA VIRUS

SV_5 is the most common contaminant of kidney cell cultures derived from newly arrived rhesus, cynomolgus, green, and patas monkeys. SV_{41}—serologically related but not identical to SV_5—is also commonly found in green monkey kidney cell cultures. No CPE can be seen in infected cultures, but RBC from various animal species readily adsorb onto SV_5-infected cell sheets (figure 29.2*E*). Thus, the hemadsorption method is commonly used to detect the presence of these viruses.

Monkey measles virus, originally known as MINIA (monkey intranuclear-inclusion agent), produces multinucleated syncytial cells in infected cultures (figure 29.2, *C1*). In stained preparations, eosinophilic inclusions located in the nuclei as well as in the cytoplasm can be readily seen (figure 29.2, *C2*).

SIMIAN FOAMY VIRUSES

These viruses often appear in aged monkey kidney cell cultures, especially in those from rhesus monkey cells. Cytopathic effect takes the form of large vacuoles (figure 29.2, *D1*), which can be distinguished morphologically from those produced by the vacuolating virus, SV_{40} (figure 29.1, *B1*). Multinucleated syncytial cells without inclusions induced by foamy viruses can be seen in stained cultures (figure 29.2, *D2*) and can easily be distinguished from those measles virus–infected cells in which intranuclear and intracytoplasmic inclusions are present (figure 29.2, *C2*).

Mixed Infections

Occasionally, a single lot of cell cultures derived from a monkey is infected with two or more virus types. Such mixed infections are recognized either by light or electron microscopy. Mixed infections of SV_{40} and measles virus are illustrated in figure 29.3. Two distinct types of intranuclear inclusions can be seen in the multinucleated giant cell, in which the eosinophilic inclusions are due to measles virus and the basophilic inclusions are due to SV_{40}. Single cells doubly infected with two viruses, for example SV_{40} and measles (figure 29.3, *B*) or SV_{40} and reovirus (figure 29.4) have been observed. Doubly infected cells are easily recognized under EM. An example is shown in figure 29.5, which shows SV_5 nucleocapsids in the cytoplasm and SV_{40} virions in the nucleus of the same cell in cell culture.

In general practice, endogenous virus infections in primary cell cultures are often not recognized. Viruses isolated from clinical specimens or virus stocks, and virus vaccines made from such infected cultures, would undoubtedly be contaminated with these endogenous agents. Precautions should therefore be taken whenever primary cell cultures, especially those derived from monkeys, are used. It is always advisable to use uninoculated control cultures in parallel for comparison.

VIRUSES OF NONPRIMATES

Latent virus infections are also found in the tissues, serum, and other animal products of a variety of nonprimate species. These include herpesvirus from

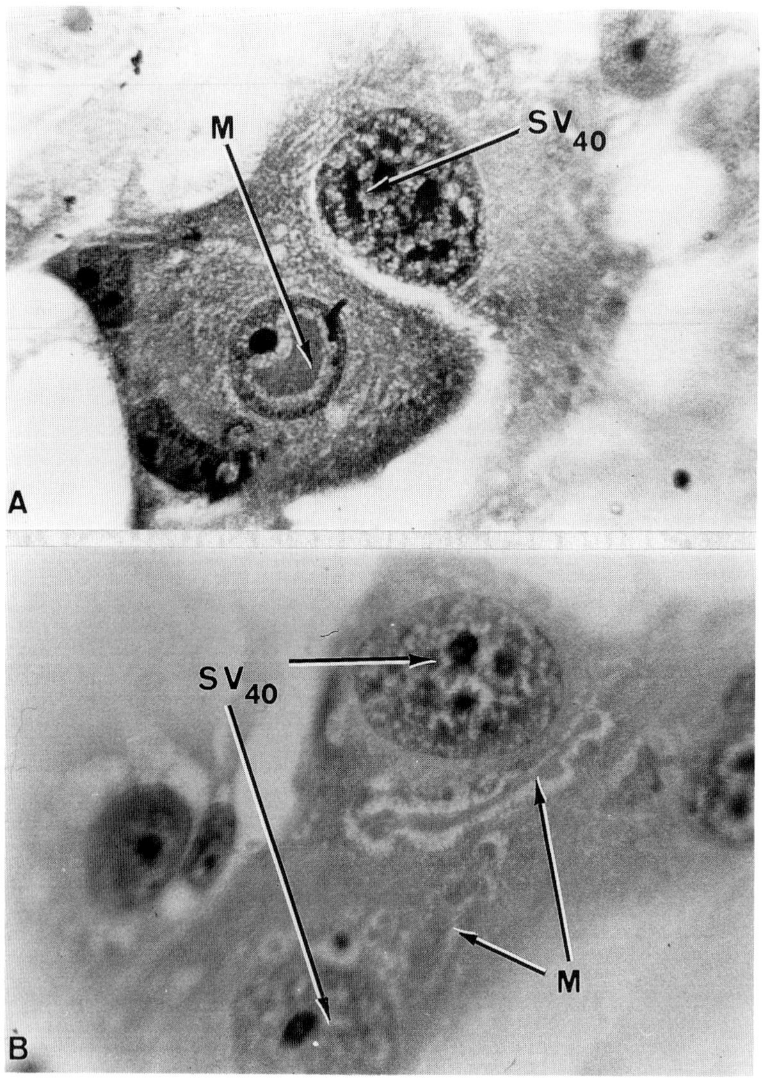

Figure 29.3 Mixed infection with measles virus and SV_{40} in a lot of AGMK cell culture (H&E, 970×) (G. D. Hsiung et al., *Proc. Soc. Exp. Bio. and Med.* 121:562 [1966]). *(A)* Measles virus (M) and SV_{40} intranuclear inclusions in two separate cells; *(B)* Measles virus (M) intracytoplasmic inclusion and SV_{40} intranuclear inclusions in the same cell

calf serum or in kidney tissues of guinea pigs, rabbits, rats, horses, dogs, and many other species; adenovirus from swine, bovine, and avian tissues; and small DNA viruses—the parvoviruses from mice, rats, and hamsters. Furthermore, certain viral agents have been recovered from animals previously considered germ-free. Thus, one cannot be assured that the organs of any animal species are virus-free. These endogenous viruses have caused many

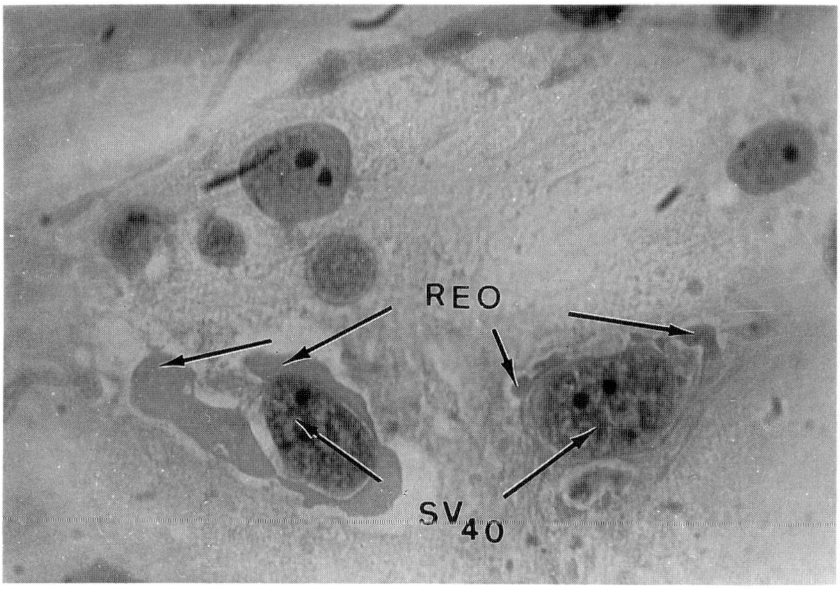

Figure 29.4 Mixed infection with reovirus and SV_{40} in a batch of AGMK cell culture (H&E, 970×) (G. D. Hsiung, *Bact. Rev.* 32:185 [1968]). Intracytoplasmic, eosinophilic inclusion at perinuclear space of reovirus (Reo) and basophilic intranuclear inclusion of SV_{40} are seen in the same cells

problems for investigators in terms of interpreting results when an isolate is obtained from a contaminated cell culture inoculated with clinical specimen. Figure 29.6 shows CPE induced in A549 cell cultures by a bovine herpesvirus (infectious bovine rhinotracheitis virus, or IBRV), which originated from a batch of contaminated bovine serum. Figure 29.7 illustrates an electron micrograph of IBRV in A549 cells; typical herpesvirus virions are noted.

Endogenous virus contaminants most commonly encountered in chicken embryo tissue culture are those that are egg-transmitted. These include avian retroviruses and Newcastle disease virus. Detection of avian leukosis virus antigen can be made by the ELISA test in cultured samples.

The chicken embryo lethal orphan (CELO) virus, an adenovirus, is one of the more easily recognized endogenous contaminants of avian tissues. Once activated, the virus produces death in developing chicken embryos. If such an agent is recognized in avian tissues intended for cell culture, it is recommended that another source for embryonated eggs be found.

Table 29.2 lists some reported cases of virus isolation from cell cultures that were prepared from the tissues of apparently healthy animals. These include both the DNA and RNA viruses; herpesviruses appear to be the most notorious. Any specimen or stock virus grown in these latently infected cultures is thus contaminated with an endogenous virus. Investigators should always keep in mind the possibility of the presence of an endogenous

Figure 29.5 Electron micrograph showing a mixed infection with SV_5 and SV_{40} in an AGMK cell (C. K. Y. Fong and G. D. Hsiung, *Virology and Rickettsiology,* 1978, vol. 1, pt. 1, 104.) SV_{40} virions are seen in the nucleus (*N*) and SV_5 helical nucleocapsids in the cytoplasm *(Cy)* (48,000×)

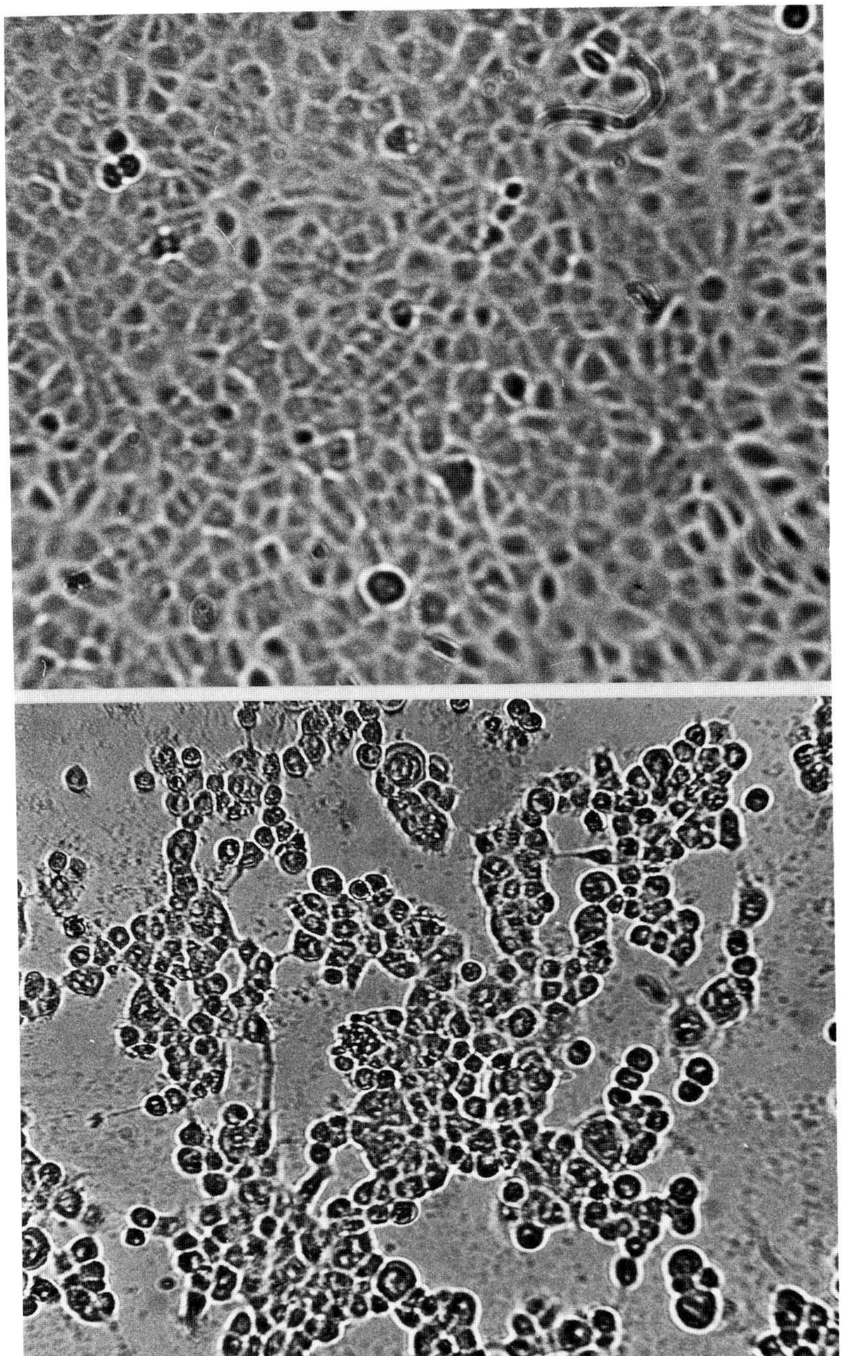

Figure 29.6 Infectious bovine rhinotracheitis virus-induced CPE in A549 cell culture (100×). *Top,* uninfected A549 cells; *bottom,* IBRV-infected cells.

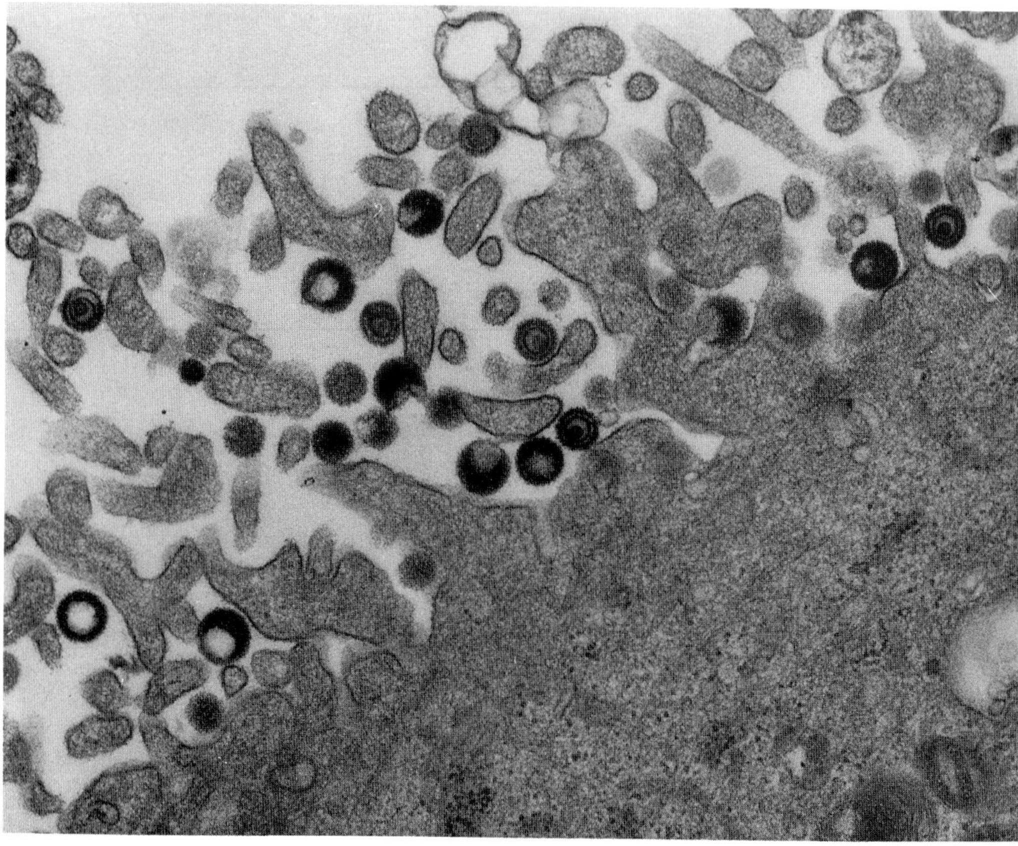

Figure 29.7. An electron micrograph showing infectious bovine rhinotracheitis herpes virions in A549 cell culture (37,500×). (Modified from C. K. Y. Fong and M. L. Landry, *J. Clin. Microbiol.* 30:1612 [1992])

virus in the test system, especially when primary cell cultures are used. Methods for recognition of these viruses are similar to those described for the simian virus groups.

MYCOPLASMA

Aside from the endogenous viruses, the most insidious chronic contaminants of cell lines are the mycoplasmas. These organisms live extracellularly and do not kill the host cells. Mycoplasma-infected cell cultures may not give rise to any overt gross morphologic changes. If any, only minimal abnormalities are observed, such as the tendency of cells to round up or simply appear granular. Therefore, the most serious problem of the chronic contamination of cell

Table 29.2 Examples of Adventitious Viruses in Cell Cultures Prepared from Tissues of Domestic and Laboratory Animals and Avian Species

Animal Species	Virus Types
Domestic animals	
Cattle	Adenovirus, herpesvirus, reovirus, parainfluenza virus, picornavirus
Horse	Herpesvirus
Pig	Adenovirus, picornavirus, parvovirus
Dog	Adenovirus, herpesvirus, parainfluenza virus, reovirus
Laboratory animals	
Rabbit	Poxvirus, papovavirus, herpesvirus
Guinea pig	Herpesvirus, parainfluenza virus, retrovirus
Rat	Parvovirus, herpesvirus, coronavirus
Hamster	Parvovirus
Mouse	Parvovirus, papovavirus, herpesvirus, retrovirus, reovirus
Avian species	
Chicken	Adenovirus, herpesvirus, myxovirus, retrovirus
Duck	Herpesvirus

lines concerns the mycoplasmas and the L forms of other organisms—organisms that are present in many animal tissues. Since the screening of bovine sera for viral and mycoplasmal contaminations is now routine and mouth pipetting has been eliminated, the contamination of cell lines with mycoplasma has been greatly reduced.

Testing for Mycoplasma Contamination

Periodically in cell culture work, cell lines must be tested for mycoplasma contamination. Several procedures are available, but DNA fluorochrome staining is sensitive, simple, and easy to perform, and the results are obtained in an hour.

DNA FLUOROCHROME (HOECHST) STAINING

Hoechst stain is a DNA fluorochrome that specifically binds to nuclear but not mitochondrial DNA. It will also stain DNA of mycoplasma, bacteria, yeast, and fungi and cause fluorescence. Mycoplasma-positive cell cultures are distinguishable from the other DNA-containing microorganisms. Briefly, the staining procedure is as follows:

1. Seed cells in shell vials using growth medium without antibiotics and allow to propagate until the cell monolayers are about 50–80% confluent.

2. Without decanting the growth medium, fix the coverslips directly by adding sufficient amount of Carnoy's fixative (1 part glacial acetic acid, 3 parts methanol, freshly prepared) and fix for 2 minutes. Then remove the medium-fixative mixture and discard.
3. Repeat the fixation procedure with 1 ml of Carnoy's fixative, allowing 5 minutes for fixation, then discard. Repeat step 3 for an additional 5 minutes fixation.
4. Air dry the fixed cells on the coverslips for 30 minutes.
5. While the fixed cells are drying, dilute the Hoechst stain as instructed in the kit (Mycoplasma Stain Kit). *Caution*: Handle the stain carefully; it is a *mutagen*.
6. Overlay 1.0 ml of the working dilution of stain onto the fixed cells on coverslips placed in a dry sterile petri dish together with the known positive and negative controls. Allow 10–30 minutes staining at room temperature. Cover the petri dish while staining to prevent drying.
7. Aspirate stain and discard.
8. Wash the stained cells with distilled water allowing 1 minute soaking time per wash. Aspirate washes and discard.
9. Air dry stained cells. Mount on slides with 2 drops of mounting fluid (provided in the kit) with cell layer side up, then add another 1 or 2 drops of mounting fluid on the cell layer and cover it with a clean coverslip.
10. Examine mounted coverslips and control slides using a fluorescence microscope under oil immersion (1,000X) and a UV-IA filter. A negative culture will show only fluorescent nuclei (figure 29.8*A*). Occasionally, micronuclei or nuclear fragments from dead or disrupted cells will fluoresce as spherical bodies, but these are larger and brighter than the mycoplasma fluorescence. A mycoplasma-positive (figure 29.8*B*) cell culture will show fluorescent, pleomorphic, pinpoint dots surrounding the cells, ranging from 0.1 to 0.3 μm in diameter, either aggregated in clusters or rows.

    Discard contaminated cell lines. New cell lines should also be tested before establishment in the laboratory and must be mycoplasma-negative before use.

MYCOPLASMA CULTURING PROCEDURE

1. Spread by streaking 0.1 ml of cell culture medium from tubes or flasks onto special mycoplasma agar (Difco) plates.
2. Incubate plates in a CO_2 incubator at 37°C for 5–7 days.
3. Examine plates under a dissecting microscope. Mycoplasma colonies are usually 0.5 mm in diameter and present as a typical "fried egg" appearance (figure 29.9).

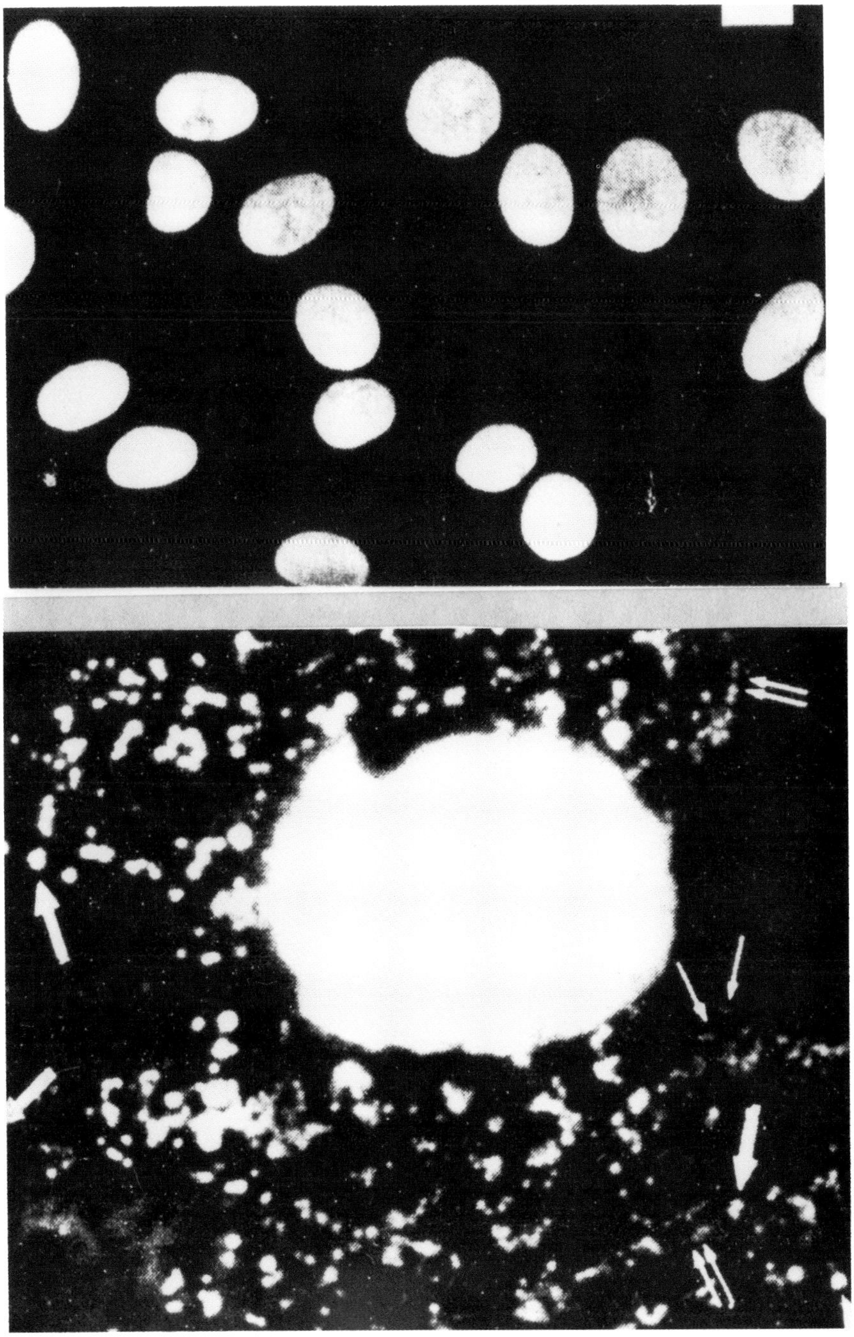

Figure 29.8 Cell cultures prepared with fluorescent Hoechst 33258 stain showing mycoplasma contamination (modified from T. R. Chen, *Exp. Cell Res.* 104:258 [1977]). *(A)* Cell culture control, only bright fluorescent nuclei are seen; *(B)* Cell culture heavily infected with mycoplasmas; arrows show mycoplasmas in doublets and chains scattered around the big cluster of cells

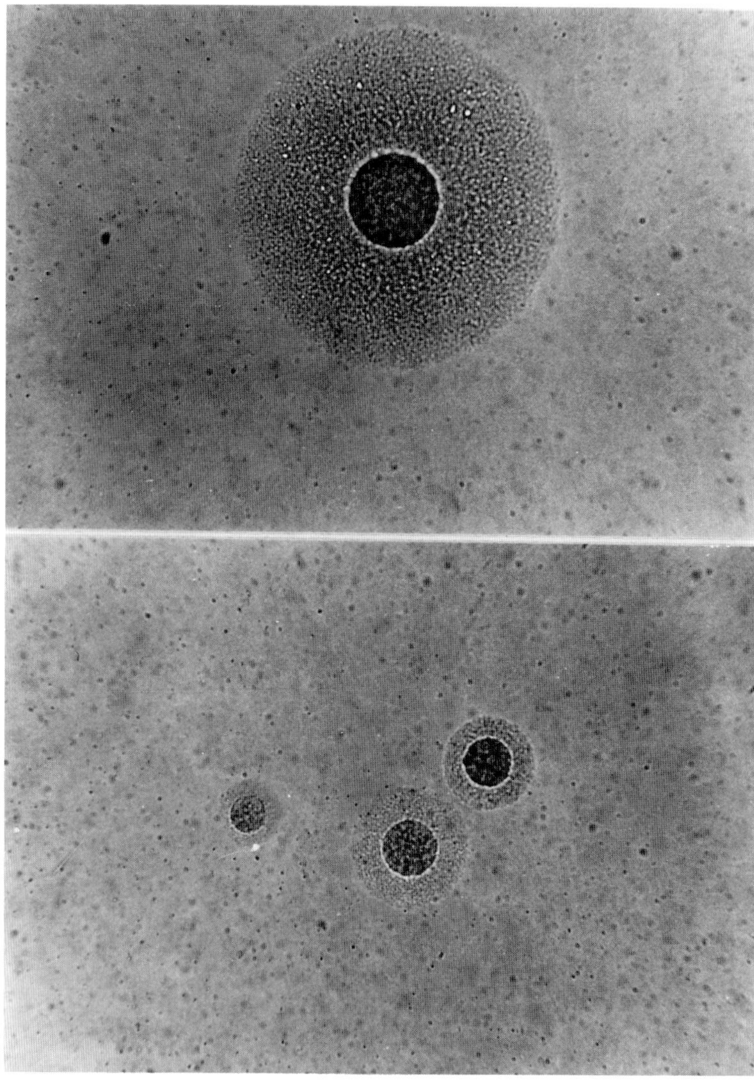

Figure 29.9 Mycoplasma colonies on agar plate

ELECTRON MICROSCOPY

Alternatively, mycoplasmas can be recognized by EM examination of thin sections of cultured cells (figure 29.10).

IMMUNOFLUORESCENT STAINING FOR IDENTIFICATION

IF using polyvalent antiserum against mycoplasmas has also been used. Species identification is also possible by IF staining of infected cells or colonies. Reagents are commercially available.

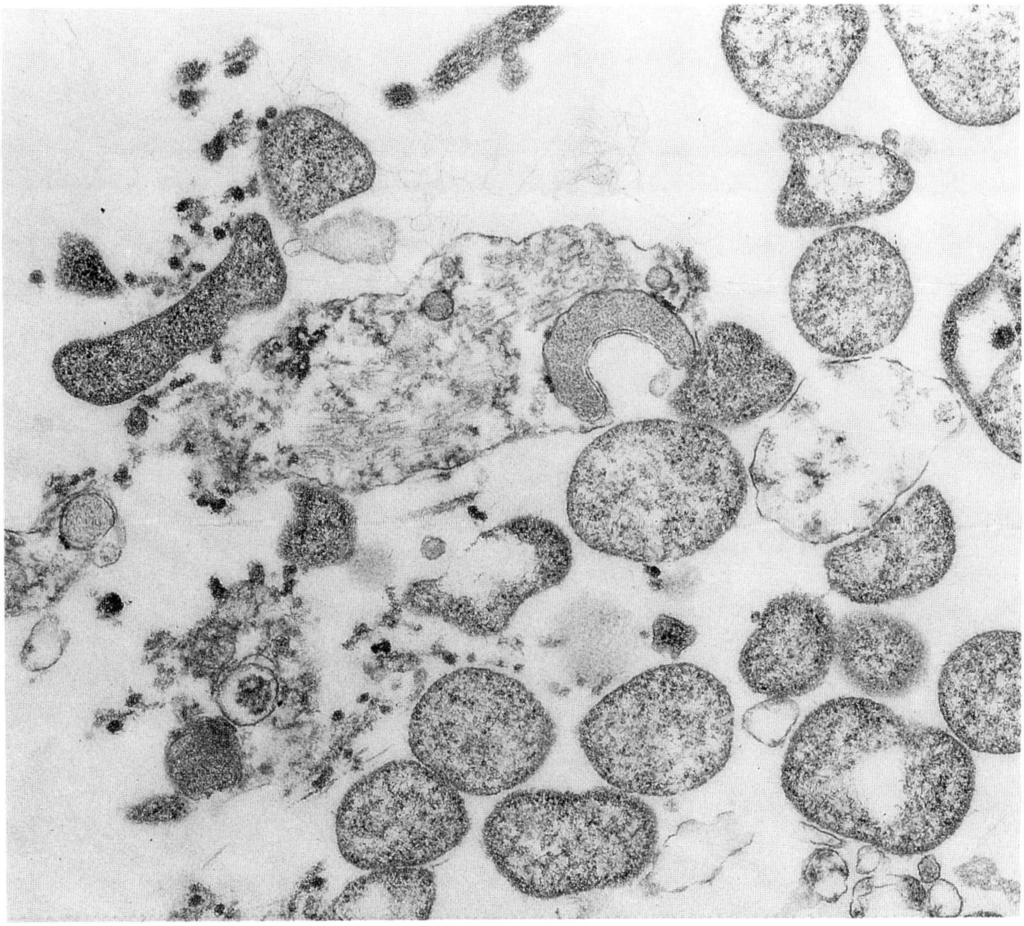

Figure 29.10 An electron micrograph showing a cell culture contaminated with mycoplasma

Control and Elimination of Mycoplasma Contamination

It is better to discard than to attempt to rescue mycoplasma-contaminated cell cultures. However, if the cell line is a valuable strain, elimination and control of mycoplasma contamination can be achieved by washing and treatment with antibiotics.

1. Remove and discard culture fluid from the cells.
2. Wash repeatedly with Hanks' BSS containing kanamycin, 200 μg/ml.
3. Detach cells by trypsin-versene solution (for formula, see chapter 30).
4. Resuspend cells in fresh growth medium (see EHS, chapter 30) containing kanamycin, 100 μg/ml.
5. Subculture at low density in a clean, sterile vessel.

6. Repeat the entire process of washing, antibiotic treatment, and subculturing at low cell density, each time in a clean, sterile vessel, for a period of 3 weeks.
7. Test for mycoplasma contamination as described above.
8. If recovery from contamination is achieved, remove the kanamycin from the maintenance medium. Maintain the cells without kanamycin for several subcultures before the cell culture is considered safe to use. (Note: It is not advisable to incorporate routinely low concentrations of kanamycin to the maintenance medium, because of the possibility of the development of resistant strains of mycoplasmas.)

30

CULTURE MEDIA, REAGENTS, AND SOLUTIONS

V. F. CHAN AND G. D. HSIUNG

Culturing cells in vitro requires conditions as close as possible to the environment existing in vivo. Environmental factors that are critical for growing cells include nutrition, temperature, hydrogen ion concentration (pH), osmotic pressure, and the matrix on which the cells grow. Of these factors, nutritional needs in the form of synthetic media commonly used in a diagnostic virology laboratory will be presented in this chapter.

Rapid progress in the development of synthetic media has been made since Eagle first reported in 1955 the minimum essential components for the growth of mammalian cells in vitro. Eagle's MEM in Hanks' or Earle's BSS is the simplest medium for general use. As the variety and number of cell cultures has increased—both primary and passaged—the number of cell culture media has also increased. Most, if not all, the commonly used media may be purchased from commercial sources. Some companies also provide custom service.

Sterile culture media may be purchased in a form ready for immediate use or for use after the addition of certain components. Many media may be purchased in concentrated form, usually 10X, so that reconstitution can be performed as needed. The latter offers advantages when storage is a problem. Powdered medium for tissue culture is convenient for large laboratories, although it requires additional filtration equipment and further testing for sterility and cell growth. In certain instances, investigators prefer to prepare their own media and solutions. The few described below are simple and can be prepared with minimal laboratory facilities. The preference of antibiotics used in culture media varies from laboratory to laboratory. The amounts listed below for penicillin and streptomycin are given in paren-

ISBN 0-300-05845-4.

Table 30.1 Antibiotic Stock Solutions

Name of Antibiotic	Stock Solution (μg/ml)	Final Concentration for Culture Medium (μg/ml)	Treatment for Clinical Specimens* (μg/ml)
Amphotericin B (Fungizone)	250	1.25	10
Gentamicin	50,000	50	500
Kanamycin	10,000	50	100
Neomycin	10,000	50	100
Penicillin G	200,000 u	200 u	500 u
Streptomicin sulfate	500,000	200	500
Tetracycline	10,000	50	100

*Antibiotic concentrations are for treatment of heavily contaminated clinical specimens like stools or rectal swabs and throat washings or throat swabs.

theses to indicate that they should be added as needed. For additional information, see table 30.1. All water used in the following formulae should be demineralized. Often, when water quality is a problem, it is better to purchase media at a 1X concentration.

PREPARATION OF MEDIA AND OTHER REAGENTS

Depending upon the manufacturer from whom media or reagents are purchased and the supplement(s) added to the media, the upkeep, storage temperature, and shelf lives of different media and reagents may vary.

Liquid Media

As a general guide, all liquid media—such as MEM, Hanks', and Earle's BSS, and 7.5% sodium bicarbonate—are best kept at 4°C and should be protected from exposure to UV light. Unsupplemented (unopened) liquid media can be stored for two years at 15°–30°C, but once opened they are usable within one month if stored at 4°C.

Fetal Bovine Serum

If in lyophilized or unopened liquid form, FBS can be stored indefinitely at −20°C. Once reconstituted or opened, FBS is best stored at 4°C and should be used within three months. Large volumes of liquid FBS should be aliquoted into smaller volumes to minimize contamination and mishandling.

L-Glutamine, Essential Vitamins, and Amino Acids

These supplements are better stored at −20°C for a longer shelf life (2 years). Like the liquid media, once thawed and opened, they should be stored at 4°C and used within one month. It is good practice to add L-glutamine to MEM if it is three to four weeks old at the time of use.

Antimicrobials

In powder form, store antimicrobials at 4°C. In liquid form, they can be stored at or lower than −20°C for extended periods, usually up to one year. Once thawed or reconstituted, antibiotics should be used. Avoid repeated freezing and thawing.

GLASSWARE

The storage of media and reagents in reusable glassware may introduce toxic substances that can kill or prevent the growth of cells in cultures. Therefore, reusable glassware for cell culture quality must be cleaned meticulously. The washing and sterilization procedures for glassware in a cell culture laboratory are as follows.

1. Glassware containing medium must be rinsed immediately with tap water once it is empty. Medium that has dried on the glassware surface will be more difficult to clean. Contaminated glassware should be autoclaved before cleaning.
2. Soak glassware overnight in chlorox (0.5% hypochlorite solution) or special glassware detergent.*
3. Brush manually or by machine as follows: Pipettes are sonicated for 1–1.5 hours in detergent nontoxic to cell cultures. Other glassware is cleaned in a washing machine with glassware detergent for 15 minutes.
4. Rinse thoroughly with 10 changes of tap water followed by three changes of single distilled water (manually). If a washing machine is available, all glassware, including pipettes, is rinsed with demineralized filtered water for a 15-minute cycle.
5. Drain glassware and dry. Cover flasks, bottles, beakers, and graduated cylinders with aluminum foil. Plug pipettes with cotton and sort them by size into canisters. Screw caps for bottles and tubes are packed in heat-resistant containers. Use autoclave indicators on aluminum foil–covered glassware, screw cap boxes, and canisters of pipettes.
6. Sterilize glassware, preferably by dry heat (hot-air oven) at 171°C for 2

*7X detergent has been found nontoxic to cell cultures.

hours. Steam sterilize (15-lb pressure for 30 minutes at 121°C). Clean glassware should not be mixed with dirty glassware during steam sterilization, since toxic substances, if present, may deposit onto clean glassware. In general, screw caps and automatic pipettes with rubber tubing are sterilized by autoclaving at 15-lb pressure for 20 minutes at 121°C.

7. Store sterilized glassware in closed, dust-free cabinets. As an additional precaution against contamination, storage cabinets can be equipped with UV lights.

CULTURE MEDIA AND REAGENTS

Hanks' Balanced Salt Solution (10×)

	Solution A	
	1 liter	5 liters
NaCl	80 g	400 g
KCl	4 g	20 g
$MgSO_4 \cdot 7H_2O$	1 g	5 g
$MgCl_2 \cdot 6H_2O$	1 g	5 g
$CaCl_2$ (anhydrous)	1.4 g	7 g
Glucose	10 g	50 g
Demineralized water	800 ml	4,000 ml
	Solution B	
$Na_2HPO_4 \cdot 2H_2O$	0.6 g	3 g
KH_2PO_4	0.6 g	3 g
Demineralized water	200 ml	1,000 ml

1. Weigh each salt in the order listed. Dissolve in demineralized water.
2. Autoclave solutions A and B separately at 15 lb for 15 minutes.
3. Mix A and B when cool. Pour B into A *slowly*, stirring constantly.
4. Dispense in 100–500 ml amounts in sterile bottles and cap tightly.
5. Store at room temperature or 4°C.

Hanks' Balanced Salt Solution (1×)

	1 liter	5 liters
Demineralized water, sterile	889 ml	4,445 ml
Hanks' BSS (10×)	100 ml	500 ml
$NaHCO_3$ (7.5%)	7 ml	35 ml
Phenol red (0.5%)	4 ml	20 ml
Penicillin 200,000 u/ml	(1 ml)	(5 ml)
Streptomycin 500,000 μg/ml	(0.4 ml)	(2 ml)

Using sterile precautions

1. Combine each solution in the order listed above; final pH 7.0–7.2.
2. Mix well and dispense in 500–1,000 ml amounts into sterile bottles and cap tightly.
3. Culture 1–5 ml in thioglycollate broth to test for sterility.
4. Store in refrigerator at 4°C.

Earle's Balanced Salt Solution (10×)

	Solution A	
	1 liter	5 liters
NaCl	68 g	340 g
KCl	4 g	20 g
$MgSO_4 \cdot 7H_2O$	2 g	10 g
$NaH_2PO_4 \cdot H_2O$	1.4 g	7 g
Glucose	10 g	50 g
Demineralized water	800 ml	4,000 ml
	Solution B	
$CaCl_2$ (anhydrous)	2 g	10 g
Demineralized water	200 ml	1,000 ml

1. Dissolve each salt in demineralized water in the order listed above.
2. Autoclave solutions A and B separately at 15 lb for 15 minutes.
3. Mix solutions A and B when cool. Pour B into A and stir *slowly*.
4. Dispense in 100–500 ml amounts in sterile bottles and cap tightly.
5. Culture 1–5 ml in thioglycollate broth to test for sterility.
6. Store at room temperature or 4°C.

Eagle's Growth Medium with 10% Fetal Bovine Serum (EHS)

	1 liter	5 liters
Demineralized water, sterile	756 ml	3,775 ml
Hanks' BSS (10X)	100 ml	500 ml
Phenol red (0.5%)	4 ml	20 ml
Amino acids (100X)*	10 ml	50 ml
NaOH (1N), adjusted to pH 7.0	(5 ml)	(25 ml)
Vitamins (100X)*	10 ml	50 ml
Fetal bovine serum	100 ml	500 ml
$NaHCO_3$ (7.5%)	4.5 ml	23 ml
Penicillin (200,000 u/ml)	(1 ml)	(5 ml)
Streptomycin (500,000 μg/ml)	(0.4 ml)	(2 ml)
L-Glutamine stock solution (2.9%)†	10 ml	50 ml

*Purchased from a commercial source.

†L-Glutamine stock solution can be sterilized by filitration and stored at −20°C. Add to the medium just before use.

Using sterile precautions:

1. Mix each solution in the order listed; final pH 7.0–7.2.
2. Dispense in 500–1,000 ml amounts into sterile bottles and cap tightly.
3. Culture 1–5 ml in thioglycollate broth to test for sterility.
4. Store at 4°C.

Eagle's Maintenance Medium with 2% Fetal Bovine Serum (EES)

	1 liter	5 liters
Demineralized water, sterile	820 ml	4,100 ml
Earle's BSS (10X)	100 ml	500 ml
Phenol red (0.5%)	4 ml	20 ml
Amino acids (100X)*	10 ml	50 ml
NaOH (1N), adjusted to pH 7.0	(5 ml)	(25 ml)
Vitamins (100X)*	10 ml	50 ml
Fetal bovine serum	20 ml	100 ml
$NaHCO_3$ (7.5%)	29.6 ml	148 ml
Penicillin, 200,000 u/ml	(1 ml)	(5 ml)
Streptomycin, 500,000 μg/ml	(0.4 ml)	(2 ml)
L-Glutamine stock solution (2.9%)†	10 ml	50 ml

*Purchased from a commercial source.

†L-Glutamine stock solution can be sterilized by filitration and stored at −20°C. Add to the medium just before use.

Using sterile precautions, follow the same procedures as for EHS. The final pH should be 7.6–7.8.

REAGENTS AND SOLUTIONS

Phosphate-Buffered Saline (PBS)

	Solution A	
	1 liter	5 liters
NaCl	8.0 g	40 g
KCl	0.2 g	1 g
$CaCl_2 \cdot 2H_2O$	0.132 g	0.66 g
$MgCl_2 \cdot 6H_2O$	0.1 g	0.5 g
Demineralized water	800 ml	4,000 ml
	Solution B	
Na_2HPO_4	1.15 g	5.75 g
KH_2PO_4	0.2 g	1 g
Demineralized water	200 ml	1,000 ml

1. Dissolve each salt in demineralized water in the order listed.
2. Autoclave solutions A and B separately at 15 lb for 15 minutes.
3. Mix A and B when cool. Pour B into A and stir *slowly*. Final pH 7.0.
4. Dispense in 500–1,000 ml amounts into sterile bottles tightly capped.
5. Culture 1–5 ml in thioglycollate broth to test for sterility.
6. Store in refrigerator at 4°C.

Trypsin Solution in PBS 2.5%

PBS without calcium and magnesium	1,000 ml
Trypsin (1:300) (Difco)	25 g

1. Add trypsin to PBS and shake to dissolve.
2. Filter through Millipore filter, 0.45-μm pore size.
3. Dispense in 200–500 ml amounts into sterile bottles tightly capped.
4. Culture 1–5 ml in thioglycollate broth to test for sterility.
5. Store at −20°C.

Versene Solution (0.02%)

	1 liter	5 liters
NaCl	8 g	40 g
KCl	0.2 g	1 g
KH_2PO_4	0.2 g	1 g
Na_2HPO_4	1.15 g	5.75 g
Disodium ethylenediamine tetraacetate (EDTA)	0.2 g	1 g
Demineralized water	1,000 ml	5,000 ml

1. Weigh each salt separately and dissolve in sequence.
2. Dispense in 50 ml amounts into bottles with loose caps.
3. Autoclave at 15 lb for 15 minutes.
4. Store at room temperature, tightly capped.

Trypsin-Versene Solution

Trypsin (2.5%)	5 ml
Versene (0.02%)	45 ml

1. Mix trypsin-versene using sterile procedures.
2. Before use, warm the mixture at 37°C in water bath.

Thioglycollate Broth for Bacteriological Sterility Tests

1. While stirring, add 24 g thioglycollate (Difco) powder to 500 ml demineralized water.
2. Add 5 g dextrose.
3. Add sufficient distilled water to make 1,000 ml.
4. Dispense in 5-ml amounts into tubes.
5. Autoclave at 15 lb for 15 minutes.
6. Store at room temperature in the dark.

Sodium Bicarbonate (7.5%)

$NaHCO_3$	75 g
Demineralized water	1,000 ml

1. Dissolve $NaHCO_3$ in sterile, demineralized water.
2. Filter through Millipore filter, 0.45-μm pore size, using *positive* pressure.
3. Dispense in 50–100 ml amounts into sterile bottles and cap tightly.
4. Culture 1–5 ml in thioglycollate broth to test for sterility.
5. Store in refrigerator or at room temperature.

Phenol Red Solution (0.5%)

Phenol red	5 g
NaOH (0.1 N)	150 ml
Demineralized water	850 ml

1. Add 0.1 N NaOH to phenol red powder by drops until dissolved.
2. Add demineralized water.
3. Dispense in 100–500 ml amounts into bottles with loose caps.
4. Autoclave at 15 lb for 15 minutes.
5. Culture 1–2 ml into thioglycollate broth to test for sterility.
6. Store at room temperature or in refrigerator, tightly capped.

Trypan Blue Solution for Cell Counts (0.4%)

Trypan blue dye (Sigma)	0.4 g
PBS	100 ml

1. Dissolve trypan blue in PBS, adjusted to pH 7.3. Filter through no. 41 ashless filter paper.
2. Keep at room temperature in a bottle tightly capped.

Alsever's Solution

	1 liter
Dextrose	20.5 g
Sodium citrate	8.0 g
Citric acid	0.55 g
NaCl	4.2 g
Demineralized water	1,000 ml

1. Weigh each item in the order listed above. Dissolve in demineralized water.
2. Dispense in 50–100 ml amounts into bottles with loose caps.
3. Autoclave at 10 lb for 10 minutes.
4. Culture 1 ml into thioglycollate broth to test for sterility.
5. Store sterile solution tightly capped at 4°C in refrigerator or at room temperature.

SUPPLEMENTARY READING, PART SIX

Chapter 28: Cell Culture Preparation

Bashor, M. M. Dispersion and disruption of tissues. *Methods Enzymol.* 58:119–31 (1979).

Bodian, D. Simplified method of dispersion of monkey kidney cells with trypsin. *Virology* 2:575–76 (1956).

Caviedes, R.; Liberona, J. L.; Hidalgo, J.; Tascon, S.; Salas, K.; and Jaimovich, E. A human skeletal muscle cell line obtained from an adult donor. *Biochem. Biophys. Acta* 1134:247–55 (1992).

Coriell, L. L. Preservation, storage and shipment. In Jakoby, W. B., and Pastan, I. H., eds., *Methods in Enzymology*, vol. 58, pp. 29–36, *Cell Culture*. New York: Academic Press, 1979.

Freshney, R. I. Culture of animal cells. In *A Manual of Basic Techniques*. New York: Alan R. Liss, 1983.

Hay, R. J. The seed stock concept and quality control for cell lines. *Anal. Biochem.* 171:225–37 (1988).

Hodinka, R. L. Cell culture techniques: Serial propagation and maintenance of monolayer cell cultures. In Clarke, L. M., sec. ed., Viruses, rickettsiae, chlamydiae, and mycoplasmas, vol. 8.20, pp. 1–11, in Isenberg, H. D., ed.-in-chief, *Clinical Microbiology Procedures Handbook*. Washington, D.C.: American Society for Microbiology, 1992.

Janada, J. M.; Desmond, E. P.; and Abbott, S. L. Animal and animal cell culture systems. In Balows, A.; Hausler, W. J., Jr.; Herrmann, K. L.; Isenberg, H. D.; and Shadomy, H. J., eds., *Manual of Clinical Microbiology*, 5th ed., pp. 137–46. Washington, D.C.: American Society for Microbiology, 1991.

Kennett, R. H. Cell fusion. In Jakoby, W. B., and Pastan, I. H. *Methods in Enzymology*, vol. 58, pp. 354–59, *Cell Culture*. New York: Academic Press, 1979.

McGarrity, G. J.; Sarama, J.; and Vanaman, V. Cell culture techniques. *ASM News* 51:170–83 (1985).

Mallampalli, R. K.; Floerchinger, C. S.; and Hunninghake, G. W. Isolation and immortalization of rat pre-type II cell lines. *In Vitro Cell. Develop. Biol.* 28:181–87 (1992).

Markham, P. D., and Salahuddin, S. Z. In vitro cultivation of human leukocytes: Methods for the expression and isolation of human viruses. *Bio. Tech.* 5:432–43 (1987).

Paul, J. *Cell and Tissue Culture*, 5th ed. London: Churchill Livingstone, 1975.

Peterson, W. D.; Simpson, W. F.; and Hukku, B. Cell culture characterization: Monitoring for cell identification. *Methods Enzymol.* 58:164–78 (1979).

Price, P.; Bush, C.; Parkins, C. S.; McMillan, T. J.; Robinson, M.; and Steel, G. G. Evaluation of cell attachment matrix (CAM) coated plates for primary culture of human tumor biopsies. *Radiother. Oncol.* 21:282–85 (1991).

Reuveney, S. Microcarriers in cell culture: Structure and applications. In Maramorosch, K., ed., *Advances in Cell Culture*, vol. 4, pp. 213–47. New York: Academic Press, 1985.

Schmidt, N. J. Cell culture procedures for diagnostic virology. In Schmidt, N. J., and Emmons, R. W., eds., *Diagnostic Procedures for Viral, Rickettsial and Chlamydial Infections*, 6th ed., pp. 51–100. Baltimore, Md.: American Public Health Association, 1989.

Younger, J. S. Monolayer tissue cultures. 1, Preparation and standardization of suspensions of trypsin-dispersed monkey kidney cells. *Proc. Soc. Exp. Biol. Med.* 85:202–5 (1954).

Chapter 29: Endogenous Viral and Mycoplasmal Contaminants in Cell Cultures

Barile, M. G. Mycoplasma contamination of cell cultures. In Hsiung, G. D., and Green, R. H., eds., *Virology and Rickettsiology,* vol. 1, pt. 2, pp. 381–406, in *Handbook Series in Clinical Laboratory Science*. West Palm Beach, Fla.: CRC Press, 1978.

Chen, T. R. Microscopic demonstration of mycoplasma contamination in cell cultures by flourescent Hoechst 33258 stain. *Exp. Cell. Res.* 104:255–62 (1977).

Chu, F. C.; Johnson, J. B.; Orr, H. C.; Probst, P. G.; and Petricciani, J. C. Bacterial virus contamination of fetal bovine sera. *In Vitro* 9:31–34 (1973).

Croghan, D. L.; Matchett, A.; and Koski, T. A. Isolation of porcine parvovirus from commercial trypsin. *Appl. Microbiol.* 26: 431–33 (1973).

Fong, C. K. Y., and Landry, M. L. An adventitious viral contaminant in commercially supplied A549 cells: Identification of infectious bovine rhinotracheitis virus and its impact on diagnosis of infection in clinical specimens. *J. Clin. Microbiol.* 30:1611–13 (1992).

Hay, R. J. The seed concept and quality control for cell lines. *Analyt. Biochem.* 171:225–37 (1988).

Hsiung, G. D. Latent virus infections in primate tissues with special reference to simian viruses. *Bact. Rev.* 32:185–205 (1968).

Hsiung, G. D.; Fong, C. K. Y.; and Bia, F. Viruses of guinea pigs: Considerations for biomedical research. *Micro. Rev.* 44:468–90 (1980).

Kniazeff, A. J. Viruses infecting cattle and their role as endogenous contaminants of cell cultures: Cell culture for virus vaccine production. *Nat. Cancer Inst. Monogr.* 29:123–32 (1968).

McGarrity, G. J. Detection of mycoplasma infection of cell cultures. In Maramorosch, K., ed., *Advances in Cell Culture,* pp. 99–131. New York: Academic Press, 1982.

McGarrity, G. J.; Steiner, T.; and Gamon, L. Prevention, detection and control of mycoplasmal infections of cell cultures. In Kilbey, B. J.; Legator, M.; Nichols, W.; and Ramel, C., eds., *Handbook of Mutagenicity Test Procedures,* 2d ed., pp. 823–39. New York: Elsevier, 1984.

Molander, C. W.; Kniazeff, A. J.; Boone, C. W.; Paley, A.; and Imagawa, D. T. Isolation and characterization of viruses from fetal calf serum. *In Vitro* 7:168–73 (1972).

Nettleton, P. F., and Rweyemamu, M. M. The association of calf serum with the contamination of BHK 21 clone 13 suspension cells by a parvovirus serologically related to the minute virus of mice (MVM). *Arch. Virol.* 64:349–74 (1980).

Paul, J. Prevention of contamination. In *Cell and Tissue Culture,* 5th ed., pp. 133–54. London: Churchill Livingstone, 1975.

Petricciani, J. C. Regulatory philosophy and acceptability of cells for the production of biologicals. *Dev. Biol. Standard.* 74:9–15 (1991).

Smith, K. O. Adventitious viruses in cell cultures. *Prog. Med. Virol.* 12:302–26 (1970).

Tsukamoto, K.; Hihara, H.; and Kono, Y. Detection of avian leukosis virus antigens by the ELISA and its use for detecting infectious virus after cultivation of samples and partial characterization of specific pathogen-free chicken lines maintained in this laboratory. *J. Vet. Med. Sci.* 53:339–408 (1991).

Chapter 30: Culture Media, Reagents, and Solutions

Bird, B. R., and Forrester, F. T. *Basic Laboratory Techniques in Cell Culture*. U.S. Department of Health and Human Services, Centers for Disease Control, Atlanta, 1981.

Eagle, H. Nutrition needs of mammalian cells in tissue culture. *Science* 122:501–4 (1955).

Freshney, R. T. Culture of animal cells. In *A Manual of Basic Techniques*. New York: Alan R. Liss, 1983.

Hay, R. J. *ATCC Quality Control Methods for Cell Lines*. 1st ed. Rockville, Md.: ATCC, 1985.

Hodinka, R. L. Cell culture techniques: Preparation of cell culture medium and reagents. In Clarke, L. M., sec. ed., Viruses, rickettsial, chlamydiae, and mycoplasmas, vol. 8.19, pp. 1–12, in Isenberg, H. D., ed.-in-chief, *Clinical Microbiology Procedures Handbook*. Washington, D.C.: American Society for Microbiology, 1992.

Morton, H. J. A survey of commercially available tissue culture media. *In Vitro* 6:89–108 (1970).

Paul, J. *Cell and Tissue Culture,* 5th ed. London: Churchill Livingstone, 1975.

Rutzky, L. P., and Pumper, R. W. Supplement to a survey of commercially available tissue culture media (1970). *In Vitro* 9:468–69 (1974).

Sato, G. H.; Pardee, A. B.; and Sirbasku, D. A. eds. *Growth of Cells in Hormonally Defined Media.* In *Cold Spring Harbor Conferences on Cell Proliferation,* vol. 9. New York: Cold Spring Harbor Laboratory, 1982.

Schafer, T. W.; Pascale, A.; Shimanaski, G.; and Carne, P. E. Evaluation of gentamicin for use in virology and tissue culture. *Appl. Microbiol.* 23: 565–70 (1972).

Schmidt, N. J. Cell culture procedures for diagnostic virology. In Schmidt, N. J., and Emmons, R. W., eds., *Diagnostic Procedures for Viral, Rickettsial and Chlamydial Infections,* 6th ed., pp. 51–100. Washington, D.C.: American Public Health Association, 1989.

Sigma Chemical Co. *Technical Information on Commonly Used Tissue Culture Techniques*. St. Louis, Mo., 1990.

Swack, N. S. Cell cultures, culture media, reagents, and endogenous viruses. In Hsiung, G. D., and Green, R. D. eds., *Virology and Rickettsiology,* vol. 1, pt. 2, p. 369, in *Handbook Series in Clinical Laboratory Science*. West Palm Beach, Fla.: CRC Press, 1978.

Tissue Culture Association, Inc. Commonly used tissue culture terms. *In Vitro Cell. Dev. Biol.* 26:97–100 (1990).

Wang, R. J. Effect of room fluorescent light on the deterioration of tissue culture medium. *In Vitro* 12:19–22 (1976).

Whittaker, M. A. *Bioproducts Use of Cultured Cells in Virology*. Walkersville, Md.: Whittaker M. A. Bioproducts, 1990.

INDEX

Page numbers are followed by *F* for figures and *T* for tables.

FRONT COVER

Cytopathic effect (CPE) induced by different viruses in various cell systems. Clockwise from the top:

1. CPE induced by enterovirus in MK cells
2. CPE induced by rhinovirus in HDF cells
3. Syncytium induced by RS virus in Hep-2 cells
4. CPE induced by adenovirus in Hep-2 cells
5. CPE induced by HSV in HDF cells
6. CPE induced by CMV in HDF cells

BACK COVER

Electron micrographs of various viruses

1. HIV in infected H9 cells (thin section)
2. Influenza virus–infected MK cells (thin section)
3. Rotavirus in stool sample
4. Enterovirus aggregates (PTA-stained)
5. Parvovirus aggregates (PTA-stained)
6. Papillomavirus aggregates (PTA-stained)
7. Adenovirus in infected HEK cells (thin section)
8. Herpes simplex virus in infected RK cells (thin section)